Mobile Satellite Communications

For a complete listing of the *Artech House Mobile Communications Library*, turn to the back of this book.

Mobile Satellite Communications

Shingo Ohmori
Hiromitsu Wakana
Seiichiro Kawase

Artech House
Boston • London

Library of Congress Cataloging-in-Publication Data
Ohmori, Shingo
   Mobile satellite communications / Shingo Ohmori, Hiromitsu Wakana, Seiichiro Kawase
      p.   cm.
   Includes bibliographical references (p.   ) and index.
   ISBN 0-89006-843-7 (alk. paper)
   1. Mobile communication systems.   2. Artificial satellites in telecommunication.   I. Wakana, Hiromitsu.   II. Kawase, Seiichiro.   II. Title.
   TK6570.M6036   1997
   621.3845—dc21   97-41708
                                                                 CIP

British Library Cataloguing in Publication Data
Ohmori, Shingo
   Mobile satellite communications
   1. Artificial satellites in telecommunication   2. Mobile communication systems
   I. Title   II. Wakana, Hiromitsu   III. Kawase, Seiichiro
   621.3'825

   ISBN 0-89006-843-7

Cover design by Jennifer L. Stuart

International Standard Book Number: 0-89006-843-7
Library of Congress Catalog Card Number: 97-41708

10 9 8 7 6 5 4 3

Contents

Preface

Mobile satellite communication, which provides communication links to ships, aircraft, and land vehicles, is a very new satellite communication system that began in the 1970s.

Mobile communications have become extremely important for modern societies because of the large number of people traveling all over the world using aircraft, ships, and land vehicles. In the 21st century, mobile satellite communication systems will play a much more important role in contributing to worldwide peace and prosperity.

Mobile Satellite Communications aims to provide the basic and latest knowledge of mobile satellite communications, including information on satellite orbits, the system design concept, vehicle antennas, digital communication technologies, Earth stations, and other related technologies. The book also provides basic knowledge of radiodetermination technologies, including information on the global positioning system (GPS) and OmniTRACS, a hybrid system of radio determination and message communications.

The authors have been engaged in the research and development of mobile satellite communications and have extensive knowledge and experience as a result of carrying out the Engineering Test Satellite Five (ETS-V) experiments. The ETS-V, an R&D satellite launched in 1987, was the world's first to conduct experiments on satellite communications for ships, aircraft, and land vehicles.

We believe that *Mobile Satellite Communications* will be very useful for undergraduate students, engineers, and managers in the mobile satellite communications fields.

Our book has 11 chapters. Chapter 1 gives an overview on mobile satellite communications and classifies satellite communication services. It also covers the history of development and international coordination of frequency allocation. Chapter 2 discusses satellite orbits from the standpoint of mobile satellite communications, including both geostationary and nongeostationary Earth orbiting satellite communication systems. Chapter 3 deals with key parameters in designing satellite links such as noise temperatures and the capabilities of radiation power and signal receiving. The basic characteristics of the transponders of satellites and frequency interference are also described in this chapter. Chapter 4 describes vehicle antennas and the basic characteristics required for their installation. Chapter 5 deals with the propagation problems, which are essential matters to be discussed in mobile satellite communications. Chapter 6 presents digital communication technologies such as modulation techniques, multiple access, error correction, and voice coding. Chapter 7 introduces some examples of operational and forthcoming mobile satellite communications, including geostationary Earth orbiting systems (INMARSAT and AMSC) and nongeostationary Earth orbiting satellite systems (Iridium and Odyssey). Chapter 8 describes mobile Earth stations, hand-carried, as well as those installed on aircraft, ships, and land vehicles. It describes their basic system configurations, antennas, and satellite tracking functions. Chapter 9 gives an overview of radio-navigation systems, including the GPS and the OmniTRACS. Chapter 10 describes the broadcasting for vehicles. Chapter 11 describes an intelligent satellite, which in the near future will provide personal and multimedia services for handy terminals. The research and development programs ACTS, ITALSAT, and COMETS are introduced.

There is a lot of leading edge research in the satellite communications field; researchers are now trying to implement commercial systems using a group of nongeostationary orbiting satellites. In the 21st century, innovative technologies in mobile satellite communications will be needed for personal communications, broadband satellite systems for multimedia services, and gigabit-class satellite systems for global information infrastructures.

We hope that this book will contribute to the comprehensive understanding of mobile satellite communications for everyone interested in these systems.

We are most grateful to the reviewers for their suggestions and comments, and to our editors for their continuous and patient encouragement. Finally, we would like to express our sincere appreciation to our colleagues at the Communications Research Laboratory for their helpful discussions.

Shingo Ohmori,
Hiromitsu Wakana,
Seiichiro Kawase,
November 1997

1

Introduction

1.1 Overview of Mobile Satellite Communications

1.1.1 What Are Mobile Satellite Communications?

"What are mobile satellite communications?" In a word, they are radiocommunication systems for "mobiles" such as ships, aircraft and land vehicles using satellites. It must be noticed that a mobile satellite does not mean a moving satellite, which has nongeostationary orbits. It is not important whether mobile satellite communication systems use geostationary satellites or nongeostationary satellites. Figure 1.1 shows the concept of mobile satellite communication systems, which basically consist of a satellite, a gateway Earth station, and various mobile Earth stations such as aircraft, ships, land vehicles, and portable terminals. A communication link from the Earth (mobile Earth stations) to a satellite is called an uplink, and that from a satellite to the Earth is called a downlink. A communication link between a gateway Earth station and a satellite is called a feeder link. In almost all mobile satellite communication systems, 1.6 and 1.5 GHz frequencies (L band) are used in uplinks and downlinks between satellites and mobiles, respectively. This is denoted as 1.6/1.5 GHz. In a feeder link, 6/4 GHz (C band), or 14/12 GHz (Ku band), is usually used in the present systems.

A gateway Earth station is sometimes called a base Earth station, where satellite communication links are connected to other communication systems such as terrestrial public switched networks. A gateway Earth station is also called a coast Earth station in the case of maritime satellite communication systems.

1

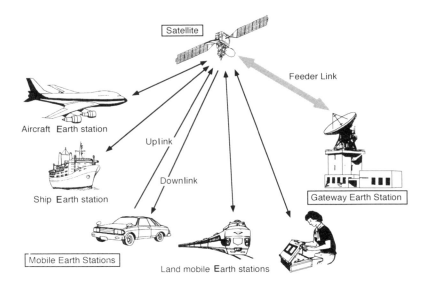

Figure 1.1 The concept of a mobile satellite communication system.

1.1.2   Mobiles Want to Know Where They Are: Radio Determination

Mobile satellite communication provides so-called bidirectional communications, such as voice, data, telex, and facsimile for mobiles. However, with advances in mobile satellite communications, mobiles not only want to communicate using telephone or facsimile, but they also want to know their positions, and inform others of these positions. This positioning function is called radio determination, which is defined as determining the position by using radiowaves. When radio determination is used for the purpose of navigation, this function is called radio navigation. Consequently, mobile satellite communications have very close relations with radiodetermination and radionavigation systems. In this book, mobile satellite communications include communication, radio determination, radio navigation, and broadcasting. The definitions of these services will be described in Section 1.2.

1.1.3   Generations of Satellite Communications—Fixed, Mobile, and Personal

"Are mobile satellite communication systems completely new concepts?" No, they are not necessarily new. They have been developed and implemented based on technologies of the first generation of satellite communications. The first generation is called *fixed satellite communication systems*. Mobile satellite communication systems are called the second generation of satellite communication systems. Further, in the 21st century, mobile satellite communications will advance to the third generation; that is, personal satellite communication systems.

The characteristics of these three satellite system generations can clearly be seen by comparing the sizes of the satellites and the antennas of Earth stations. Figure 1.2 shows the relationship between the growth of satellites (the INTELSAT, International Telecommunications Satellite Organization satellites are shown as an example) and the antenna sizes of Earth stations. In satellite communication systems, the antenna sizes of Earth stations will mainly be decided by satellite capabilities such as transmission power and the size of the antenna.

In order to provide communication services for small terminals with small antennas, the satellite is required to transmit sufficient power to be able to establish communication links between the satellite and Earth stations. In general, the more transmission power required, the bigger the satellite will become.

In the early stages of satellite communications (from the middle of the 1960s to the middle of the 1970s) satellites were too small, so they transmitted very weak power to the Earth stations. This meant that the physical dimensions of the antennas of Earth stations had to very large in order to be able to receive very weak signals from the satellite, and they had to be able to transmit strong signals that could be received by a "weak satellite." For example, the first INTELSAT satellite (known as the Early Bird, which was launched in 1965) weighed only about 40 kg in a geostationary orbit. The antennas of the Earth

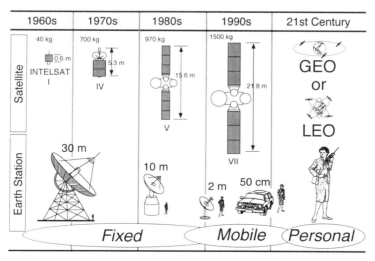

Figure 1.2 Three generations of satellite communication systems: fixed, mobile, and personal.

stations were about 30m in diameter. The Earth stations had to be fixed on the ground because of their large antennas and related facilities including terminals for communications. In a similar sense to mobile satellite communications, *fixed* means fixed Earth stations or fixed points on the ground. It is not important whether the system uses geostationary or nongeostationary satellites.

With technical developments in the 1980s, satellites became larger and more powerful, which made it possible to decrease the antenna size of ground stations to around 2m. The INTELSAT-VII in operation weighs about 1,500 kg in orbit, and the antenna sizes of Earth stations are less than 2m in diameter. These satellite communications with "very small" antennas are called VSATs (very small aperture terminals) systems. Despite this, these smaller antennas are still too large to install on mobiles such as ships, aircraft, and land vehicles.

From the middle of the 1970s to the beginning of the 1990s, many countries and organizations throughout the world actively carried out research and development programs into mobile satellite communication systems, such as the MSAT-X in the United States [1], the MSAT in Canada [2], the PROSAT in Europe [3], and the ETS-V in Japan [4]. In 1982, the International Maritime Satellite Organization (INMARSAT) started global commercial maritime satellite communication services in the Atlantic, Pacific, and Indian Oceans [5]. It has been expanding services to aircraft and land vehicles. In the late 1990s, domestic mobile satellite communication systems have been implemented on a commercial basis such as the Mobilesat in Australia, MSAT in Canada, and AMSC in the United States.

In the 21st century, with technical innovations, these mobile satellite communication systems will be upgraded to personal satellite communication systems in which individuals can directly access the satellite to establish communication channels with very small handheld terminals [6].

These personal satellite systems will be defined as the third generation of satellite communications.

The main characteristics of each generation are summarized as follows:

1. *First-generation fixed satellite communication systems*: Fixed satellite systems provide radiocommunication services between fixed Earth stations through satellite links. A system consists of a satellite and gateway Earth stations, which are very large and expensive requiring complex facilities with large antennas. In almost all cases, geostationary satellites have been used. A typical fixed satellite communication system is the INTELSAT system.

2. *Second-generation mobile satellite communication systems*: Mobile satellite communication systems provide radiocommunication services

between mobiles and a gateway Earth station through satellite links. The system consists of a satellite, a gateway Earth station, and mobiles such as ships, aircraft and land vehicles. Direct communication channels between mobiles cannot be serviced mainly because of a lack of satellite capabilities such as transmission power and channel switching functions. The communication channels between mobiles are established through a gateway Earth station. In almost all cases of mobile satellite systems, geostationary satellites have been used, or will be used in forthcoming systems. Typical systems are the INMARSAT, which has been providing worldwide commercial services for ships, aircraft, and land mobiles.

3. *Third-generation personal satellite communication systems:* The concept of a personal satellite service will be defined as one that provides radiocommunication services between very small handheld terminals using one or more satellites. A system consists of a satellite and handheld terminals, which are small enough for personal use. A personal terminal can directly access the satellite to establish communication links without connection to a gateway Earth station. Therefore, the satellites may have sophisticated functions onboard such as channel switching, networking, and signal processing.

Although personal satellite communication systems have not yet been realized, many countries and organizations have started research and development programs into advanced satellite communication systems including personal satellite communication systems, which can be operated in higher frequency bands such as 30/20 GHz (Ka band) and millimeter waves. Typical programs are the Advanced Communication Technology Satellite (ACTS) in the United States [7] and the Communication and Broadcasting Engineering Test Satellite (COMETS) in Japan [8]. The ACTS, launched in 1993, has Ka-band missions, and has been used for many kinds of advanced communications experiments. The COMETS, which will be launched in 1997, has both Ka-band and millimeter-wave (47/44 GHz) missions.

1.1.4 Toward Personal Communications—GEO or LEO?

As mentioned in the previous sections, mobile satellite communications will be replaced by personal satellite communications, which use very compact handy terminals. In the past, to achieve handy terminals with compact antennas, satellites needed to be enormous to have sufficient transmission power in the case of systems that use geostationary Earth orbit (GEO). However, in recent years, new concepts of using nongeostationary orbiting satellites have been proposed by several private sectors in the United States and by several other countries. The new systems are generally called low Earth orbiting (LEO) satellites, which use a group of low-altitude satellites. "Low" means that the orbit is lower than the geostationary orbit of about 36,000 km above the Earth in the equatorial plane. Typical examples of LEO satellites systems are the Iridium [9], Odyssey [10], and Globalstar [11], which plan to use 66, 12, and 48 satellites in circular low Earth orbits of 780 km, 10,354 km, and 1,400 km in altitudes, respectively. Odyssey is sometimes called a medium Earth orbiting (MEO) system or an intermediate circular orbiting (ICO) system because of its relatively high altitude of about 10,000 km. Although there are no strict definitions, the altitude of highly inclined elliptical orbit (HEO) systems ranges from 500 km to several thousands kilometers, while that of MEOs ranges from several thousands kilometers to 20,000 km. In this book, the term LEO generally includes MEO, however, sometimes MEO is used when needed.

There are other non-GEO systems, which use (HEO). A typical HEO system is the Molniya, which has operated since 1965 in the former USSR mainly for domestic communications. The apogee and perigee of Molniya's orbit are 40,000 km and 500 km, respectively. Molniya was initially the name of a satellite, but more recently Molniya has given its name to the elliptical orbit first used by the Molniya system. Figure 1.3 shows the orbits of GEO, LEO, and HEO satellites. Table 1.1 shows the main characteristics of GEO, LEO, and HEO orbits.

The present mobile satellite systems use, and forthcoming ones will use, GEO satellites. However, in designing future personal communication systems, it is a very difficult to choose between GEO or LEO system. A comparison of the main characteristics of GEO and LEO systems for personal satellite communication are shown in Table 1.2.

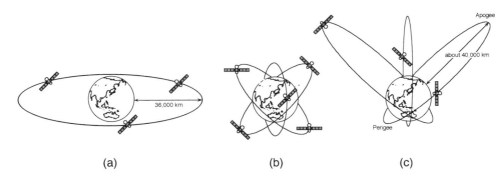

(a)   (b)   (c)

Figure 1.3 Orbits of (a) GEO, (b) LEO, and (c) HEO.

Table 1.1
Main Characteristics of GEO, LEO, and HEO Orbits

	GEO	*LEO*	*HEO*
Typical System	*Inmarsat*	*Iridium*	*Molniya*
Orbit			
Type	Circular	Circular	Oval
Number of orbits	1	6	4
Altitude	36,000 km	780 km	
Apogee			40,000 km
Perigee			500 km
Period	24 hours	1 hour 40 minutes	12 hours
Satellite			
Weight	about 1,500 kg	about 700 kg	about 1,000 kg
Numbers	3	66 (11/orbit)	12 (3/orbit)
Minimum ele. angle	5 degrees	8 degrees	80 degrees
Visible time	24 hours	10 minutes	8 hours

Several studies on advanced systems using geostationary satellites with large multibeam antennas to serve handheld voice terminals have been carried out. If GEO satellites are adopted, high-frequency bands such as the Ka band and the millimeter wave have to obtain sufficient frequency bandwidths to accommodate numerous users. The trend toward using the Ka band for such missions has been purely driven by frequency congestion. In these systems, considerable frequency reuse is possible. By using a high frequency, the terminal can be small; however, the satellites have to be gigantic in order to have sufficient transmission power. Further, the antenna of the user terminal has to be directional in order to be able to track satellites because of its high gain. The use of the Ka band, of course, brings additional difficulties in terms of high rain fade losses and is not optimal in a number of respects. On the other hand, if LEO satellites are adopted, many satellites and orbits will be required in order to be able to achieve seamless coverage, although such sophisticated problems may be resolved by channel switching between different satellites. However, the LEO systems have the advantage of being able to transmit sufficient power to allow an omnidirectional antenna to be used for user terminals, because of the small value of free-space propagation loss caused by low altitude and low frequencies such as L and S bands. Further, LEO systems have the significant advantage of lower propagation delay. This fact is very important for mobile systems, especially for real-time voice communications systems.

Table 1.2
Comparison of GEO and LEO Satcom Systems

		GEO	LEO
System	Frequency	Ka band, Millimeter waves	800 MHz, L band, S band
	Coverage	Global coverage except polar regions	Global coverage with polar regions
	Elevation Angle	Over 5 degrees	10–20 degrees
	Delay Time	270 msec	10–30 msec
	Backup Satellites	Spare in orbit	Active satellites with mutual backup
	Switching of Satellites	None	May be required
Satellite	Bus Size	Very Large (over 2 tons)	Many small satellites
	Onboard Processing	Necessary	Sophisticated
	Onboard Antenna	10–30 meters	Under 5 meters
	EIRP	High	Low
	Orbit	Geostationary (36,000 km)	Arbitrary (Iridium; 750 km)
User Terminal	Volume	Handy terminal	Handheld terminal
	Antenna	Directional with tracking	Omnidirectional without tracking
	Countermeasures against Doppler Shift	Necessary (High frequency used)	Required (moving satellites used)
Notes	Radiodetermination	Two satellites required	Essentially suitable
	Coverage Beams	Fixed	Moving (Iridium; about 5 km/sec)
Typical Examples		ACTS (USA), COMETS (Japan)	Iridium, Globalstar, Odyssey (USA)

1.2 Classification of Satellite Communications

1.2.1 Satellite Services

Depending on the purpose, satellite communications can be classified into several services such as radio communication, radio navigation, radio determination, broadcasting, meteorological, standard frequency and time signal, and amateur. These services, terminology, and allocated frequency bands have been defined by the Radio Regulations (RR) of the International Telecommunication Union (ITU) [12].

The satellite services, which are related to mobile satellite communications, are listed in Table 1.3. Mobile satellite services are classified into the maritime mobile satellite service (MMSS), aeronautical mobile satellite service (AMSS), and land mobile satellite service (LMSS).

1.2.2 Frequency Designations

The radio frequency is divided into nine frequency bands as shown in Table 1.4. In satellite communication fields, frequency bands are often denoted with alphabetical symbols such as C, L, S, Ku, and Ka bands, as shown in Table 1.4. Band numbers and band names are defined by the Radio Regulation, and alphabetic symbols such as L and S are defined by the IEEE Standard Radar Definitions [13].

Typical satellite services and their designated frequency bands are shown in Table 1.5. The operational systems, planned systems, and research and

Table 1.3
Classification of Services Related to Mobile Satellite Communications

Fixed Satellite Service (FSS)	
Mobile Satellite Service (MSS)	
	Maritime mobile satellite service (MMSS)
	Aeronautical mobile satellite service (AMSS)
	Land mobile satellite service (LMSS)
Radiodetermination Satellite Service (RDSS)	
Radionavigation Satellite Service (RNSS)	
	Aeronautical radionavigation service
	Maritime radionavigation satellite service
Broadcasting Satellite Service (BSS)	

Table 1.4
Designation of Frequency Bands

Band Number	Band Name	Alphabetic Symbol	Frequency
4	VLF		3–30 kHz
5	LF		30–300 kHz
6	MF		300–3 GHz
7	HF		3–30 MHz
8	VHF		30–300 MHz
9	UHF		300 MHz–3 GHz
		L band	1–2 GHz
		S band	2–4 GHz
10	SHF		3–30 GHz
		C band	4–8 GHz
		X band	8–12 GHz
		Ku band	12–18 GHz
		K band	18–27 GHz
11	EHF		30–300 GHz
		Ka band	27–40 GHz
		Millimeter waves	40–300 GHz
12		Submillimeter waves	300–3,000 GHz

Note: "Band Number N" extends from 0.3×10^N Hz to 3×10^N Hz

development programs are denoted by thick bold, bold-solid, and solid lines, respectively.

Table 1.6 shows part of the frequency bands, which have been allocated for mobile satellite communication services.

1.2.3   Fixed Satellite Services (FSS)

A typical example of FSS is the INTELSAT system. The first generation of INTELSAT systems operated in the C band (6/4 GHz). At present, many domestic systems such as OPTUS [14] in Australia and JCSAT in Japan operate in the Ku band (14/12 GHz). A few systems in the world are using the Ka band (30/20 GHz), such as OLYMPUS [15] and CS [16], which provide coverage throughout most of Europe, and Japan, respectively.

1.2.4   Mobile Satellite Service (MSS)

Mobile satellite services are categorized into three services: maritime, aeronautical, and land mobile communications. A typical example is the INMARSAT

Table 1.5
Satellite Services and Frequency Allocations

Service	1 GHz · L	2 · S	4 · C · 8 · X	12 · Ku · 18	K (27)	Ka (40 GHz)	MM-wave	300 GHz Laser · Sub-MM
Fixed (FSS)			Early INTELSAT (6/4 GHz)	Many domestic systems (14/12 GHz)		Few domestic systems CS (Japan) Olympus (Italy) (30/20 GHz); ACTS (USA) (30/20 GHz)	COMETS (JPN) (30/20, 47,44 GHz)	
Mobile (MSS)	INMARSAT AMSC (USA) MSAT (CND) OPTUS (AUS) (1.6/1.5 GHz)	NSTAR (JPN) (2.5/2.0 GHz); IRIDIUM (2.5/1.6 GHz)		OmniTRACS (14/12 GHz)				
Radio Determination (RDSS)								
Radio Navigation (RNSS)	GPS (1.6 GHz) (1.2 GHz)							
Broadcasting (BSS)	Digital Audio 1.5 GHz-World 2.3 GHz-USA 2.5 GHz-JPN			TV (12 GHz)	COMETS (JPN) (21 GHz)			
Intersatellite (ISS)		TDRS (USA); ETS-VI (JPN)		TDRS (USA); ARTEMIS (ESTEC)		ETS-VI (JPN) COMETS (JPN)	ETS-VI	ETS-VI ARTEMIS

Present systems   Planned systems   R&D programs

Table 1.6
Frequency Allocation for Mobile Satcom (L Band)

Downlink (Space-Earth)

	1,215 MHz	1,260	1,492	1,525	1,530	1,535	1,544	1,545	1,555	1,559	1,610 MHz
Region 1	RNSS (S-E)				MMSS LMSS (S-E)		Distress and safety (S-E)	AMSS (S-E)	LMSS (S-E)	RNSS (S-E)	
Region 2	RNSS (S-E)			MSS (S-E)	MMSS LMSS (S-E)		Distress and safety (S-E)	AMSS (S-E)	LMSS (S-E)	RNSS (S-E)	
Region 3	RNSS (S-E)			MSS (S-E)	MMSS LMSS (S-E)		Distress and safety (S-E)	AMSS (S-E)	LMSS (S-E)	RNSS (S-E)	

Uplink (Earth-Space)

	1,610 MHz	1,626.5	1,631.5	1,645.5	1,646.5	1,656.5	1,660 MHz
Region 1	MSS (E-S)	MMSS LMSS (E-S)	MMSS LMSS (E-S)	Distress and safety (E-S)	AMSS (E-S)	LMSS (E-S)	
Region 2	RDSS MSS (E-S)	MSS (E-S)	MMSS LMSS (E-S)	Distress and safety (E-S)	AMSS (E-S)	LMSS (E-S)	
Region 3	RDSS MSS (E-S)	MSS (E-S)	MMSS LMSS (E-S)	Distress and safety (E-S)	AMSS (E-S)	LMSS (E-S)	

Table 1.6 (Continued)

Frequency Allocation for Mobile Satcom (S Band)

	1,930 MHz	1,980	2,010	2,120	2,170	2,200	2,483.5	2,500	2,520	2,670	2,690 MHz
Region 1		MSS (E-S)									
Region 2		MSS (E-S)			MSS (S-E)		RDSS MSS (S-E)	MSS (S-E)		MSS (E-S)	
Region 3											

Note (1) Region 1: Europe, Russia, Africa   Region 2: North & South America   Region 3: Asia, Australia

Note (2) MSS: Mobile satellite service   MMSS: Martime mobile satellite service
RDSS: Radio-determination satellite service   AMSS: Aeronautical mobile satellite service
RNSS: Radio-navigation satellite service   LMSS: Land mobile satellite service

system. The INMARSAT has been operated in the L band (1.6/1.5 GHz), and almost all domestic mobile satellite communication systems, such as American Mobile Satellite Corporation (AMSC) in the United States [17] and MSAT in Canada, have also been operated in the L band. For the first time in the world, the OPTUS system in Australia started domestic services mainly for land mobiles in the L band by launching the satellite OPTUS-B in 1993. Japan started mobile communication services in the S band (2.5/2.0 GHz) for domestic maritime and land mobiles by launching the N-STAR satellite in 1997. The OmniTRACS system in the United States [18] is a very unique system using the Ku band (14/12 GHz), which has been allocated to FSS, not to MSS. It provides message communication services mainly for long-distance tracks using satellites that have been designed for FSS. OmniTRACS is also providing a self-consistent radiodetermination service using two geostationary satellites. The OmniTRACS systems are also operated by JCSAT in Japan, and by Eutelsat in Europe under the name EutelTracs.

1.2.5 Radiodetermination Satellite Service (RDSS)

A typical system will be Iridium, which will provide voice communication and positioning services in a self-consistent system. Radio determination has a wider meaning than radio navigation. In a radiodetermination system, not only I (a mobile), but also others (for example, fixed Earth stations) can know my position. The RDSS system is a bidirectional system. On the other hand, in a radionavigation system, although a mobile can know its position, others cannot. The radionavigation system is a one-directional system, from a satellite to a mobile or from a mobile to a satellite.

1.2.6 Radionavigation Satellite Service (RNSS)

Typical examples of radionavigation systems are the Navy Navigation Satellite System (NNSS), sometimes better known as the TRANSIT [19], and the Navigation System with Time and Ranging/Global Positioning System (NAVSTAR/GPS) [20]. The NNSS was the first radionavigation system in the world. It was originally developed as a U.S. Navy military satellite system, but has been open to civilian use since 1967. The position of a mobile terminal is determined by measuring the Doppler frequency shift in 150-MHz and 400-MHz signals transmitted from the NNSS satellite in polar orbits. The NAVSTAR/GPS, operated in two frequencies of 1.6 GHz and 1.2 GHz, is a second-generation radionavigation system and is the most widely used radionavigation system in the world. Tsikada and GLONASS have been used in Russia, which are equivalent to NNSS and GPS.

One of the most important points is that radionavigation systems such as NNSS and GPS are one-directional systems from a satellite to the Earth. Mobile terminals only receive signals from the satellite and never transmit signals to the satellite. These systems are called passive radionavigation systems. In passive systems, it must be noted that a mobile can know its own position, but it can never inform others of it. Radionavigation systems have usually adopted MEO satellites, because MEOs have many advantages in obtaining information on position. More detail will be provided in Chapter 9.

1.2.7 Broadcasting Satellite Service (BSS)

This service broadcasts TV and radio programs via a satellite to Earth stations. Present broadcasting satellite systems, operated at 12 GHz, are designed for community reception (fixed terminals with large antennas). If a satellite has enough power to transmit signals to be received by small antennas suitable for individual reception (fixed terminal with small antennas), the system is called a direct broadcasting satellite (DBS) system. Although present systems are designed for fixed terminals and not for mobiles, some mobiles such as large ships, trains, and buses have received TV programs from direct broadcasting satellites (such as the BS satellite in Japan) while moving. In Europe, the United States, and Japan, direct digital audio broadcasting (DAB) satellite systems in the L and S bands have been investigated to allow mobiles with high-quality programs comparable to compact disks to be developed. An advanced broadcasting system using the Ka band (21 GHz) has already been studied for the COMETS program. More detail will be described in Chapter 10.

1.2.8 Intersatellite Service (ISS)

There are two types of ISS from the standpoint of satellite orbits. The first is to establish links between GEO-GEO satellites, and the second is to establish links between GEO-LEO satellites. At present, the Tracking and Data Relay Satellite Systems (TDRSS) is the only system operating in the world [21]. The TDRSS provides data links between GEO and LEO satellites in the S (2.3 GHz) and Ku (15/13 GHz) bands. The ETS-VI satellite, which was launched in 1994, had three intersatellite communication missions: the Ka band, millimeter waves, and laser. Laser communications between satellites uses very challenging technologies [22].

From the standpoint of mobile satellite communications, the Iridium system is going to carry out intersatellite communications in the Ka band between its adjacent four satellites.

1.3 History of Mobile Satellite Communications

1.3.1 Dawn of Mobile Satcom—1960s

Table 1.7 shows a history of satellite communications in terms maritime, aeronautical, and land mobiles. The world's first experimental mobile satellite communications were carried out in aeronautical communications. In 1964, NASA and Pan Am airlines succeeded in achieving aeronautical satellite communications using the Syncom-III satellite, which was the first geostationary satellite in the world. The frequency used for the experiments was the VHF band (117.975 to 136 MHz), which had been allocated for aeronautical mobile service (R). The (R) denotes civil aviation routes used for international flights.

1.3.2 The Start of Commercial Services and R&D Programs—1970s

In 1971, the International Civil Aviation Community (ICAO) recommended an international program of research and development and system evaluation. In the same year, the L band was allocated for Distress and Safety, MMSS, and AMSS(R) by the World Administrative Radio Conference held in 1971 (WARC '71). In 1974, according to the recommendations, Canada, the Federal Aviation Administration (FAA) of the United States, and the European Space Agency (ESA) signed a memorandum of understanding to develop the AEROSAT system, which would be operated in the VHF and L bands. Although AEROSAT was scheduled to be launched in 1979, the program was canceled in 1982, mainly because of the withdrawal of financial support by the United States.

At present, however, maritime satellite communication is more advanced than aeronautical or land mobile communication. In 1976, Comsat General in the United States started commercial maritime satellite communication services by using MARISAT satellites (which were launched in 1976). The MARISAT satellite were designed for the U.S. Navy, and they had UHF transponders onboard. A transponder is a set of transmitters and responders onboard the satellite that receives and transmits signals through frequency conversion and signal amplification. Because there was sufficient margin for additional weight and volume, an L-band transponder was able to be installed on the MARISAT satellites to provide commercial services for maritime communications.

On July 16, 1979, INMARSAT was inaugurated as an international organization to provide commercial maritime satellite communications. To offer maritime satellite communication services as quick as possible, the INMARSAT leased three MARISAT satellites from Comsat General and two Maritime European Communication Satellites (MARECS) from the ESA.

Table 1.7
History of Mobile Satellite Communications

	Maritime	Aeronautical	Land
1960s		64 World's first Mobile Satcom Experiment with Syncom-III, NASA/Pan Am	
1970s		71 Allocation of L band (WARC'71) 74 Start of AEROSAT program (Canada, ESA, FAA)	70 Start of MUSAT program (Canada)
1980s	76 Start of Marisat system (USA) 77 Start of AMES R & D program (Maritime & Aero; Japan) 79 Inauguration of INMARSAT 82 Start of INMARSAT global service 82 Start of PROSAT R&D program (Maritime, Aero & Land; ESA) 84 Start of ETS-V/EMSS R&D program (Maritime, Aero and Land; Japan) 87 Launch of ETS-V for mobile satcom experiments (Maritime, Aero & Land; Japan)	84 Racal Decca satcom trials in the AMSS(R) band 85 Start of AvSAT program (Arinc, USA) 85 Revision of INMARSAT treaty for AMSS 87.6 PRODAT Phase 1 satcom trials 87.10 Satcom trials using commercial flights (JAL, Japan) 87 AMSS(R) was opened to public satcom (WARC'87)	79 Allocation of 800 MHz (WARC'79) 80 Start of MSAT program (Canada) 81 Inauguration of AUSSAT (Australia) 87 Allocation of L band (WARC'87) 88 Inauguration of AMSC (USA) 88 Inauguration of TMI (Canada) 89 Revision of Inmarsat treaty of LMSS
1990s	89 Start of Inmarsat-C 90 Launch of INMARSAT 2nd generation 92 Allocation of L & S bands for LEO mobile satcom (WARC '92) 93 Launch of ACTS for advanced satcom experiments using Ka band (USA) 97 Launch of COMETS for advanced satcom experiments using Ka band and MM waves (Japan)	91 Start of INMARSAT-Aero (Pacific)	91 Inauguration of Optus (Australia) 92 Launch of Optus-B1 (Australia) 93 Start of Optus system (Australia) 95 Launch of AMSC & MSAT satellites

In the 1970s, along with the starting of commercial maritime satellite communications of INMARSAT, many research and development programs started in various countries to study the feasibility of mobile satellite communications, not only for large ships but also for small ships, aircraft, and land mobiles. In 1970, Canada started the MUSAT program for land mobile satellite communication in the VHF band (200 to 400 MHz). In Japan, the Aeronautical and Maritime Engineering Satellite (AMES) program was started in 1975 to develop an aeronautical and maritime communication system mainly for small fishing vessels. Canada promoted the MUSAT program, which carried out aeronautical communication experiments mainly for military purposes using the ATS-6 satellite in the VHF frequency band.

1.3.3   Acceleration of Mobile Satcom—1980s

The MUSAT program in Canada was changed to the MSAT program in 1980 because the 800-MHz frequency band was allocated to land mobile satellite communications by the WARC '79. In Europe, ESA started the PROSAT program in 1982 to promote industrial technologies in Europe and to contribute to the INMARSAT system. The INMARSAT system officially went into operation on February 1, 1982 with worldwide services in the Pacific, Atlantic, and Indian Oceans.

The middle of the 1980s has significant meaning both for aeronautical and land mobile satellite communications. In 1985, Aeronautical Radio Inc. (ARINC) in the United States proposed the Aviation Satellite (AvSAT) program to provide integrated voice and data communications between air and ground services throughout the world. INMARSAT, on the other hand, was approved by the 4th assembly in October 1985 to provide aeronautical satellite communications. The controversy and competition between INMARSAT and AvSAT programs stimulated the early introduction of aeronautical satellite communications in the avionics community. In October 1987, the world's first in-flight telephone calls from a commercial jetliner over transoceanic flight routes were successfully carried out from a Japan Air Lines Boeing 747 using the INMARSAT satellite. The experiment was jointly carried out by Kokusai Denshin Denwa (KDD), Japan Air Lines, and the Communications Research Laboratory (CRL) of the Ministry of Posts and Telecommunications. This experiment was based on the Engineering Test Satellite-Five/Experimental Mobile Satellite System (ETS-V/EMSS) research and development program of CRL, which was reorganized from AMES in 1984. There were also other significant aerosatcom voice trials such as the INMARSAT/British Telecom/British Airways/Racal trials, which started in May 1988 using fully avionics-compliant packaging (located in the avionics bay, not in the passenger cabin).

In 1987, the AMSS(R) L-band frequency, which had been exclusively allocated for air traffic control (ATC) and aeronautical company communications (AOC), was opened to aeronautical public communications (APC) at WARC '87. At the same conference, the L band was allocated to LMSS. These trends made the L band the most suitable frequency for integrated mobile satellite communications, which provide worldwide services to ships, aircraft, and land mobiles. In 1989, INMARSAT expanded its services to land mobiles by introducing the INMARSAT-C system, which provides low bit rate (600 bits/sec and 1200 bits/sec) message and data communications with low-cost portable terminals.

1.3.4 Liftoff of Mobile Satcom—1990s

In 1991, American Mobile Satellite Corporation (AMSC) in the United States and Telecast Mobile Incorporated (TMI) in Canada, which were inaugurated in 1988, started low bit rate message communication services, mainly for long-distance tracks, by leasing transponders from MARISAT and INMARSAT, respectively. These services were almost the same as the INMARSAT-C service, which was introduced in 1989. The first AMSC satellite and the first MSAT satellite were launched in 1995 and 1996, respectively. Both satellites have been mutual backup systems. Australia became a new domestic mobile satellite communication provider by launching the OPTUS-B1 (the former AUSSAT-B1) satellite in 1992.

  With further advances in mobile satellite communications, strong demand for positioning services occurred for mobiles and base stations. As mentioned previously, communication and navigation systems had been developed and implemented separately, and mobiles usually used the GPS system to locate their positions and report these to the base station through communication links using the INMARSAT-C system. Only the OmniTRACS, founded by Qualcomm in 1985, has provided a self-consistent positioning service using two GEO satellites since May 1990. This positioning system is called Qualcomm's Automatic Satellite Position Reporting (QASPR).

  At the beginning of the 1990s, several U.S. private firms proposed new concepts for mobile satellite communications using a group of LEO (or MEO) satellites. Typical proposed systems are, as previously mentioned, Iridium, Odyssey, and Globalstar. In 1991, INMARSAT also proposed the Project 21/INMARSAT-P, which would provide global personal satellite communication services using non-GEO orbits. After feasibility studies, the INMARSAT-P is now going to use the intermediate circular orbit (ICO), which is the same as MEO. WARC '92 responded to these activities and allocated the L (1,626.5 to 1,631.5 GHz) and S (2,483.5 to 2,500 GHz) bands for mobile satellite services using LEO satellites.

1.4 International Coordination

1.4.1 International Organizations

International coordination in satellite communications have been carried out by the International Telecommunication Union (ITU) of the United Nations. The ITU was inaugurated in 1932 and reorganized in 1992. The terminology, definition of satellite services, technical standards, and frequency allocations are defined in the Radio Regulations (RR), which have been drafted by the World Administrative Radio Conference (WARC) of ITU. In 1992, the WARC was reorganized as the World Radiocommunication Conferences (WRC) of ITU. The basic concept of the present RRs relevant to satellite communications was drafted by the WARC-ST (WARC-satellite) in 1971. The satellite communication systems have been internationally authorized by the International Frequency Registration Board (IFRB) of ITU, the present ITU-R (ITU-Radio), with registration of such system parameters as frequency and the orbit to be used.

1.4.2 Regulatory Procedures to Operate Satellite Systems

The administration of a country that intends to establish satellite communication systems, operated in any GEO or LEO orbit, shall send to the ITU-R, not earlier than five years before the date of introducing the service, information on each satellite network of the planned system, information such as the frequency bands to be used, the modulation types of signals and the radiation characteristics of antennas, satellites, and Earth stations. The ITU-R shall publish the information in a weekly circular distributed to all administrators in the world. If any administration is of the opinion that interference will be caused to its existing or planned space radiocommunication services, it shall, within four months after the publication of the weekly circular, send its comments to the administration concerned. If no such comments are received from any administration within the period mentioned above, it may safely be assumed that no administration has basic objections to the planned satellite network for the system on which details have been published. An organization to receive comments shall endeavor to resolve any difficulties that may arise and shall provide any additional information. After completing all coordination, the planned system will be internationally authorized by recording it in the Master International Frequency Register of frequency assignments.

References

[1]   A series of MSAT-X Quarterly, Jet Propulsion Laboratory, No. 1 (1984) to No. 25 (1990).

[2]   *Proc. 2nd Int. Mobile Satellite Conf.*, Ottawa, Canada, June 1990.

[3]   "PROSAT Phase I Report," European Space Agency, ESA STR-216, May 1986.

[4]   Hamamoto, N., S. Ohmori, and K. Kondo, "Results on CRL's Mobile Satellite Communication Experiments Using ETS-V Satellite," *Space Communications*, 1990, pp. 483–493.

[5]   "Never Beyond Reach-The World of Mobile Satellite Communications," INMARSAT, 1989.

[6]   Stabrook, P., et al., "A 20/30 GHz Personal Access Satellite System Design," *Int. Conf. on Communications '89*, Boston, MA, June 1989.

[7]   Richard, R., "Advanced Communications Technology Satellite (ACTS)," *Int. Conf. on Communications*, ICC'89, 52.1–12, 1989.

[8]   Ohmori, S., et al., "Advanced Mobile Satellite Communications Using COMETS Satellite in MM-wave and Ka-band," *3rd Int. Mobile Satellite Conf.*, Pasadena, CA, June 1993, pp. 549–553.

[9]   Raymond, J. L., "Low-Earth Orbiting Satellite System," Workshop on Personal Satellite Communication Systems in the *1st Int. Conf. on Universal Personal Communications, ICUPC'92*, Dallas, TX, Sept. 1992.

[10]   Horstein, H., "Odyssey-A Satellite-Based Personal Communication Systems," *2nd Int. Conf. on Universal Personal Communication, ICUP'93*, Ottawa, Canada, 1993.

[11]   Louie, M., and P. Monte, "Globalstar Communications Payload for Global Mobile Communications," *14th Int. Communication Systems Conf. and Exhibit*, AIAA-92-1953-CP, Washington, DC, March 1992.

[12]   Radio Regulations of International Telecommunication Union.

[13]   "IEEE Standard Radar Definitions," *The Institute of Electrical and Electronics Engineers, Inc.*, 1982.

[14]   Cooper, P., and L. Crawley, "Mobilesat: A World First," *Proc. of the Int. Mobile Satellite Conf., IMSC'93*, Pasadena, CA, June 1993, pp. 273–277.

[15]   Bonhomme, R., L. Herdan, and R. Steels, "Development and Application of New Technologies in the ESA Olympus Programme," *10th American Institute of Aeronautics and Astronautics (AIAA) Int. Communication Satellite Systems Conf.*, AIAA-84-0706, 1984.

[16]   "Japan's CS (Sakura) Communications Satellite Experiments," Special edition, *IEEE Trans. on Aerospace and Electronic Systems*, Vol. AES-22, No. 3, May 1986.

[17]   Garener, W., "Description of the AMSC Mobile Satellite System," *13th American Institute of Aeronautics and Astronautics (AIAA) Int. Communication Satellite Systems Conf.*, AIAA-90-0872-CP, Los Angeles, CA, March 1990.

[18] Jacobs, I. M., "An Overview of the OmniTRACS: The First Operational Two-way Mobile Ku-band Satellite Communications System," *Space Communications*, No. 7, 1989, pp. 25–35.

[19] Stansell, T. A., "The Navy Navigation Satellite System," *NAVIGATION*, Fall 1968.

[20] Easton R. L., "The Navigation Technology Program," *NAVIGATION*, Vol. 25, No. 2, 1978.

[21] Spearing R. E., "Tracking and Data Relay Satellite System (TDRSS)," *IAF'85*, 1985.

[22] Arimoto, Y., et al., "Preliminary Result on Laser Communications Experiment Using ETS-VI," *Photonics West '95, Free-Space Lasercommunication Technologies VII*, 2381, Feb. 1995, pp. 151–158.

2

Mobile Satellite Orbits

2.1 Overview

The orbital dynamics support a communications satellite project, usually in two phases—orbital design and orbital operation. The orbital design, or the mission analysis as it is often referred to, begins at the early stage of the satellite project, where various types of orbits are examined so as to identify the best suitable orbit for the objective communications service. The orbital operation starts immediately after the satellite launch, where satellite tracking, orbit determination, and orbit maintenance will be carried out periodically over the lifetime of the satellite. The orbital operation is based on precise knowledge of rather short-term orbital motions of the satellite, while the orbital design is based on a generic survey of orbits, with particular interest in the long-term stability of the orbits.

In this chapter, we will study introductory orbital dynamics from the orbital design's point of view. Starting from the most basic equations of motion, and avoiding the use of sophisticated mathematics, we will examine the fundamental properties of circular orbits and elliptical orbits, and briefly discuss the concept of the orbital formation of a number of satellites. While there is growing interest in nongeostationary orbits and their formations, the geostationary orbit still retains its practical importance, so the basic property of this important orbit will be figured out in a compact formulation. This enables, in particular, a straightforward analysis of the stability of a geostationary communication link.

In this way, we will learn the basic knowledge for understanding the orbital design of communication statellites, particularly for mobile services.

2.2 Circular Orbit

2.2.1 Basic Formulations

The dynamics of circular orbits may appear rather simple, in such a way that the "gravitational and centrifugal forces are in balance." But if the forces really balance out so that there are no forces acting on the satellite, why doesn't the satellite make a rectilinear motion to fly away from the Earth? Answering such a naïve question requires a study of the orbit with basic equations of motion, and this provides at the same time a good starting point for studying elliptical and geostationary orbits.

Since satellites go around the Earth, their motions are conveniently described in polar coordinates. Referring to Figure 2.1, let us derive the equations of motion for polar coordinates (r, θ) in place of rectangular coordinates (x, y). From the identity equations of

$$x = r\cos \theta, \qquad y = r\sin \theta$$

we can immediately write down the following equations:

$$\ddot{x} = \ddot{r}\cos \theta - 2\dot{r}\dot{\theta} \sin \theta - r\dot{\theta}^2 \cos\theta - r\ddot{\theta} \sin \theta \qquad (2.1)$$

$$\ddot{y} = \ddot{r}\sin \theta + 2\dot{r}\dot{\theta} \cos \theta - r\dot{\theta}^2 \sin\theta + r\ddot{\theta} \cos \theta \qquad (2.2)$$

The force f acting on the satellite is measured in the acceleration that the force has caused to the satellite (that is, force is measured per unit mass of the satellite), so that the equations of motion in the original rectangular coordinates appear simple as

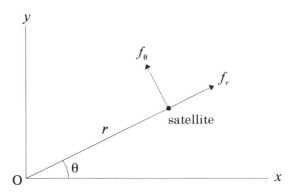

Figure 2.1 Rectangular and polar coordinates.

$$f_x = \ddot{x}, \; f_y = \ddot{y} \tag{2.3}$$

If we write the force f in radial and tangential components (f_r, f_θ), they are related with x and y components as

$$f_r = f_x \cos \theta + f_y \sin \theta \tag{2.4}$$

$$f_\theta = -f_x \sin \theta + f_y \cos \theta \tag{2.5}$$

Substituting (2.1) and (2.2) into (2.3) and then (2.3) into (2.4) and (2.5) yields

$$\ddot{r} - r\dot{\theta}^2 = f_r \tag{2.6}$$

$$2\dot{r}\dot{\theta} + r\ddot{\theta} = f_\theta \tag{2.7}$$

which are the equations of motion in polar coordinates.

If the Earth is ideally spherical and its internal mass is distributed in spherical symmetry, we are allowed to calculate the Earth's gravity by supposing that all the Earth's mass has been concentrated at its center point. This point of mass placed at the origin O in Figure 2.1 will attract the satellite, with the inverse-square gravitational force, as

$$f_r = -\mu/r^2, \quad f_\theta = 0$$

where $\mu = 398600.5 \text{ km}^3/\text{s}^2$ is the constant of the Earth's gravity. The satellite motion then obeys the equations

$$\ddot{r} - r\dot{\theta}^2 = -\mu/r^2 \tag{2.8}$$

$$2\dot{r}\dot{\theta} + r\ddot{\theta} = 0 \tag{2.9}$$

which will serve as the basis of all our subsequent discussions. Considering the satellite motion as such—that we have simply the point-mass Earth and the satellite—is referred to as a *two-body problem*.

Now, to make a circular orbit, put $r = \text{const}$. Then, from (2.9), it must be such that $\dot{\theta} = \text{const} = \Psi$, and then (2.8) requires that

$$r\Psi^2 = \mu/r^2$$

This equation certainly states that the centrifugal force (left-hand side) balances the gravitational force (right-hand side). Note that the centrifugal force appears as long as we observe the satellite motion along the r axis that rotates with the satellite. If one wishes to define the circular orbit in terms of the balance of forces, one should state first that the centrifugal and gravitational forces acting on the satellite are both constant in magnitude, and then that the two forces balance each other.

Given the orbital radius r, the orbital motion is characterized by the angular rate of the orbital revolution

$$\Psi = \sqrt{\mu/r^3} \qquad (2.10)$$

the satellite orbital velocity

$$\nu = r\Psi = \sqrt{\mu/r} \qquad (2.11)$$

and the orbital period

$$P = 2\pi/\Psi = 2\pi\sqrt{r^3/\mu} \qquad (2.12)$$

Problem 2.1

In order that the orbital period should be equal to the Earth's rotation period which is 23h 56m 4s (or 86,164s), what should be the orbital radius?

We have so far been discussing the satellite motion as being restricted in our (x, y)-plane, or alternatively, in our (r, θ)-plane. This was allowed because the gravitational force had no out-of-plane component, so that our plane—the *orbital plane*—keeps its orientation unchanged in the inertial space. In order to define this orientation in the inertial space, two parameters are used (see Figure 2.2). One is the angle that the orbital plane makes against the equatorial plane, which is called the *inclination*. The other is defined as follows: The points at which the satellite crosses the equatorial plane are called *nodes,* of which that of crossing from south to north is called the *ascending node.* The orientation of the ascending node, as measured from some fixed reference direction along the equatorial plane, is called the *right-ascension of the ascending node.* Theoretically, the reference direction can be any if it is fixed in the inertial space. In practice we select (due to the historical consequence) this reference direction to the direction at which the sun is found at the time of the vernal equinox—this direction points to a particular corner of the zodiac constellation of Pisces. Thus the orientation of the orbital plane is specified by i the inclination and Ω the right-ascension of the ascending node.

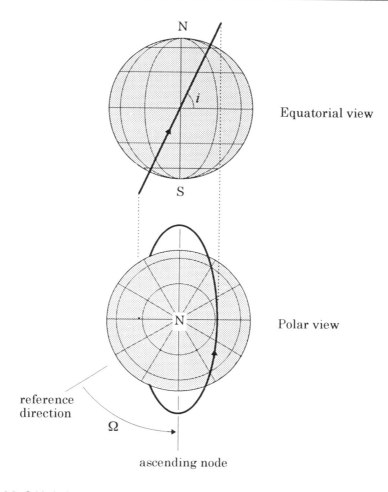

Equatorial view

Polar view

reference
direction

Ω

ascending node

Figure 2.2 Orbital plane orientation: equatorial view and polar view.

2.2.2 Perturbation of the Orbital Plane

Though we have so far assumed that the Earth is spherically symmetric, it is
not precisely true. The Earth is slightly oblate, so that its equatorial radius is
greater than its polar radius by 21 km. This oblateness causes, as we will soon
see, the orbital plane to gradually change its orientation in the inertial space.
Such a slow change arising in the orbital motion is called a *perturbation.*
Though the oblateness-caused perturbation of the orbital plane is small, taking
it into account is essential for discussing how to construct an orbital formation
of a number of satellites for communication services.

   The gravitational force of the oblate Earth appears as illustrated in
Figure 2.3, where the near-equator bulging part of the Earth's mass attracts

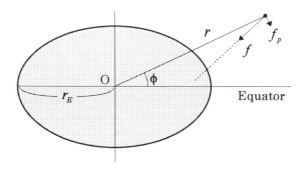

Figure 2.3 Earth's oblateness deflecting the gravitational force.

the satellite so as to make the pointing direction of the gravitational force f slightly turn aside from the Earth's center O. The force f will then have a small component f_p perpendicular to the orbital plane and oriented towards the equatorial plane. This component, for a satellite at radius r and latitude ϕ, is known to be equal to

$$f_p = 3\mu J_2(r_E^2/r^4)\sin \phi \cos \phi \qquad (2.13)$$

where $r_E = 6378.15$ km is the Earth's equatorial radius and the constant $J_2 = 0.0010823$ is the measure of the Earth's oblateness. Let us show that this force component produces a perturbation in the right ascension of the ascending node.

Assume that our satellite S has crossed the ascending node and traveled an angle θ of its revolution, as illustrated in Figure 2.4. (Note here that we are doing spherical trigonometry with this figure, with triangles all being formed with great circles.) During a short time period Δt, the out-of-plane force component f_p will produce a slight velocity change $f_p \Delta t$ pointed towards the equator. This velocity change has a component perpendicular to the satellite velocity, which is

$$\Delta v = f_p \Delta t \cdot \sin \alpha$$

where α is the azimuth of the satellite heading direction. Due to this velocity change of Δv, the satellite heading direction will change as much as $\Delta \psi = (\Delta v)/v$. Then, by spherical trigonometry, a negative change in Ω arises as

$$\Delta \Omega = -\frac{\sin \theta}{\sin i}\Delta \psi = -\frac{\sin \theta}{\sin i}\frac{\sin \alpha}{v}f_p \Delta t \qquad (2.14)$$

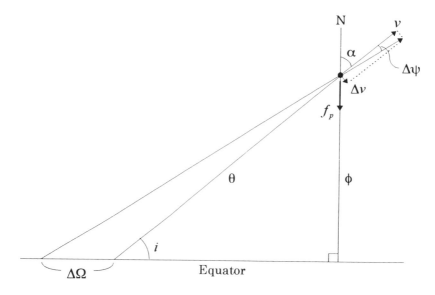

Figure 2.4 Perturbation in Ω.

Therefore the change in Ω per one orbital revolution becomes

$$\Delta\Omega = -\int_0^P \frac{\sin\theta}{\sin i} \frac{\sin\alpha}{v} f_p \, dt \qquad (2.15)$$

Now, replace dt with $d\theta$ by the relation $d\theta = (v dt)/r$ and use the spherical trigonometry formulas

$$\sin\alpha = \cos i/\cos\phi, \quad \sin\phi = \sin i \sin\theta$$

and (2.11); then (2.15) becomes

$$\Delta\Omega = -3\frac{J_2 r_E^2}{r^2}\cos i \int_0^{2\pi} \sin^2\theta \, d\theta$$

The definite integral in this equation is equal to π, so that we obtain

$$\Delta\Omega = -3\pi J_2 (r_E/r)^2 \cos i \qquad (2.16)$$

to estimate the perturbation of the node in radian per one revolution. Practically, this equals $\Delta\Omega = -9.96(r_E/r)^{3.5} \cos i$ deg per day. If the orbit is a polar orbit ($i = \pm 90$ degrees), its node does not move. For an orbit with an inclination between +90 degrees and −90 degrees, the node moves in the opposite sense against the direction of the satellite motion, which is often referred to as *nodal regression*.

While the inclination i also seems likely to change in Figure 2.4, it can be shown that this change will vanish after taking the integration over one revolution, so there will be no long-term change in the inclination.

A satellite revolving in a circular orbit has a constant angular momentum around the Earth's center. So, the behavior of an orbital plane is analogous to the motion of a spinning top. Imagine that you have a thin solid disc made of uniform material, and that this disc is supported at its center with a thin long thread suspended from the ceiling (see Figure 2.5). Given a proper initial motion, the disc will spin in a stationary plane at a constant inclination against the horizontal—regard this here as the equatorial—plane. If the thread supports the disc right at its center of mass, the spin plane will not move (that is, no perturbation). Let us then attach a short rod to the disc, just like an umbrella's handle, and put a small weight at the rod's end, as shown in the figure. This will operate so as to drive the disc's plane towards the horizontal plane, and so this is analogous to the oblateness-caused perturbing force that drives the orbital plane towards the equatorial plane (Figure 2.3). The spinning disc will then show a steady motion of the node, with its inclination being kept unchanged, and this represents exactly the perturbation of the orbital plane.

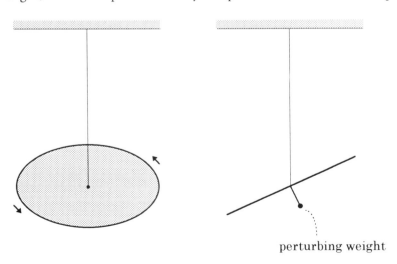

perturbing weight

Figure 2.5 Modeling the orbital plane perturbation.

Due to this analogy, the perturbation in Ω is sometimes referred to as the *precession* of node.

Problem 2.2

In our top analogy, the node will precess if the disc spins in a vertical plane, while actually the node of a polar orbit does not move. From where does this discrepancy originate?

One should not be perplexed by the node change for a near-equatorial orbit. Assume, in Figure 2.6, that our satellite has crossed node N. Equation (2.16) states that the node will be at some N' after one revolution, but this does not mean that the satellite should be found at N' after one revolution; actually, the satellite will be at S after one orbital period. For an equatorial orbit, particularly note that the node loses its definition so that one should forget about the node precession. More precisely speaking, the oblateness-caused perturbing force has other components than (2.13), and this makes the statellite reach some S' after one orbital period defined by (2.12). That is to say, the orbital period is also perturbed, and this perturbation depends on the inclination. The relationship between orbital radius and orbital period as defined by (2.12) thus needs a slight correction due to the perturbation. This may seem to cause a difficult situation—you select a particular orbital period for your communications objective, and then use (2.12) to find the orbital radius, and thus the found radius will differ from what it should actually be. Fortunately, this difference in orbital radius is so small as to be negligible from the viewpoint of communications services. So, what we should bear in mind in our orbital design is, after all, the node-precession due to the oblate Earth.

2.2.3 Multiple Satellite Formation

Using circular orbits for mobile communications proves its worth when a number of satellites are employed so as to provide a wide service coverage. The satellites have to be arranged in a proper orbital formation so as to assure continuous services over the coverage. Designing such a formation will be, if conceptually outlined, like the following.

The design starts with defining ϵ, the minimum satellite elevation angle that should be guaranteed for the communication service, and h, the satellite

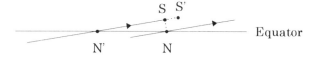

Figure 2.6 Nodal regression for near-equatorial orbit.

altitude. Given ϵ and h, the satellite has a ground service area that is bounded by a circle centered at the subsatellite point. We try to cover a wide region—presumably global—with a number of such service circles revolving round the Earth. Note that we may forget about the Earth's rotation if we consider a global coverage. Consider a circular orbit of altitude h and place the necessary number of satellites in this orbit at an even spacing so as to make a train of mutually overlapping service circles, as illustrated in Figure 2.7. That the satellites have identical altitudes is essential because the service circles then move at identical velocities of revolution according to (2.10) where $r = r_E + h$, and then a fixed overlapping geometry can be maintained of the service circles. Thus we have a service belt surrounding the Earth surface along a great circle. Making up the global coverage requires a number of such service belts. Since any two service belts cross each other at two places, redundancy in coverage-making is inevitable, while efforts can be made to minimize this redundancy by an adequate formation of the orbital planes. So that this orbital plane formation be conserved, the orbital plane perturbations due to (2.16) must appear in unison; therefore, the orbital inclinations must all be identical. Since the service belts become sparsest near the equator, a sufficient number of belts must be placed so as not to leave any space between them near the equator. Note that the necessary minimum separation between adjacent belts depends on whether the belts are corevolving (separation wider) or counter-revolving (separation narrower), because of the uneven width of the service belt.

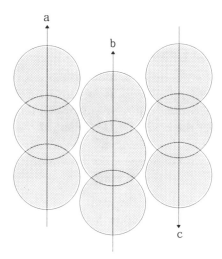

Figure 2.7 Corevolving (a,b) and counter-revolving (b,c) service belts.

Selection of the satellite altitude needs thorough consideration. Lower altitudes would be advantageous because of the lower propagation loss and the lower delay time for the communications, while they cause a drastic increase in the total number of satellites because of the smaller service-circle size. If the altitude becomes less than 1,000 km and approaches 500 km, the effect of the atmospheric drag will cause difficulty in orbital maintanenace. Due to the existence of the van Allen radiation belt, the altitude zones from 1,500 to 5,000 km and from 13,000 to 20,000 km should be avoided. Thinking out a formation complying with these conditions is very much a matter of trial and error, requiring some skill and imagination. However, no matter how complicated a multiple-satellite orbital formation may appear, its physical principle never goes beyond the reach of (2.10) and (2.16).

2.3 Elliptical Orbit

2.3.1 Orbital Shape

The satellite in a circular orbit undergoes its revolution at a fixed altitude and a fixed velocity, thereby providing communication services in a uniform manner along its orbit. In contrast, a satellite in an elliptical orbit can drastically vary its altitude and velocity during one revolution. Such a dynamism can be useful for developing an enhanced communication service for a particular ground area. In this section, we will study the dynamics of this potentially useful orbit, using the basic equations that have been prepared for circular orbits.

To start with, assume in Figure 2.8 that we have a satellite in a circular orbit A of radius r_0 and orbital velocity v_0. Now, imagine that the satellite

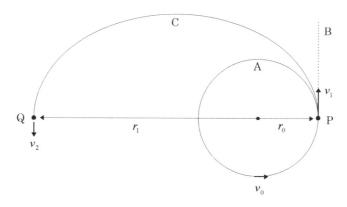

Figure 2.8 Orbital change by velocity boost.

velocity is boosted from v_0 up to v_1 instantaneously at some point P—then what would the new orbit be like? If v_1 is very large, the satellite will fly away along line B as if there were no Earth. On the contrary, if $v_1 = v_0$, the satellite simply remains in the initial orbit A. So, for a moderate magnitude of the boosted v_1, the new orbit should be something between A and B, more like C, with a finite maximum radius r_1. Prior to showing that C is an elliptical orbit, let us first examine how much the velocity boost will raise the maximum radius of the new orbit.

Between the basic equations (2.8) and (2.9) of the two-body problem, we first make use of (2.9). Its left-hand side multiplied by r makes $2r\dot{r}\dot{\theta} + r^2\ddot{\theta}$, which is identical to $d(r^2\dot{\theta})/dt$. So, we have

$$d(r^2\dot{\theta})/dt = 0$$

from which we can write

$$r^2\dot{\theta} = \text{const} = p \tag{2.17}$$

This quantity p has an important physical meaning. In Figure 2.9, a satellite is moving at a velocity v. During a short time interval dt, the satellite travels as far as $v\,dt$ and its component perpendicular to the radius vector of the satellite is $r\dot{\theta}dt$. Then, during this dt, the radius vector sweeps out an area equals to $(1/2)r \cdot r\dot{\theta}\,dt = (p/2)dt$. So, $p/2$ indicates the area swept out by the satellite radius vector per unit time. We will call this the *area-sweeping rate,* and (2.17) states that the area-sweeping rate of a satellite should remain a constant throughout the orbital revolution.

Going back to Figure 2.8, the area-sweeping rate for orbit C is evaluated at P; it is equal to $(1/2)r_0 v_1$. If Q is the point of maximum orbital radius— though we do not yet know where and when this Q takes place—the velocity v_2 at Q must be perpendicular to the radius r_1, so that the area-sweeping rate at Q is written as $(1/2)r_1 v_2$. Since these must be equal, we have

$$r_0 v_1 = r_1 v_2 \tag{2.18}$$

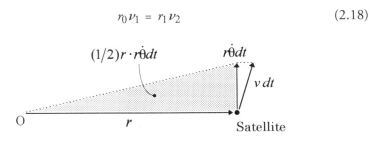

Figure 2.9 Area-sweeping rate.

Next, we apply the law of conservation of energy. Evaluating the energy (i.e., kinetic plus potential: $v^2/2 - \mu/r$) at P and at Q, and equating them, we have

$$\frac{v_1^2}{2} - \frac{\mu}{r_0} = \frac{v_2^2}{2} - \frac{\mu}{r_1} \qquad (2.19)$$

Solving (2.18) and (2.19) for r_1 and v_2 results in

$$r_1 = \frac{r_0}{2(v_0/v_1)^2 - 1} \qquad (2.20)$$

$$v_2 = v_1[2(v_0/v_1)^2 - 1] \qquad (2.21)$$

As the boosted velocity v_1 approaches $\sqrt{2}\, v_0$, the maximum radius r_1 from (2.20) goes to infinity, so that the satellite escapes away from the Earth. If v_1 is slightly below this *escape velocity*, orbit C will have a raised but finite maximum radius, and the satellite will stay longer near the maximum radius point because of a small v_2 by virtue of (2.21). Simply owing to this mechanism, communication services using elliptical orbits can have a prolonged service time at higher elevation angles.

Problem 2.3

Assuming a circular orbit of 1,000-km altitude, find the magnitude of velocity boost that is necessary for raising the maximum radius to the geostationary radius of 42,164 km.

Let us now examine the exact shape of the new orbit C. Use $\dot{\theta} = p/r^2$ from (2.17) and substitute this into (2.8), and we have

$$\ddot{r} = -\frac{\mu}{r^2} + \frac{p^2}{r^3} \qquad (2.22)$$

Solving this equation needs some change in technique. Introduce a fictitious quantity u by $r = 1/u$, where $u > 0$. Express $\ddot{r}$ in terms of u, by using (2.17), as

$$\frac{dr}{dt} = -\frac{1}{u^2}\frac{du}{dt} = -\frac{1}{u^2}\frac{du}{d\theta}\frac{d\theta}{dt} = -p\frac{du}{d\theta}$$

and

$$\frac{d^2 r}{dt^2} = -p\frac{d^2 u}{d\theta^2}\frac{d\theta}{dt} = -p^2 u^2\frac{d^2 u}{d\theta^2}$$

Then, (2.22) is rewritten as

$$\frac{d^2 u}{d\theta^2} + u = \frac{\mu}{p^2} \qquad (2.23)$$

Thus the motion of u against θ is found to be a harmonic oscillation, which can be solved as

$$u = \frac{\mu}{p^2} + q\cos(\theta + \omega)$$

with arbitrary constants q and ω. One can assume that $q > 0$ by properly writing ω. Then,

$$0 \le q < \mu/p^2 \qquad (2.24)$$

must hold in order that $u > 0$ should hold. Coming back from u to r, the orbital shape is now described by

$$r = \frac{p^2/\mu}{1 + (p^2/\mu)q\cos(\theta + \omega)} \qquad (2.25)$$

Let us show that this makes an ellipse.

If satellite S plots an ellipse, as illustrated in Figure 2.10, S must move in such a way that the sum of the distances from S to two focal points is kept constant, as

$$r + r' = 2a \qquad (2.26)$$

where $2a$ is the major axis of the ellipse and the focal points are separated from each other by $2ae$. Set $r' = 2a - r$ from (2.26) and substitute this into the trigonometric relation of

$$(r')^2 = r^2 + (2ae)^2 + 2r(2ae)\cos\theta$$

Then, we have

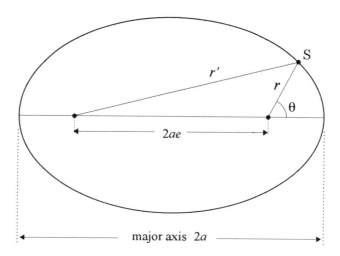

Figure 2.10 Definition of ellipse; $r + r' = 2a$.

$$r = \frac{a(1 - e^2)}{1 + e \cos \theta} \tag{2.27}$$

to formulate an ellipse with one of the focal points at the origin. This equation is identical to (2.25) if we allow for ω as defining the orientation of the major axis and if we put

$$e = (p^2/\mu)q \tag{2.28}$$

and

$$a(1 - e^2) = p^2/\mu \tag{2.29}$$

Note that $0 \le e < 1$ by (2.24). The radius r from (2.27) takes its minimum and maximum of

$$r_{min} = a(1 - e), \; r_{max} = a(1 + e) \tag{2.30}$$

at $\theta = 0$ (perigee) and at $\theta = \pi$ (apogee), respectively. So, the parameter e indicates to what degree the orbital shape differs from a circle, while a, being equal to $(r_{min} + r_{max})/2$, measures the size of the orbit. Thus a and e, respectively called the *semimajor-axis* and *eccentricity*, define the size and shape of an elliptical orbit. The parameter ω is called the *argument of perigee*, and it defines the orientation of the perigee, normally measured from the ascending node. Thus an elliptical orbit defined by a, e, and ω will be arranged in its orbital plane, for example as illustrated in Figure 2.11, where the orbital plane intersects with the equatorial plane at the line connecting the ascending-descending nodes. The orientation of the orbital plane is defined again by i and Ω in the same manner as for the circular orbit. So, we have the parameters a, e, i, Ω, and ω specifying the size, shape, and orientation of an elliptical orbit in the inertial space.

Problem 2.4

Confirm that the above discussion about elliptical orbits reduces to that of circular orbits if $e = 0$.

2.3.2   Satellite Position as a Function of Time

Though we have defined the exact shape of the orbit, still we do not know the satellite motion in this orbit as a function of time. This is because the time argument t has disappeared from (2.23) while we were solving the equations. In what way should we know the satellite motion?

First, finding out the oribital period is not difficult. An ellipse specified by a and e has the area of

$$A = \pi a^2 \sqrt{1 - e^2}$$

From (2.29), the area-sweeping rate of the satellite is

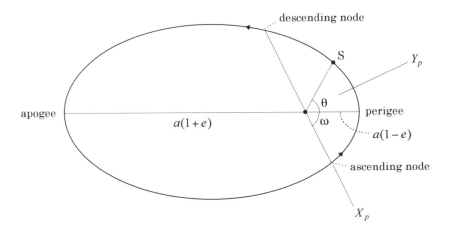

Figure 2.11 Elliptical orbit in its orbital plane.

$$p/2 = \sqrt{\mu a(1 - e^2)}/2$$

The orbital period, which is the time spent for sweeping out the whole elliptical area, is therefore

$$P = A/(p/2) = 2\pi\sqrt{a^3/\mu} \qquad (2.31)$$

Comparing this with (2.12) clarifies that the semimajor-axis a determines the orbital period in the same way as the radius r did for the circular orbit.

Now let us figure out the satellite position-time relationship. In Figure 2.11, the angle θ indicates the satellite motion of revolution as measured from the perigee—this angle has a particular name of *true anomaly*. We want to express the true anomaly θ as a function of time t. The standard procedure for doing this, as most textbooks teach, is like the following: First convert t into a fictitious quantity m, which is called the *mean anomaly*. The mean anomaly m varies from zero to 2π linearly with the lapse of time after the perigee passing over one orbital period. From this m, find another fictitious quantity E called the *eccentric anomaly*, by the equation

$$m = E - e\sin E$$

This equation, known as *Kepler's equation,* cannot be solved analytically whenever $e \neq 0$, so that numerical techniques must be employed for its solution. Then, from the solved E, some trigonometric manipulations yield θ, thus establishing the rigorous relationship between t and θ. The actual procedure can be found, for example, in [1] or [2].

The computing code for the above-cited standard procedure will amount, if printed out, to not much more than a page or so. This would, however, attract few readers to this book. If our interest lies mainly in the conceptual design of orbital systems, a more understandable and easily applicable procedure would be welcome to replace this rigorous but cumbersome procedure. So as to meet this requirement, a simplified procedure is presented in the following.

Given an orbit with a and e specified, let us derive the relationship $\theta(t)$ directly from the law of conservation of area-sweeping rate. Consider, in Figure 2.12, satellite position samples (r_i, θ_i) for $i = 0, 1, \ldots, N$ over one revolution, where the θ_i values are equally separated in true anomaly so that $\theta_i = i\Delta\theta$ with $\Delta\theta = 2\pi/N$. The sectorial segment of the ellipse between $\theta = \theta_{i-1}$ and $\theta = \theta_i$ has an area that approximates to

$$A_i = (1/2)\, r_{i-1}\, r_i \Delta\theta$$

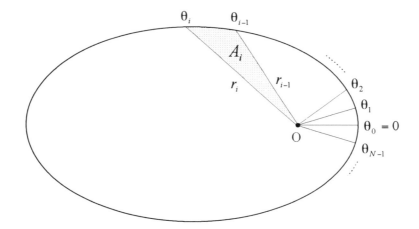

Figure 2.12 Finding the θ-t relationship.

where the r values are calculated with (2.27). Sum up all the sectorial areas to evaluate the total area

$$A_T = \sum_{i=1}^{N} A_i$$

from which we evaluate the area-sweeping rate

$$(p/2) = A_T/P$$

using the orbital period of (2.31). Then, the time needed for the satellite to travel from θ_{i-1} to θ_i is

$$t_{i-1,i} = A_i/(p/2) = (A_i/A_T)P$$

So, if the satellite passed the perigee at $t = t_p$, the time t_i of passing the point $\theta = \theta_i$ will be

$$t_i = \sum_{k=1}^{i} t_{k-1,k} + t_p = P\sum_{k=1}^{i} A_k/A_T + t_p$$

Thus, a "timetable" for one orbital revolution is obtained as (t_i, θ_i), $i = 1, 2,$ ..., N. When the satellite goes into next revolution, the time t_i is, of course, replaced by $t_i + P$. For an arbitrarily specified time, the satellite position will be interpolated out of the timetable. The coding for our procedure is simple enough, since it is nothing more than summing up and normalizing the areas A_i. Note, however, that the accuracy of our timetable depends on the number of the sample points N and on the eccentricity e. Table 2.1 evaluates the timetable error, where the θ_i values have been checked against their rigorous values at t_is. Given e, a proper N must be selected so as to meet the required accuracy.

Our timetable has a direct application if one wishes to view a satellite's motion on a computer screen. Display one position of the satellite for $\theta = \theta_{i-1}$. Place the wait time of $t_{i-1,i}$, and then go to the display of next position for $\theta = \theta_i$, and so proceed. The satellite on the screen will then show physically exact orbital motions.

It has been pointed out qualitatively, from the discussions over Figure 2.8, that a satellite in an elliptical orbit should stay longer in the near-apogee region. This can be analyzed now with our timetable approach. Find, from the timetable, the length of the period of time that true anomaly falls within $\pm\theta$ of the apogee. This period of time is then evaluated in ratio to one orbital period, and its result becomes as shown in Figure 2.13. Increasing the eccentricity thus actually produces the near-apogee concentration of the time of stay. Our timetable is particularly suitable for this kind of analysis because the sample points θ_i are uniformly distributed.

Now that we have exactly figured out the orbital motion, was our former discussion with Figure 2.8 redundant with a dead end? Definitely not. What we discussed over Figure 2.8 provides us with a clear image that every elliptical orbit comes from its original circular "parking" orbit through a velocity boost, and this image helps us to understand the satellite motion near perigee. Many books and materials on orbital mechanics show figures of elliptical orbits, and not infrequently these figures turn out to be incorrect if looked at closely, in that minimum radius does not actually take place at perigee. Bearing in your mind the image of Figure 2.8 that the orbital ellipse is osculating at its perigee

Table 2.1
Mean Anomaly Timetable Error (millidegree)

Sample Points N	Eccentricity e			
	0.5	0.6	0.7	0.8
200	6.5	9.2	14	22
400	1.6	2.3	3.4	5.6
600	0.7	1.0	1.5	2.5
800	0.4	0.6	0.8	1.4

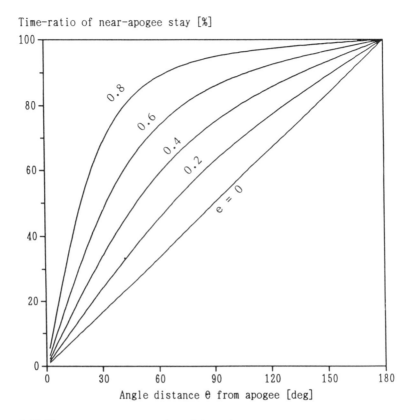

Figure 2.13 Near-apogee concentration of time of stay.

to its original circular parking orbit, you will automatically draw a correct figure of the elliptical orbit, thereby correctly imagining the near-perigee motion of the satellite.

To end this section, a comment on *orbital elements* will be helpful. We have so far learned five parameters to specify the orbit: a, e, i, Ω, and ω, while there should be six orbital elements. At most sites of orbital operations, the sixth element is the mean anomaly m, while this is simply due to historical consequence. Specifying the mean anomaly is equivalent to stating when the satellite passed, or will pass, the perigee. So, it will be suitable here to define the orbital elements as a, e, i, Ω, ω, and t_p, with t_p, the perigee passing time.

2.3.3 Perturbations

The Earth's oblateness also causes perturbations to elliptical orbits. Its analysis is, in principle, the same as for the circular orbit. During a short time dt, a perturbing force f_p produces a slight change to an orbital element, for example as

$$d\Omega = \alpha f_p dt$$

through a sensitivity coefficient α. The time argument t is then replaced by the true anomaly θ, so that the integral over one revolution becomes

$$\Delta\Omega = \int_0^{2\pi} \alpha(\theta) f_p(\theta) \frac{r}{\nu} d\theta$$

The perturbing force must be expressed in three (i.e., r, θ, and out-of-plane) components, with each having its own sensitivity coefficient, and the integrations for the three components are then summed up. This procedure for elliptical orbits is quite cumbersome, so we will see only the results. The perturbations that appear in the long term are

$$\Delta\Omega = -3\pi J_2 \frac{r_E^2}{a^2(1 - e^2)^2} \cos i$$

$$\Delta\omega = 3\pi J_2 \frac{r_E^2}{a^2(1 - e^2)^2}\left(2 - \frac{5}{2}\sin^2 i\right)$$

in radians per revolution. Thus the long-term perturbation appears only in node and perigee, so that the orbital shape and size are both kept unchanged. That the perigee moves—and so does the apogee—will cause difficulty to communication services. It is fortunate, however, that the perigee perturbation vanishes at the particular inclination of $i = 63.4$ degrees. Note that this particular inclination does not depend on the size, mass, and oblateness of the Earth. If, in the future, we discover an unknown planet, and if this planet has a satellite in an elliptical orbit with the particular inclination of 63.4 degrees, then most probably the satellite is an artificial satellite providing regional services to the planet's surface.

2.3.4 Multiple Satellite Formation

No matter how the near-apogee stay time may be prolonged, a single satellite in an elliptical orbit cannot provide a continuous communication service to a ground user. So, similar to the case of circular orbit, the need arises to employ a number of satellites to construct an orbital formation, and how to design such a formation will be seen from the following example.

Selecting the particular inclination of 63.4 degrees so as to make the apogee-perigee orientation fixed is a prerequisite for the formation design, provided that we are not going to set $i = 0$ so as to serve equatorial regions. This does not mean, however, that the apogee must come above the geographical latitude of 63.4 degrees. An orbit with a fixed size and shape with the fixed $i = 63.4$ degrees can still vary its geometry by changing its perigee orientation ω. Selecting ω to either 90 degrees or 270 degrees brings the apogee latitude to ±63.4 degrees, while ω selected to 0 or 180 degrees brings the apogee onto the equator. So, the apogee latitude can be any within ±63.4 degrees.

A number of orbits of identical a, e, i, and ω are then arranged with different Ω values—a typical example for this is illustrated schematically in Figure 2.14. In this example, three satellites—A, B, and C—make a continuous communication service to a ground area represented by point G. (Here the numbers affixed indicate the motions of satellites and ground area). While G

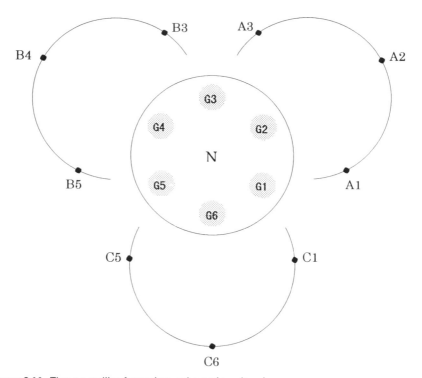

Figure 2.14 Three-satellite formation; schematic polar view.

moves from G1 to G3 as the Earth rotates, satellite A undergoes its orbital motion from A1 to A3 via apogee A2. By a proper selection of the perigee orientation ω, the apogee A2 will be above G2, so that the users around G enjoy the merit of the satellite at near apogee. When satellite A goes down to arrive at A3, it will hand over its communication service to satellite B, which has just come up to B3; then satellite B, during its motion from B3 to B5, will serve G, which then moves from G3 to G5; and so it proceeds. In this manner the communication service for G will continue without break, with each satellite operating in turn for 8 hours.

Some conditions are necessary for such a system to be workable. Each time G comes to G2, satellite A must come up to the apogee A2. So the orbital period should be either P = one day, 1/2 day, 1/3 day, and so on (here "day" is precisely for a sidereal day, which is 23h 56m 4s). Correspondingly, by (2.31), the semimajor-axis must be of a_1 = 42,164 km, a_2 = 26,562 km, a_3 = 20,270 km, and so on. Let h_p and h_a be the height of perigee and apogee, respectively. A lower perigee is advantageous for obtaining a larger eccentricity, while too low a perigee is impractical due to the presence of atmospheric drag; so, assume here h_p = 1,000 km. Then, from the relation

$$a = \frac{r_{max} + r_{min}}{2} = \frac{h_p + h_a}{2} + r_E$$

apogee height must be h_a = 70,572 km (for a_1), 39,367 km (for a_2), 26,781 km (for a_3), and so on. The first one appears too high, so select the second h_a = 39,367 km as reasonable. This specifies the eccentricity, through (2.30), as

$$e = (r_{max} - r_{min})/(2a) = (h_a - h_p)/(2a) = 0.722$$

and the revolution period is 11h 58m. The time of stay of the satellite in the operational orbital arc from A1 through A3 is 8h, which is 67% of one revolution. Then, owing to the near-apogee concentration indicated by Figure 2.13, the operational orbital arc falls within ±37 degrees of the apogee, and this will make the operational arc appear well above the horizon at G (how to calculate actually the satellite elevation will be shown at the end of this chapter). However, near the switching points of A3 and B3, and at other switching points as well, the satellite elevation will unavoidably go down. This becomes less if the apogee latitude and the service area latitude are both high enough. So, this kind of orbital formation is suitable particularly for a service to high-latitude areas, and such uses of elliptical orbits are referred to as *Molniya orbits*.

If the apogee latitude is selected to be its maximum of 63.4 degrees, with ω = 90 degrees or 270 degrees, the formation can offer a special advantage. A slight adjustment of the eccentricity around the above-referred e = 0.722 makes the satellite-switching points A3 and B3 coincide with each other, and all other pairs of switching points as well. This means that the satellite line-of-sight direction and the satellite range are both continuous at the moment of satellite switching, thus enabling good maintenance of communication links.

While the ω values were all identical in the above-discussed formation, other kinds of formations with unequal ω values may be possible to provide a wider coverage. Even more sophisticated formations employing circular and/or geostationary orbits in combination can theoretically exist—thinking out such orbital formations is similar to the circular orbit's case (inventive work relying on imagination).

2.4 Geostationary Orbit

2.4.1 Communication Link Stability

We have observed that a stable communication link cannot be provided by satellites in circular and elliptical orbits since the link must be switched regularly from one satellite to another. Providing a stable link is the unique merit of the geostationary satellite owing to its fixed position relative to the Earth. However, no satellites can be absolutely stationary—perturbations make the orbit gradually lose its initial stationary condition, thus making the satellite drift away. Consequently, the satellite needs orbital corrections from time to time so as to maintain its stationary position, which is referred to as *stationkeeping*. The established practice for a communication satellite is to keep its position within the limit of deviation of 0.1 degree in latitude and longitude from its assigned nominal position. A satellite motion taking place inside this limit of stationkeeping is negligible if viewed from a mobile user's antenna with broad directivity, while it will be of interest if it causes range variations that may affect the signal transmission (particularly digital). Examining the stability of range needs proper modeling of the satellite motion, and this will be answered in the following study.

2.4.2 Near-Geostationary Orbital Motion

If we had an ideally geostationary satellite, its orbit must be circular, equatorial, and synchronous to the Earth's rotational period. This period is 23h 56m 4s, so that the orbital radius must be, by (2.12), r_0 = 42,164.2 km. Consider such an ideal satellite as fictitiously marking the nominal stationary position, and consider our actual satellite as moving in close proximity around the nominal position, as illustrated in Figure 2.15. Our satellite's position is then measured relative to the nominal position, radially in R and tangentially in L. In the Earth-centered polar coordinates, the nominal position is located at

$$r = r_0, \quad \theta = \Psi_0 t$$

where $\Psi_0 = 7.292115 \times 10^{-5}$ (rad/sec) is the Earth's rotation rate, with $r_0 \Psi_0^2 = \mu/r_0^2$. The actual satellite is then located, for small R and L, at

$$r = r_0 + R, \quad \theta = \Psi_0 t + L/r_0 \tag{2.32}$$

Now consider the motion of this (r, θ) as obeying the two-body motion equations (2.8) and (2.9). First, substituting (2.32) into (2.8) makes

$$\ddot{R} - (r_0 + R)(\Psi_0 + \dot{L}/r_0)^2 = -\mu/(r_0 + R)^2 \tag{2.33}$$

Assume here that time derivatives of R and L are also small; then, higher order small terms such as $R\dot{L}$ are neglected, and the right-hand side is approximated as

$$-\mu(r_0 + R)^2 \approx -\mu/r_0^2 + 2(\mu/r_0^3)R = -\mu/r_0^2 + 2\Psi_0^2 R$$

so that we have

$$\ddot{R} - 2\Psi_0 \dot{L} - 3\Psi_0^2 R = 0 \tag{2.34}$$

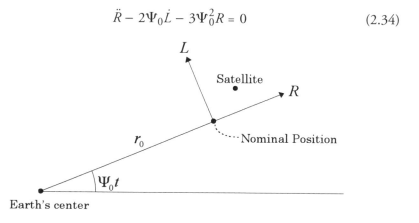

Figure 2.15 Satellite position relative to its nominal position, in equatorial plane.

In the same manner, substituting (2.32) into (2.9) yields

$$\ddot{L} + 2\Psi_0 \dot{R} = 0 \tag{2.35}$$

These results are for the motions taking place in the equatorial plane. Actually however, the satellite may slightly displace out of this plane, and this motion is measured along the Z axis, which is perpendicular to (R, L) plane, as illustrated in Figure 2.16. Consequently, the right-hand side of (2.33) must change to $-\mu/[(r_0 + R)^2 + Z^2]$; but Z is small compared to r_0 so that Z^2 may be discarded as a higher order term. Therefore (2.34) and (2.35) need not change their forms if we consider Z motions. The Earth's gravitational force has its Z component, which approximates

$$-\frac{\mu}{(r_0 + R)^2 + Z^2} \frac{Z}{\sqrt{(r_0 + R)^2 + Z^2}} \approx -(\mu/r_0^3)Z = -\Psi_0^2 Z$$

so that the motion along Z will obey

$$\ddot{Z} + \Psi_0^2 Z = 0 \tag{2.36}$$

Thus the satellite motion in (R, L, Z) coordinates is governed by (2.34–36) for a two-body problem. If we consider perturbing forces in (R, L, Z) components,

$$\ddot{R} - 2\Psi_0 \dot{L} - 3\Psi_0^2 R = f_R \tag{2.37}$$

$$\ddot{L} + 2\Psi_0 \dot{R} = f_L \tag{2.38}$$

$$\ddot{Z} + \Psi_0^2 Z = f_Z \tag{2.39}$$

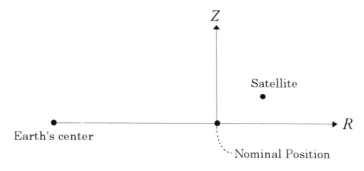

Figure 2.16 Out-of-plane motion.

are the motion equations. (Originally, these equations were derived for studying rendezvous guidance and control of two satellites—see for example [2].)

Now, assume a two-body problem; the satellite motion is then solved as

$$R = -\frac{2}{3\Psi}D + E_x \cos\Psi_0 t + E_y \sin \Psi_0 t \tag{2.40}$$

$$L = L_0 + Dt - 2E_x \sin \Psi_0 t + 2E_y \cos\Psi_0 t \tag{2.41}$$

$$Z = I_x \cos\Psi_0 t + I_y \sin \Psi_0 t \tag{2.42}$$

where L_0, D, E_x, E_y, I_x, I_y are arbitrary constants that specify the orbital motion as follows: L_0 and D indicate a linear drift motion in L, and this motion arises if the radius is offset from the synchronous radius—see (2.40). E_x and E_y indicate an elliptical motion in the (R, L) plane, and this is related to a small eccentricity of the orbit—see how the radius R oscillates by (2.40), and compare it with (2.30), to find the relation

$$\sqrt{E_x^2 + E_y^2} = r_0 e \tag{2.43}$$

The values I_x and I_y indicate a sinusoidal motion along Z, which corresponds to a small inclination, with the relation

$$\sqrt{I_x^2 + I_y^2} = r_0 i \tag{2.44}$$

Thus the six parameters $(L_0, D, E_x, E_y, I_x, I_y)$ serve as orbital elements for a near-geostationary satellite.

Problem 2.5

What is the advantage of using $(L_0, D, E_x, E_y, I_x, I_y)$ over the use of Kepler's orbital elements for specifying near-geostationary orbits?

2.4.3 Motion of a Station-Kept Satellite

Geostationary satellites suffer various perturbations. Gravitational forces of the Sun and the Moon produce a diurnally sinusoidal force f_Z in (2.39), which makes Z grow in its oscillation amplitude, thus increasing the inclination. The Earth's equatorial cross-section is slightly distorted from being a circle; this produces a constant force f_L in (2.38), and this force then gives rise to an accelerated drift in L. The solar radiation pressure produces diurnally sinusoidal forces f_R and f_L in (2.37) and (2.38), so as to make R and L grow in their oscillation amplitudes, thus increasing the eccentricity. Consequently, the orbital parameters will change gradually with time, so that the satellite drifts away from its initially established nominal position. In order to counteract these perturbations, impulsive forces f_L and f_Z are applied to the satellite from time to time, with proper planning so as to maintain a condition

$$|L|, |Z| \le B \tag{2.45}$$

where

$$B = r_0 \sin 0.1 \text{ deg} = 73.6 \text{ km}$$

for the standard 0.1 degree keeping. In order to maintain this condition of stationkeeping, the L oscillation in (2.41) must be such that $2\sqrt{E_x^2 + E_y^2} < B$. Then, by (2.43), $e < B/(2r_0) = 0.00087$ is necessary for the 0.1 degree keeping. In practice, eccentricity must be kept enough smaller than this so as to allow for an L drift motion due to L_0 and D, while inclination can be any within 0.1 degree.

How to carry out the actual stationkeeping is explained in the references—concisely in [3] or thoroughly in [4]—so we will go on to examine how the range of a station-kept satellite varies with time.

2.4.4 Range Rate and Its Variation

Consider a ground user U, as illustrated in Figure 2.17, located by the offset angles α and β from the Earth's center as viewed from the nominal satellite position. As the satellite displaces from its nominal position to a point (R, L, Z), the satellite range at U will deviate from its nominal ρ_0 to

$$\rho = \rho_0 + c_R R + c_L L + c_Z Z$$

where

$$c_R = \cos \beta \cos \alpha, \quad c_L = -\cos \beta \sin \alpha, \quad c_Z = -\sin \beta$$

are range-deviation sensitivity coefficients. Then, range rate is expressed by

$$\dot{\rho} = c_R \dot{R} + c_L \dot{L} + c_Z \dot{Z}$$

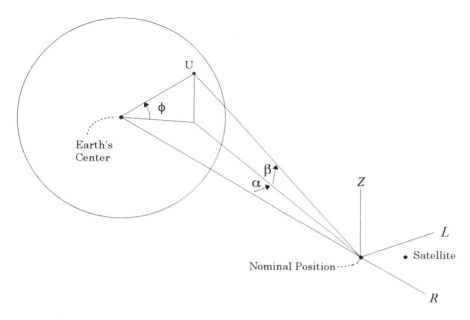

Figure 2.17 Satellite/user geometry. Angles α and β are measured along and vertically to the equatorial plane for locating the user U.

and into this equation we substitute (2.40–42). After arranging the terms, we have

$$\dot{\rho} = c_L D + \sqrt{c_R^2 + 4c_L^2}\, \Psi_0 \sqrt{E_x^2 + E_y^2} \sin(\Psi_0 t + \theta_1)$$

$$+ c_Z \Psi_0 \sqrt{I_x^2 + I_y^2} \sin(\Psi_0 t + \theta_2) \qquad (2.46)$$

where θ_1 depends on E_x and E_y, and θ_2 on I_x and I_y. Though the orbital parameters are gradually changing by perturbations, they can be thought of as constants during a day. The range rate by (2.46) then comprises one constant and two sinusoidal terms. We will examine the peak-to-peak width of the variation in range rate; so, discard the constant term. Since α and β are both within ± 8.7 degrees, approximately $\sqrt{c_R^2 + 4c_L^2} \approx 1$ and $c_Z \approx r_E \sin \phi / r_0$, where r_E is the Earth's radius and ϕ the latitude of U. Then substitute (2.43) and (2.44) into (2.46) to obtain

$$\dot{\rho} = e r_0 \Psi_0 \sin(\Psi_0 t + \theta_1) + i r_E \Psi_0 \sin \phi \sin(\Psi_0 t + \theta_2)$$

The relationship between θ_1 and θ_2 is hidden in the policy of stationkeeping that proceeds at the satellite control center, so that we cannot generally know whether the two sinusoidal terms add or cancel each other; we can simply estimate the possible minimum and maximum of the $\dot{\rho}$ value's oscillation amplitude. Hence we conclude the amplitude of the diurnal range-rate oscillation due to the satellite motion is greater than $\Psi_0|er_0 - ir_E \sin|\phi||$ and less than $\Psi_0|er_0 + ir_E \sin|\phi||$.

Problem 2.6

Assuming $e = 0.0001$, $i = 0.1$ degree, and $\phi = 45$ degrees, estimate the diurnal range-rate oscillation amplitude.

    If the user is in motion at a velocity ν towards an arbitrary heading direction, it can cause a range rate of as much as $\pm\nu \cos \epsilon$, where ϵ is the satellite elevation. Thus we know, referring to the answer to Problem 2.6: If the user's motion is faster than a usual walking speed, then it is not the satellite's motion but the user's that will specify the stability of the satellite range—so stable is the geostationary transmission link.

2.5 Calculating the Range, Azimuth, and Elevation

While we have learned by now the most basic concepts of orbital dynamics for discussing mobile communication services, some readers may have been interested in trying an orbital design. So, finally, how to calculate, practically, the range, azimuth, and elevation of a satellite is shown in the following, and it is a simplified procedure that assumes a two-body problem and uses a compact angle calculation algorithm.

    Let $(a, e, i, \Omega, \omega, t_p)$ be the orbital elements of the satellite, with t_p the perigee-passing time. Time is measured from an arbitrarily defined origin of $t = 0$. First, calculate the orbital period by

$$P = 2\pi\sqrt{a^3/\mu} \ [\text{sec}]$$

with $\mu = 398,600.5 \text{ km}^3/\text{s}^2$, and next prepare the true anomaly timetable from (a, e). Then, for each sample point (θ, t) of this timetable, the following steps are applied.

Step 1

Locate the satellite in "orbital-plane coordinates" (X_P, Y_P)—see Figure 2.11. The X_P axis points towards the ascending node, and the Y_P axis is perpendicular to X_P axis. The satellite location is

$$X_P = r\cos(\theta + \omega), \; Y_P + r\sin(\theta + \omega)$$

with $r = a(1 - e^2)/(1 + e\cos\theta)$

Step 2

Convert (X_P, Y_P) to "node-equatorial coordinates" (X', Y', Z')—see Figure 2.18. The X' axis is identical to the X_P axis, while (X', Y')-plane is equatorial and the Z' axis passes through the Earth's north pole. When viewed along the X' axis, the Y_P axis makes an inclination i against the Y' axis. Hence the conversion is

$$X' = X_P, \; Y' = Y_P\cos i, \; Z' = Y_P\sin i$$

Step 3

Convert (X', Y', Z') to *Earth-fixed coordinates* (X, Y, Z)—see Figure 2.19. The X and Y axes are fixed to the Earth so that the (X, Y) plane is the equatorial plane. The Z axis is identical to the Z'-axis, and the (X, Z)-plane contains Greenwich-meridian. Let ψ denote the Earth rotation angle; that is, the orientation angle of the X axis as measured from the vernal equinox. This angle varies with time, as

$$\psi = \Psi_0 t + \psi_0$$

where $\Psi_0 = 7.292115 \times 10^{-5}$ [rad/sec] is the Earth rotation rate. The constant ψ_0, which denotes the Earth orientation angle at $t = 0$, can be chosen arbitrarily; so, this is a design parameter specifying an orbit Earth geometry. The conversion proceeds as

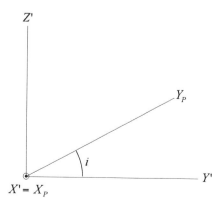

Figure 2.18 Node-equatorial coordinates (X', Y', Z').

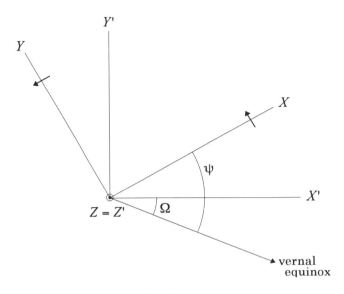

Figure 2.19 Earth-fixed coordinates (X, Y, Z).

$$X = X' \cos(\psi - \Omega) + Y' \sin(\psi - \Omega)$$

$$Y = -X'\sin(\psi - \Omega) + Y' \cos(\psi - \Omega)$$

$$Z = Z'$$

If the satellite is stationary at longitude λ_s, then forget all that has been stated and simply set the satellite position to

$$X = r_0 \cos \lambda_s, \ Y = r_0 \sin \lambda_s, \ Z = 0$$

with $r_0 = 42,164$ km—the sychronous radius.

Step 4

Consider a ground user at latitude ϕ and longitude λ. The range to a satellite at (X, Y, Z) is

$$\rho = \sqrt{(X - r_E \cos \phi \cos \lambda)^2 + (Y - r_E \cos \phi \sin \lambda)^2 + (Z - r_E \sin \phi)^2} \tag{2.47}$$

where $r_E = 6,378$ km is the Earth's radius. The Earth is assumed to be spherical; this approximation may cause an error in range less than a few tens of kilometers and errors in azimuth and elevation both less than 0.1 degree.

Step 5

Calculating the azimuth and elevation normally requires still more work with coordinate conversions, but this can be avoided by the following simplified approach.

Consider (2.47) as defining the range as a function of r_E, ϕ, and λ, as

$$\rho = f(r_E, \phi, \lambda)$$

Consider the user's reference frame: Its origin is at the user and the three axes point towards the east, north, and up. For each reference axis, calculate the satellite's direction cosine by means of numerical differentiation as

$$\text{east: } l_E = \frac{f(r_E, \phi, \lambda - \Delta\lambda) - f(r_E, \phi, \lambda + \Delta\lambda)}{2r_E\Delta\lambda\cos\phi} \tag{2.48}$$

$$\text{north: } l_N = \frac{f(r_E, \phi - \Delta\phi, \lambda) - f(r_E, \phi + \Delta\phi, \lambda)}{2r_E\Delta\phi} \tag{2.49}$$

$$\text{up: } l_U = \frac{f(r_E - \Delta r_E, \phi, \lambda) - f(r_E + \Delta r_E, \phi, \lambda)}{2\Delta r_E} \tag{2.50}$$

What these equations mean will become clear from an example: Suppose the satellite is due above the user and examine (2.50). The user's up-down displacements of $\pm\Delta r_E$ will make the range change as much as $\pm\Delta r_E$, so that the up-direction cosine equals one, while the east and north-direction cosines are zero. The dividing factors $r_E\Delta\lambda\cos\phi$, $r_E\Delta\phi$, and Δr_E in (2.48–50) must be properly chosen in their magnitudes, typically a few hundred meters.

Step 6

Thus given the direction cosines, calculate

$$\text{elevation} = \pi/2 - \cos^{-1}l_U, \quad \text{azimuth} = \tan^{-1}(x = l_N, y = l_E)$$

where azimuth is defined to be zero for north and $\pi/2$ for east.

The above-presented procedure should not be applied to a long period calculation of many orbital revolutions, since the perturbation has been neglected. For more precise orbital calculations taking account of relevant perturbations, visit the Web site

http://www.crl.go.jp/ut/orbit

where visitors can do various kinds of orbit calculations.

References

[1]   Bate, R. R., D. D. Mueller, and J. E. White, *Fundamentals of Astrodynamics*, Chapter 4, New York: Dover, 1971.

[2]   Prussing, J. E., and B. A. Conway, *Orbital Mechanics*, Chapter 2 and Chapter 8, New York, NY: Oxford, 1993.

[3]   Agrawal, B. N., *Design of Geosynchronous Spacecraft*, Chapter 2, Englewood Cliffs, NJ: Prentice-Hall, 1986.

[4]   Soop, E. M., *Handbook of Geostationary Orbits*, Norwell, MA: Kluwer, 1994.

3

System Design

3.1 System Configuration

Figure 3.1 shows the basic configuration of mobile satellite communication systems. The system consists of three basic segments: a satellite, a gateway Earth station and a mobile Earth station (Figure 1.1). From the standpoint of system design, a propagation path has to be added as the fourth segment. In mobile satellite communication systems, the propagation path is a very important factor that mainly affects the channel quality of the communication system. In land mobile satellite communications, the most serious propagation problem is the effect of blocking caused by buildings and surrounding objects, which cause signals from the satellite to shut down completely. The second problem is shadowing caused by trees and foliage, which results in signal attenuation. The third is multipath fading, which is mainly caused by buildings. However, this effect can usually be ignored because of the use of directional antennas and the great attenuation of reflected signals. In maritime satellite communications, fading caused by sea surface reflection is the most serious propagation problem. Rain attenuation has to be considered in higher frequency bands such as the Ka band and millimeter-wave band. However, it can be neglected in the L band.

A gateway and a mobile Earth station can be broken down into an antenna, a diplexer (DIP), a set of upconverters and downconverters (U/C and D/C), a high-power amplifier (HPA) and a low-noise amplifier (LNA), and a set of a modulator (MOD) and a demodulator (DEM). The configuration for a satellite is almost the same as for gateway and mobile Earth stations and can be broken down into an antenna and a set of upconverters and downconverters, and a set of this onboard equipment is called a transponder. Almost all of the

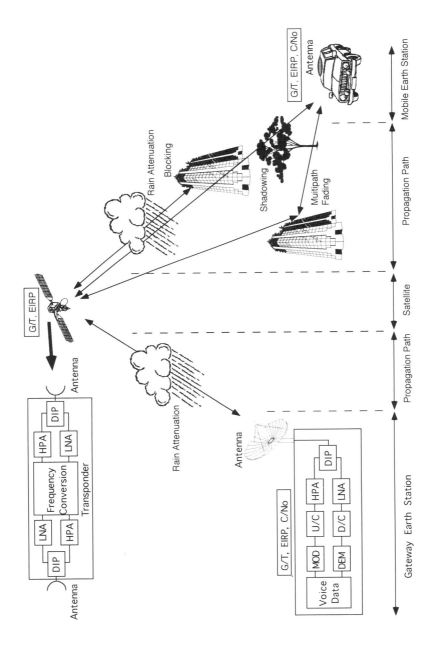

Figure 3.1 System configuration of mobile satellite communications.

present commercial satellites do not have a set of modulators and demodulators. They can only transmit a signal after converting the frequency and amplify received weak signals. This type of transponder used on the present commercial satellites is called a bent pipe transponder or a transparent transponder.

The main parameters that characterize the performance of the three segments—namely, satellites and gateway and mobile Earth stations—are G/T (the ratio of antenna gain to system noise temperature or a figure of merit), effective isotropically radiated power (EIRP) and C/N_0 (the ratio of carrier power to noise power density). The G/T and EIRP are frequently used concepts in satellite communications, and they denote the receiving and transmitting capabilities, respectively, of a satellite, a gateway Earth station, and a mobile terminal. The C/N_0 denotes the quality of the communication channel. These parameters will be described in detail in the following sections.

3.2 Main Parameters in Link Budget

3.2.1 Terminal Noise

The noise performance of communication systems can be described by a term for noise temperature. The use of a fictitious temperature stems from the fact that the basic source of noise in electrical circuits is the thermal agitation of electrons in resistive circuit components. The open-circuit root mean square (rms) noise voltage Vn generated in a resistance of value R ohms (Ω) at an absolute temperature T of kelvin (K) is given by Nyquist [1].

$$V_n = \sqrt{4\kappa TRB} \quad \text{(volts)} \tag{3.1}$$

where κ is a Boltzman's constant (1.38×10^{-23} watt/sec/K), and B is the frequency bandwidth (Hz) in which the noise voltage is measured. It is well known that maximum power can be delivered to an external load by a generator with a given internal load when the impedance of the external load is a complex conjugate of the source impedance. From this, it can easily be shown that thermal noise power Pn delivered to this optimum load by the thermal noise source of resistance R at temperature T is given by

$$P_n = \frac{V_n^2}{4R} = \kappa TB \quad \text{(watts)} \tag{3.2}$$

It must be noted that noise power does not depend on a particular value of resistance but only on absolute temperature T and frequency bandwidth B.

Hence, noise power density N_0 per unit frequency bandwidth (1 Hz) is given by

$$N_0 = \kappa T \quad \text{(watts/Hz)} \tag{3.3}$$

It is convenient to use decibel (dB) expressions in calculating the parameters of mobile satellite communications such as antenna gain, noise power, free-space propagation attenuation, and so on. In this book, log(A) is denoted by the symbol [A]. Hence, N_0 is denoted in decibels as follows:

$$
\begin{aligned}
[N_0] &= [\kappa] + [T] \\
&= 10 \cdot \log(\kappa) + 10 \cdot \log(T) \\
&= 10 \cdot \log(1.38 \times 10^{-23}) + 10 \cdot \log(T) \\
&= -228.6 + 10 \cdot \log(T) \quad \text{(dBW/Hz)}
\end{aligned} \tag{3.4}
$$

Example 3.1

Noise power density generated by a resistor at a temperature of 27°C is calculated by (3.4) as

$$
\begin{aligned}
[N_0] &= -228.6 + 10\log(273 + 27) \\
&= -228.6 + 24.8 \\
&= -203.8 \quad \text{(dBW/Hz)}
\end{aligned}
$$

3.2.2 Noise Figure

The performance of electrical circuits or components are evaluated by the parameters of a noise figure (NF), which is defined by

$$
\begin{aligned}
NF &= \frac{\dfrac{S_{in}}{N_{in}}}{\dfrac{S_{out}}{N_{out}}} \\[2mm]
&= \frac{\dfrac{S_{in}}{\kappa T_0 B}}{\dfrac{G S_{in}}{G(\kappa T_0 B + \kappa T_{in} B)}} \\[2mm]
&= 1 + \frac{T_{in}}{T_0}
\end{aligned} \tag{3.5}
$$

where S_{in} and N_{in} denote the power of signal and noise, respectively, at the input port of the circuit and S_{out} and N_{out} denote the same at the output port. The letters G and B denote the gain and frequency bandwidth of a circuit, as shown in Figure 3.2.

The T_0 denotes the physical temperature of circumstances in which the circuit is immersed, and T_{in} denotes the equivalent input noise temperature, which is an equivalent value of noise temperature at the input port of thermal noise generated in the circuit. The noise figure is frequently described in decibel units as follows:

$$[NF] = 10 \cdot \log\left(1 + \frac{T_{in}}{T_0}\right) \quad (dB) \tag{3.6}$$

When a noise figure is given, Tin is calculated as

$$T_{in} = T_0(10^{\frac{[NF]}{10}} - 1) \quad (K) \tag{3.7}$$

Example 3.2

1. When T_{in} = 400K, and T_0 = 300K, NF is given by (3.6) as follows:

$$[NF] = 10\log\left(1 + \frac{400}{300}\right) = 3.7 \ (dB)$$

2. When NF = 4 dB, and T_0 = 290K, T_{in} is given by (3.7) as follows:

$$T_{in} = 290(10^{\frac{4}{10}} - 1) = 438.4 \ (K)$$

Figure 3.3 shows the relation between noise figures in decibels and equivalent noise temperatures when T_0 = 300K.

In the same manner as the noise figure, if a circuit or a feed line has loss L_f as shown in Figure 3.4, L_f can be defined by

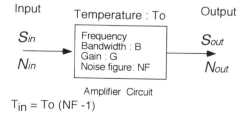

Figure 3.2 Noise figure of an amplifier circuit.

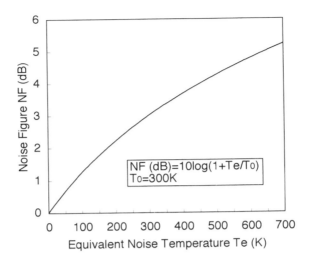

Figure 3.3 Noise figure and equivalent noise temperature.

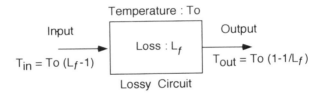

Figure 3.4 Equivalent noise temperature for a loss circuit.

$$L_f = \frac{\dfrac{S_{in}}{N_{in}}}{\dfrac{S_{out}}{N_{out}}}$$

$$= \frac{\dfrac{S_{in}}{\kappa T_0 B}}{\dfrac{\dfrac{1}{L_f} S_{in}}{\dfrac{1}{L_f}(\kappa T_0 B + \kappa T_{in} B)}}$$

$$= \frac{T_0 + T_{in}}{T_0} \tag{3.8}$$

$$\therefore \ T_{in} = T_0(L_f - 1) \tag{3.9}$$

Therefore, an equivalent noise temperature at output port T_{out} can be obtained by dividing T_{in} by L_f as follows:

$$T_{out} = T_0\left(1 - \frac{1}{L_f}\right) \tag{3.10}$$

Example 3.3

When a circuit has a loss of 3 dB and $T_0 = 300K$, equivalent input and output noise temperatures can be given by (3.9) and (3.10), respectively, as follows:

$$T_{in} = 300 \cdot \left(10^{\frac{3}{10}} - 1\right) = 298.6 \text{ (K)}$$

$$T_{out} = 300 \cdot \left(1 - \frac{1}{10^{\frac{3}{10}}}\right) = 149.6 \text{ (K)}$$

3.2.3 Noise Temperature of a Receiver

In general, a receiving system has cascade connection of loss and amplification circuits as shown in Figure 3.5. A signal from a satellite is received by an antenna with a gain of G, and the equivalent antenna noise temperature at an

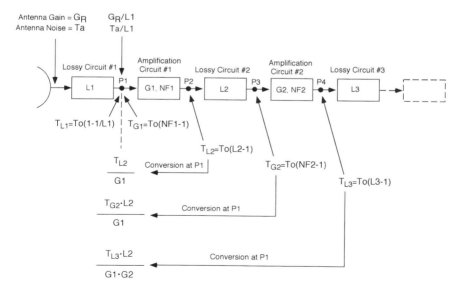

Figure 3.5 Cascade connection of loss and amplifying circuits.

output port of the antenna is denoted as T_a. Losses in loss circuits #n (n is 1, 2, 3, . . .) are denoted as L_n, and gains and noise figures of amplification circuits #n are denoted as G_n and NF_n, respectively.

In Figure 3.5, equivalent noise temperature T_{L1} at output port P1 of loss circuit #1 is described as $T_{L1} = T_0(1 - 1/L1)$. When amplification circuit #1 is concerned, equivalent noise temperature T_{G1} at input port P1 of circuit #1 can be described as $T_{G1} = T_0(NF - 1)$. Next, when loss circuit #2 is concerned, equivalent noise temperature T_{L2} at input port P2 of circuit #2 can be described as $T_{L2} = T_0(L2 - 1)$. In the same manner, equivalent noise temperature at the input port of each circuit can be described as shown in Figure 3.5. If equivalent noise temperature T_{L2} is measured at input port P1 of amplification circuit #1, this can be expressed by dividing it by gain $G1$ of circuit #1. In the same manner, equivalent noise temperature at the input port of each circuit can be converted to *the equivalent input noise temperature at the input port of amplification circuit #1* as shown in Figure 3.5.

Hence, the equivalent input noise temperature T_{in} of the whole circuit (the receiver) at input port P1 to amplifier #1 can be described as follows:

$$T_i = T_{L1} + T_{G2} + \frac{T_{L2}}{G_1} + \frac{T_{G2} \cdot L_2}{G_1} + \frac{T_{L3} \cdot L_2}{G_1 \cdot G_2} + \ldots \quad (3.11)$$

If $G_1 \gg 1$, all terms after the third term can be neglected compared to the first and second terms of (3.11). Therefore, the noise performance of the first-stage amplifier and loss circuit is found to dominate the performance of the receiver.

Overall equivalent input noise temperature Ts at the input port of the receiver can be expressed as

$$T_s = \frac{T_a}{L_1} + T_{L1} + T_{G1}$$

$$= \frac{T_a}{L_f} + T_0\left(1 - \frac{1}{L_f}\right) + T_R \quad (3.12)$$

where T_R denotes the equivalent input noise temperature of the first-stage amplifier of the receiver, which is usually called a low-noise amplifier (LNA), and L_f denotes the loss of a feed line between the antenna and the LNA. The value T_a denotes the equivalent antenna noise temperature and T_s is the system noise temperature. It must be noted that T_s depends on the measured point, and it can be usually expressed at the input port to the LNA.

3.2.4 Figure of Merit (*G/T*)

Regarding the antenna, gain G_R at receiving frequencies and equivalent input noise temperature T_a can be denoted as G_R/L_1 (= Gs) and T_a/L_1, respectively, at the input port of amplification circuit #1. The value Gs means system gain at the input port to the LNA. Consequently, the ratio of antenna gain to noise temperature at the input port to the LNA can be written as

$$\frac{G_s}{T_s} = \frac{\dfrac{G_R}{L_f}}{\dfrac{T_a}{L_f} + T_0\left(1 - \dfrac{1}{L_f}\right) + T_R}$$

$$= \frac{G_R}{T_a + T_0(L_f - 1) + T_R L_f} \tag{3.13}$$

where Gs/T_S is sometimes simply described as G/T (G over T). The G/T is an essential parameter of a receiver.

Figure 3.6 shows the relation between G/T and feeder loss in a 15-dBi antenna, which presents typical antenna gain in upcoming mobile satellite communication systems. Although T_a depends on factors such as frequency and beamwidth, a typical value is about 80K to 100K in the L band. The value L_F is a total loss of feed lines and components such as diplexers, cables, and phase shifters if a phased array antenna is used.

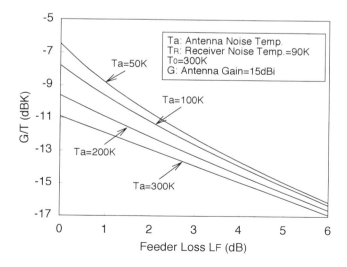

Figure 3.6 Relation between *G/T* and feeder loss. Antenna gain is 15 dBi.

3.3   Relation Between Transmitted and Received Power

The previous section showed that the sensitivity of a receiver is determined by G/T. Next, we will consider what amount of power is available at the receiver.

Figure 3.7 shows the relation between transmitted and received power. Although a perfect omnidirectional pattern in three dimensions can never be achieved, the concept of such an ideal antenna is very useful in theoretical analysis. If a transmitting antenna has an ideal isotropic radiation pattern in three dimensions, the power density on the spherical surface is

$$P_D = \frac{P_T}{4\pi d^2} \qquad (\text{watts/m}^2) \tag{3.14}$$

where P_T and d denote the transmitted power and the distance between the transmitting and receiving antennas. If the transmitting antenna has a gain of G_T, the power density of (3.14) can be written as

$$P_D = \frac{G_T \cdot P_T}{4\pi d^2} \qquad (\text{watts/m}^2) \tag{3.15}$$

where $G_T \cdot P_T$ are considered to be the radiation power transmitted by an ideal omnidirectional antenna. Therefore, this term is considered as an effective (or equivalent) isotropically radiated power (EIRP), and it is expressed as follows in antilogarithm and decibel expressions, respectively:

$$\text{EIRP} = G_T \cdot P_T \qquad (\text{watts}) \tag{3.16}$$

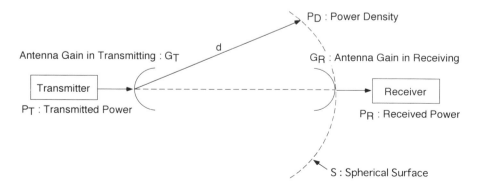

Figure 3.7 Concept of relation between transmitted and received power.

$$[\text{EIRP}] = [G_T] + [P_T] \qquad (\text{dBW}) \qquad (3.17)$$

EIRP is frequently used, and is an important concept in satellite communication systems to show the capabilities of transmission.

Then, power P_R received by the receiving antenna, which has physical aperture area A and aperture efficiency η, is

$$P_R = \frac{G_T \cdot P_T}{4\pi d^2} \cdot A \cdot \eta$$

$$= \frac{\lambda^2}{(4\pi d)^2} \cdot (G_T \cdot P_T) \cdot G_R$$

$$= \frac{(G_T \cdot P_T) \cdot G_R}{\left(\frac{4\pi d}{\lambda}\right)^2}$$

$$= \frac{\text{EIRP} \cdot G_R}{L_P} \qquad (3.18)$$

Consequently, $A\eta \cdot$ denotes the effective aperture area of the antenna, which is related to G_R and wavelength λ of the frequency obtained by the following equation [2]:

$$A \cdot \eta = \frac{\lambda^2}{4\pi} \cdot G_R \qquad (3.19)$$

Example 3.4

The frequency is 1,500 MHz. The gain of an antenna whose diameter is 100 cm and aperture efficiency is 0.6 is given by (3.19) as follows:

$$G_R = \frac{4\pi}{\lambda^2} A \cdot \eta$$

$$= 4\pi \times \left(\frac{1500 \times 10^6}{3 \times 10^8}\right)^2 \times \pi\left(\frac{1}{2}\right)^2 \times 0.6 = 148.0$$

$$[G_R] = 10\log(148.0) = 21.7 \ (\text{dB})$$

Free-space propagation loss L_P is caused by geometrical attenuation in propagation from the transmitter to the receiver. Figure 3.8 shows free-space propagation loss in the decibel scale at 1.5 GHz (L band), 4 GHz (C band),

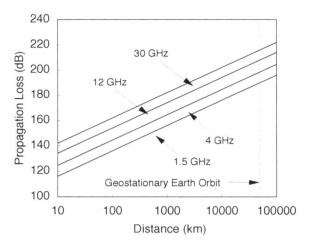

Figure 3.8 Free-space propagation loss.

12 GHz (Ku band), and 30 GHz (Ka band). The GEO has a geostationary orbit, which is about 36,000 km above the equator.

$$L_P = \left(\frac{4\pi d}{\lambda}\right)^2 \tag{3.20}$$

Example 3.5

Free-space propagation loss at 1,500 MHz from the geostationary satellite to the equator immediately below is calculated as follows:

$$L_P = \left(\frac{4\pi \times 36500 \times 10^3}{0.2}\right)^2 = (7.3\pi \times 10^8)^2$$

$$[L_P] = 10\log(7.3\pi \times 10^8)^2 = 20\log(7.3\pi \times 10^8) = 20(8 + 1.36) = 187.2 \text{ (dB)}$$

3.4 Signal-to-Noise Ratio (C/N_0) in Satellite Communication Links

The radiofrequency stages of an Earth station and a satellite, in general, consist of an antenna, a feed line, a diplexer, a high-power amplifier (HPA), and a low-noise amplifier (LNA), as shown in Figure 3.9. In Figure 3.9, G_T and G_R denote antenna gains in transmitting and receiving, respectively, and P_{out} and P_T denote the output power of an LNA and input power to an antenna,

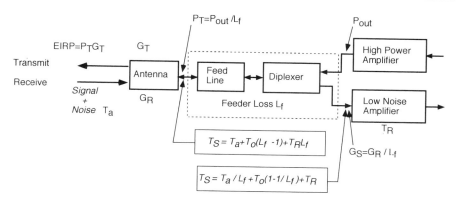

Figure 3.9 Block diagram of the RF stage of an Earth station.

respectively. The other notations are the same as those shown in the previous section.

The ratio of input signal power (C) to noise power (N) at the input point to the antenna can be written as follows using (3.2) and (3.16):

$$
\begin{aligned}
\frac{C}{N} &= \frac{\left(\dfrac{P_{out}}{L_f} \cdot G_T\right) \cdot G_R}{\dfrac{L_P}{\kappa T_S B}} \\[2em]
&= \frac{\dfrac{(P_T \cdot G_T) \cdot G_R}{L_P}}{\kappa T_S B} \\[2em]
&= \frac{\dfrac{\text{EIRP} \cdot G_R}{L_P}}{\kappa T_S B} \\[2em]
&= \frac{\text{EIRP}}{L_P}\left(\frac{G_R}{T_S}\right)\frac{1}{\kappa B}
\end{aligned}
\tag{3.21}
$$

When noise power density (C/N_0) is considered, (3.19) can be written as

$$
\frac{C}{N_0} = \frac{\text{EIRP}}{L_P}\left(\frac{G_R}{T_S}\right)\frac{1}{\kappa}
\tag{3.22}
$$

Equations (3.21) and (3.22) are basic equations that show the quality of receiving signals from a satellite to an Earth station, whose path is called a downlink.

Equation (3.20) can be written as follows in decibel expression:

$$
\begin{aligned}
\left[\frac{C}{N_0}\right] &= [P_{out}] - [L_f] + [G_T] - [L_P] + [G_R] - [T_S] - [\kappa] \\
&= [EIRP] - [L_P] + [G_R] - [T_S] + 228.6 \qquad (3.23) \\
&= [EIRP] - [L_P] + \left[\frac{G_R}{T_S}\right] + 228.6 \ (dBHz)
\end{aligned}
$$

Equation (3.23) gives us an insight into the channel quality of a downlink. The transmitted power (EIRP) is attenuated by free-space propagation (L_p) from a satellite to the Earth, amplified by receiving antenna gain (G_R), and attenuated by system noise (T_S). The channel quality of an uplink from the Earth to a satellite is expressed the same as (3.21–23).

In digital communications, the required C/N_0 is determined by the bit error rate (BER) of the required quality of communication channels. The relation between BER and C/N_0 will be discussed in detail in Chapter 6.

Example 3.6

A geostationary satellite transmits a signal at 1,500 MHz to a mobile Earth station on the equator immediately under it. The parameters are as follows:

Satellite transmission power (1W)	0	dBW
Satellite antenna gain (D = 1m)	21.7 dBi	(Example 3.4)
Propagation loss (d = 36,000 km)	187.2 dB	(Example 3.5)
Mobile antenna gain (D = 40 cm, η = 0.8)	15.0 dBi	
System noise temperature of a mobile	24.8 dBK	
Earth station (about 300K)		

Here, C/N_0 can be given by (3.21) as follows:

$$C/N_0 = 0 + 217 - 187.2 + 15 - 24.8 + 228.6 = 53.3 \ (dBHz)$$

In the above, we have considered channel quality in a downlink and an uplink separately. What about total channel quality from the base station to the mobile Earth station through the satellite? This can be easily understood by knowing that thermal noise, which is generated in an uplink and a downlink, is linearly added step by step. In general, interference noise, which is generated in the system from other systems, is added to the thermal noise. The total $(C/N_0)_T$ is given by

$$\left(\frac{C}{N_0}\right)_T = \frac{C}{(N_0)_U + (N_0)_D + I_0}$$ (3.24)

$$= \left\{ \frac{1}{\left(\dfrac{C}{N_0}\right)_U} + \frac{1}{\left(\dfrac{C}{N_0}\right)_D} + \frac{1}{\left(\dfrac{C}{I_0}\right)} \right\}^{-1}$$

where I_0 denotes the power density of interference noise and U and D denote the uplink and downlink, respectively.

In (3.24), if value (C/N_0) for one of the links is sufficiently small compared to the other values, for example $(C/N_0)_D \ll (C/N_0)_U$ and $(C/N_0)_D \ll (C/I_0)$, total quality $(C/N_0)_T$ can be approximately be given by $(C/N_0)_D$. This means that the total quality of a communication channel will be dominated by the poorest quality communication link. Figure 3.10 shows a calculated example for the total (C/N_0) depending on the uplink $(C/N_0)_s$. The parameters are downlink $(C/N_0)_s$ of 50 dBHz, 55 dBHz, and 60 dBHz. It can easily be understood in this case that the total channel quality is dominated by the poor downlink and the total channel quality never exceeds the downlink quality no matter how much uplink quality is increased.

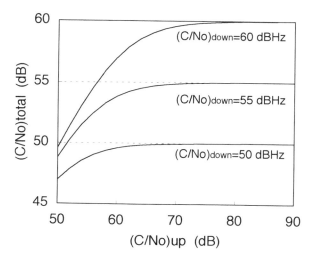

Figure 3.10 The relation between total (C/N_0) and uplink/downlink $(C/N_0)_s$.

Example 3.7

Consider the case of aeronautical satellite communication experiments using the ETS-V satellite [3]. The total C/N_0 for the forward communication link from a gateway Earth station (GES at Kashima) to an aircraft Earth station (AES at Anchorage) via satellite can be calculated by (3. 24). Interference noise has not been considered. Frequencies of 6 GHz and 1.5 GHz were used between the base Earth station and the satellite, and the satellite and the aircraft, respectively.

From a GES to a satellite (uplink)

GES EIRP   60.7 dBW

Propagation loss (6 GHz)   199.4 dB   (d = 37,270 km)

Satellite antenna gain   21.7 dBi

Feeder loss   3.0 dB

     Uplink total C = 60.7 − 199.4 + 21.7 − 3.0 = −120.0 (dBW)

     $[N_0]$ = −228.6 + 10 log(300) = −203.8 (dBHz)

     ∴ uplink $(C/N_0)_U$ = −120.0 + 203.8 = 83.8 (dBHz)

From a satellite to an AES (downlink)

Satellite EIRP   30.5 dBW

Propagation loss (1.5 GHz)   188.5 dB   (d = 41,097 km)

AES antenna gain   14.0 dBi

Antenna tracking error   0.5 dB

Feeder loss   3.0 dB

     Downlink total C = 30.5 − 188.5 + 14.0 − 0.5 − 3.0 = −147.5 (dBW)

     $[N_0]$ = −228.6 + 10 log(300) = 203.8 (dBHz)

     ∴ downlink $(C/N_0)_D$ = −147.5 + 203.8 = 56.3 (dBHz)

Therefore, total $(C/N_0)_T$ is calculated as

$$\therefore \left(\frac{C}{N_0}\right)_T = \frac{1}{\dfrac{1}{\left(\dfrac{C}{N_0}\right)_U} + \dfrac{1}{\left(\dfrac{C}{N_0}\right)_D}} = \frac{1}{\dfrac{1}{10^{8.38}} + \dfrac{1}{10^{5.63}}} = 425{,}822.28$$

$$\therefore \left[\left(\frac{C}{N_0}\right)_T\right] = 10 \log(425{,}822.28) = 56.3 \text{ (dBHz)}$$

It is confirmed that the total quality of communication channels is dominated by the poorest communication link, which is a downlink in this example.

A more detailed calculation of the example link budget is in Table 3.1.

Table 3.1
Example of a Forward-Link Budget for Aeronautical Satellite Communications [3]

From GES to Satellite (Uplink)			
Gateway Earth Station: Kashima (140.7 E, 37.0 N)			
HPA output power		dBW	10.0
Feeder loss		dB	3.0
Antenna gain		dBi	53.7
	Tx frequency	GHz	6.0
	Antenna diameter	m	10.0
EIRP		dBW	60.7
Propagation loss		dB	199.4
	Distance	km	37,270.0
Satellite:ETS-V (150 E 0)			
Antenna gain		dBi	21.7
	Antenna diameter	m	0.3
Feeder loss		dB	3.0
Received power		dBW	−120.0
System noise temp.		K	439.9
	Antenna noise temp.	K	200.0
	LNA noise temp.	K	190.0
	Environmental temp.	K	300.0
G/T		dBK	−7.7
Uplink C		dBW	−120.0
No		dBW/Hz	−203.8
Uplink C/N_0		dBHz	83.8

3.5 Key Technologies for Mobile Satellite Communication Systems

3.5.1 System Requirements and Key Technologies

Key technologies to realize mobile satellite communications are decided by such system requirements as small, lightweight terminals; enough capacity to accommodate a large number of mobile users, and high quality and reliability in communication even under the severe propagation environments of moving mobiles. These system requirements help to clarify functions such as counter-measures against fading, blocking, and shadowing caused by the propagation environment of moving mobiles. For example, in order to accommodate a large number of mobile users in an allocated frequency band, frequency band-width per channel must be as narrow as possible and the factor of frequency reuse must be increased. These required functions demand the use of key technologies in satellites, information, and mobile terminals. Figure 3.11 shows

Table 3.1 (Continued)
Example of a Forward-Link Budget for Aeronautical Satellite Communications [3]

From Satellite to AES (Downlink)			
Satellite:ETS-V (150 E, 0)			
Transponder gain		dB	128.0
HPA output power		dBW	8.0
Feeder loss		dB	3.0
Antenna gain		dB	25.5
	Tx frequency	GHz	1.5
	Antenna diameter	m	1.5
EIRP		dBW	30.5
Propagation loss		dB	188.5
	Distance	km	41,097.0
Aircraft Earth Station: Anchorage (150 W, 61.2 N)			
Antenna tracking error		dB	0.5
Antenna gain		dB	14.0
	Antenna diameter	m	0.4
Feeder loss		dB	3.0
Received power		dBW	−147.5
System noise temp.		K	294.8
	Antenna noise temp.	K	110.0
	LNA noise temp.	K	90.0
	Environmental temp.	K	300.0
G/T		dBK	−13.7
Downlink C		dBW	−147.5
No		dBW/Hz	−203.8
Downlink C/N_0		dBHz	56.3
Total C/N_0		dBHz	56.3

the mutual relations between the system requirements and key technologies of mobile satellite communications.

3.5.2 Satellite Technology

Communication equipment must be small and lightweight. Small and lightweight terminals can be achieved by technical developments in electrical components such as antennas, power amplifiers, integrated circuits, and power supplies in order to be installed on mobiles. However, the most essential factor is the capability of the satellite, which should be able to transmit sufficient power to be received by small terminals having small antennas. In GEO satellites, satellites must be large with large power amplifiers and antennas to compensate for the large attenuation in signals because of free-space propagation loss along a distance of about 36,000 km. In LEO systems, the required transmission

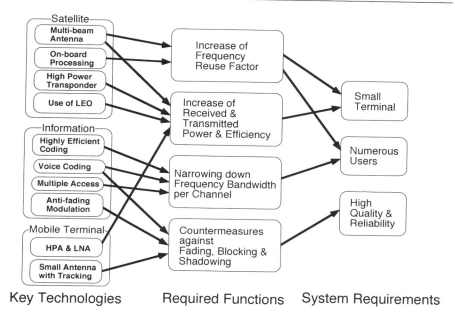

Key Technologies   Required Functions   System Requirements

Figure 3.11 Relation between system requirements and key technologies of mobile satellite communications.

power of satellites is reduced because of their low altitude. In both GEO and LEO systems, multibeam antennas onboard the satellites will greatly contribute to high power flux density on the Earth's surface and also improve frequency reuse factors by achieving multibeam on the Earth's surface. The adoption of multibeam antennas may require such intelligent functions as beam switching with onboard signal-processing capabilities.

3.5.3 Information Technology

High quality and reliability are essential factors in any communication system. However, in mobile satellite communication systems, communication quality is greatly influenced by the environmental conditions in which the mobiles are moving. In maritime satellite communications, fading caused by sea reflection greatly affects the quality of communications. In land mobile communications, signals from satellites are attenuated and obstructed by shadowing and blocking caused by obstacles such as trees and buildings. In order to overcome these, modulation and demodulation have to have antifading capabilities. In realizing such antifading capabilities and a narrow frequency bandwidth per communication channel, signal processing and voice coding using digital technologies is very important.

3.5.4 Terminal Technology

Compactness and light weight are self-evident requirements for mobile terminals in addition to mechanical strength and easy installation. A mobile antenna is one of the most important key technologies in mobile satellite communication systems. Mobile antennas must have the capability to track the satellite continuously under such severe conditions as fading, blocking, and shadowing. With handy terminals, radiation is a serious problem needing discussion because of the high-power radio wave transmissions close to the head.

3.6 Mobile Satellite Transponders

A *transponder* is a set made up of a transmitter and responder onboard the satellite. It receives and transmits signals by frequency conversion and signal amplification. Antennas for communications onboard the satellite are considered to be part of the transponder. Figure 3.12 and Figure 3.13 show transponders for the INMARSAT-2 satellite. INMARSAT operates a total of four INMARSAT-2 satellites, which were launched from 1990–1992. Each satellite is three-axis stabilized with a ten-year design life, and its initial weight in orbit and power rating are 800 kg and 1200W, respectively. Each satellite has two transponders, providing forward (C to L band) and return (L to C band) communication links that have a capacity equivalent to about 250 INMARSAT-A voice channels. The effective L-band isotropic radiated power (EIRP) is 39 dBW. Each satellite has a global beam that covers roughly one-third of the Earth's surface.

3.6.1 C/L Transponders

The C/L-band transponder receives uplink signals in the C band (6.4 GHz) from base stations and retransmits downlink signals in the L band (1.5 GHz) after frequency conversion and signal amplification by a high-power amplifier (HPA). The signals received by a C-band antenna are fed via a bandpass filter to a downconverter. A signal channel filter is followed by automatic level control (ALC) equipment, which limits the level of the signals to the amplifier. The high-power amplifier consists of six traveling-wave tubes and their associated power supplies. In front of each TWTA, there is a driver/linearizer to compensate for the nonlinear RF properties of the TWT. The signal divider supplies an equal drive signal to each of the four (out of the six) TWTAs that are active at any time, and the other two can be activated for backup if the operating TWTAs malfunction. The active TWTAs are selected by 2/3 and

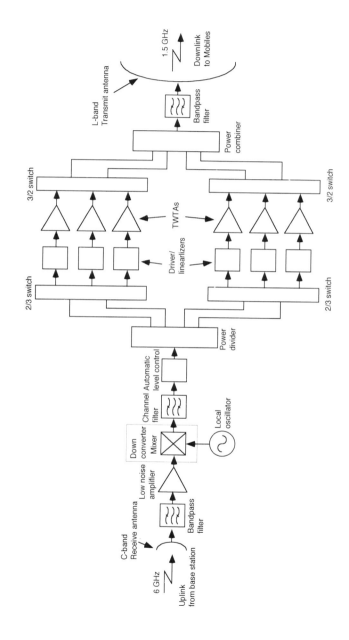

Figure 3.12 C/L-band transponder of INMARSAT-2 satellite.

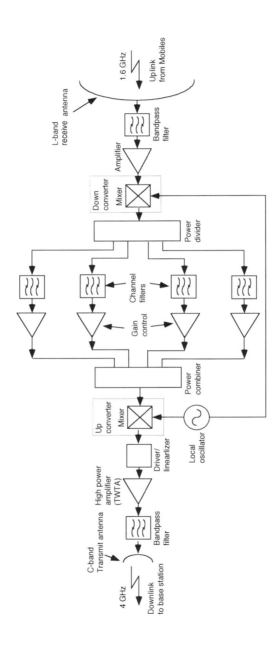

Figure 3.13 L/C-band transponder of INMARSAT-2 satellite.

3/2 switches, and their output powers are combined by a power combiner. The total power is fed to an L-band transmission antenna via a bandpass filter.

3.6.2　L/C Transponders

The L/C-band transponder receives uplink signals in the L band (1.6 GHz) from mobiles and retransmits downlink signals in the C band (3.6 GHz) after frequency conversion and signal amplification by the HPA. The signals received by an L-band antenna are fed to a downconverter via a bandpass filter and low-noise amplifier. At the downconverter, signals are converted into 60-MHz intermediate frequencies. A filter assembly then provides the required characteristics for the four channels. After the power for four channels is combined, signals are upconverted from 60 MHz to 3.8 GHz by activated TWTAs and fed to a C-band transmission antenna for the downlink to Earth.

In April 1996, the first of four INMARSAT-3 satellites was launched, and the fourth satellite was launched in July 1997. The INMARSAT-3 satellite has used the latest spotbeam technology for worldwide voice and data communications services to further decrease the size of mobile terminals, enabling pocket-size messaging units on ships, aircraft, and land vehicles. The INMARSAT-3 satellite can deliver an EIRP of up to 48 dBW in the L band. This value is eight times higher than that of the INMARSAT-2 satellite, and this can provide eight times more voice channel capacity than the INMARSAT-2.

3.7　Frequency Interference

3.7.1　Interference Coordination

As mentioned in Chapter 1, frequency bands have been internationally allocated to all radiocommunication services, not only satellite communications but also terrestrial services. Most frequency bands allocated to satellite services have been allocated on a shared basis to terrestrial communication systems. Interference is classified into six types as follows:

Between satellite and satellite systems
(1)　Satellite 1 to Earth station 2
(2)　Satellite 1 from Earth station 2
Between satellite and terrestrial systems
(3)　Satellite to terrestrial station
(4)　Satellite from terrestrial station
(5)　Earth station to terrestrial station
(6)　Earth station from terrestrial station

In order to be able to share the same frequency bands and to guarantee coexistence between different systems, many provisions such as coordination procedures and technical requirements have been introduced in the Radio Regulations of the ITU [4]. Table 3.2 shows the provisions required for satellite and terrestrial stations that are operated in the same frequency band over 1 GHz.

3.7.2 Evaluation of Interference

Interference is considered to be undesirable signals from other systems, and its effect is evaluated by an increase in an equivalent noise temperature ΔT, which is caused by interference. This method can be applied to any satellite communication system independent of modulations and exact frequencies. If $(\Delta T/T) \geq 6\%$, interference may be caused by other systems. As a result, frequency coordination has to be done between system operators according to ITU procedures.

The interference between system 1 and system 2 has been considered. As shown in Figure 3.14, system 1 consists of satellite 1 (S1) and Earth station 1 (E1), and system 2 consists of satellite 2 (S2) and Earth station 2 (E2). System 2 creates interference in system 1 in the uplink from E2 to S1 and in the downlink from S2 to E2.

The increases in equivalent received noise temperatures $\Delta T_{S12\Delta}$ and ΔT_{E12} of S1 and E1, respectively, have been calculated as follows:

$$\Delta T_{S12} = \frac{P_{E2} \cdot G_{T2}(\theta_{21}) \cdot g_{r1}(\phi_{12})}{\kappa L_{S1E2}} \qquad (3.25)$$

$$\Delta T_{E12} = \frac{P_{S2} \cdot g_{T2}(\phi_{21}) \cdot G_{R1}(\theta_{12})}{\kappa L_{S2E1}} \qquad (3.26)$$

where the notations are as follows:

- P_{E2} : Maximum power density (W/Hz) put into antenna of E2;
- P_{S2} : Maximum power density (W/Hz) put into antenna of S2;
- $G_{r2}(\theta_{21})$: Transmission antenna gain of E2 toward S1;
- $G_{R1}(\theta_{12})$: Receiving antenna gain of E1 toward S2;
- $g_{r1}(\phi_{12})$: Receiving antenna gain of S1 toward E2;
- $g_{r2}(\phi_{21})$: Transmission antenna gain of S2 toward E1;
- κ : Boltzmann's constant (1.38×10^{-23} watt/sec/K);
- L_{S1E2} : Propagation loss between S1 and E2;
- L_{S2E1} : Propagation loss between S2 and E1.

Table 3.2

Provisions for Terrestrial, Satellite, and Earth Stations Operated in the Same Frequency Band Over 1 GHz

Applied Stations	Items	Limitations		Remarks
Terrestrial Station	EIRP	> 1 GHz	55 dBW	Max. value	A = angle to the satellite with respect to the boresight direction of the antenna
		1–10 GHz	35 dBW	A > 2 deg	
		10–15 GHz	45 dBW	A > 1.5 deg	
	Input power to an antenna	1–10 GHz	13 dBW	Max. value	
		> 10 GHz	10 dBW		

	Frequency band	0 < E < 5	5 < E < 25	25 < E < 90	Unit	Satellite service
			Elevation Angle (degrees)			
Power flux density on the Earth surface	1.69–1.7 GHz	−133	−133	−133	dBW/m²/1.5MHz	Meteorology
	1.525–2.5 GHz	−154	−155 + 0.5(E − 5)	−144	dBW/m²/4 kHz	Research operation
	2.5–2.69 GHz	−152	−155 + 0.75(E − 5)	−137	dBW/m²/4 kHz	Broadcast, fixed
	3.4–7.75 GHz	−152	−155 + 0.5(E − 5)	−142	dBW/m²/4 kHz	Fixed, mobile Meteorology Research
Space Station (GEO Satellite)	8.025–11.7 GHz	−150	−150 + 0.5(E − 5)	−140	dBW/m²/4 kHz	Fixed, research Earth probing
	12.2–12.75 GHz	−148	−148 + 0.5(E − 5)	−138	dBW/m²/4 kHz	Fixed
	17.7–19.7 GHz	−115	−115 + 0.5(E − 5)	−195	dBW/m²/1MHz	Fixed
	31.0–40.5 GHz	−115	−115 + 0.5(E − 5)	−195	dBW/m²/1 MHz	Earth probing Fixed, mobile Research

Maintaining position	Nominal +/− 0.1 deg	Experimental satellite
Accuracy of beam direction	Nominal +/− 0.5 deg	
	Within Max (0.3 deg, 10% of HPBW) to nominal value	HPBW = half-power beam width
Cessation of transmission	Remote operation function is required	

Table 3.2 (Continued)
Provisions for Terrestrial, Satellite, and Earth Stations Operated in the Same Frequency Band Over 1 GHz

Applied Stations	Items	Limitations	Remarks
Earth station	EIRP limitation	1–15 GHz 40 dBW/4 kHz 40 + 3E dBW/4kHz >15 GHz 64 dBW/1 MHz 64 + 3E dBW/1MHz	$E < 0$ deg E = elevation angle to the satellite $0 < E < 5$ deg No Limitations over 5 deg $E < 0$ deg $0 < E < 5$ deg
	Minimum elev. angles	3 deg.	
	Antenna radiation pattern	$D/\lambda > 100$ (max. gain > about 48 dB) $G(\phi) = G_{max} - 2.5 \times 10^{-3}(D_\phi/\lambda)^2$ (dBi) $G(\phi) = G_1$ (dBi) $G(\phi) = 32 - 25 \log\phi$ (dBi) $G(\phi) = -10$ (dBi) $D/\lambda < 100$ (max. gain < about 48 dB) $G(\phi) = G_{max} - 2.5 \times 10^{-3}(D\phi/\lambda)^2$ (dBi) $G(\phi) = G_1$ (dBi) $G(\phi) = 52 - 10 \log(D/\lambda) - 25\log(\phi)$ (dBi) $G(\phi) = 10 - 10 \log(D/\lambda)$ (dBi)	$0 < \phi < \phi_m$ $\phi_m < \phi < \phi_r$ $\phi_r < \phi < 48$ $48 < \phi < 180$ $0 < \phi < \phi_m$ $\phi_m < \phi < \phi_r$ $\phi_r < \phi < 48$ $48 < \phi < 180$ D = antenna diameter λ = wave length ϕ = beam direction (degrees) G_1 = gain of the 1st sidelobe $\phi_m = -G_1 + 20\lambda\text{SQRT}(G_{max})/D$ $\phi_r = 15.85(D/\lambda)^{-0.6}$

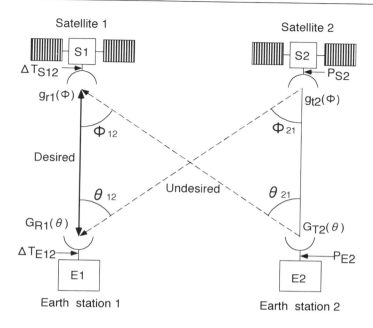

Figure 3.14 Interference between two satellite systems.

The increase in equivalent noise temperature ΔT_{12} in system 1 caused by the interference of system 2 is given by

$$\Delta T_{12} = r\Delta T_{S12} + \Delta T_{E_{12}} \qquad (3.27)$$

where r denotes the total gain from an output port of the receiving antenna for S1 to an output port of the receiving antenna for E1 in system 1.

If polarizations are taken into consideration, (3.27) can be written as follows by introducing coefficients of polarization discrimination:

$$\Delta T_{12} = \frac{r\Delta T_{S12}}{Y_{\text{up}}} + \frac{\Delta T_{E_{12}}}{Y_{\text{down}}} \qquad (3.28)$$

Table 3.3 shows coefficients of polarization discrimination.

In (3.28), Y_{up} and Y_{down} denote coefficients of polarization discrimination in uplinks and downlinks, respectively. In almost all cases, the coefficient is generally 1, because interference has been evaluated as the worst case.

Example 3.8

Two satellite systems are considered. System 1 consists of satellite 1 (S1) and Earth station 1 (E1) and system 2 consists of satellite 2 (S2) and Earth station

Table 3.3
Coefficients of Polarization Discrimination

Polarization		Polarization Discrimination Coefficient Y
System 1	System 2	
LHC	RHC	4
LHC	L	1.4
RHC	L	1.4
LHC	LHC	1
RHC	RHC	1
L	L	1

Note: LHC: left-handed circular polarization; RHC: right-handed circular polarization; L: linear polarization.

2 (E2). Both systems are operated in C-band frequencies (6 GHz) and only an uplink of system 2 can create interference in system 1. A downlink from S2 does not cause interference in E1 of system 1. In Figure 3.14, an uplink from E2 may cause interference in satellite 1. The total noise temperature in the C-C (up-down) link of system 1 is 80K.

The parameters, whose notations are the same as in Figure 3.14, used for calculation are as follows:

System 1
   Satellite 1 (S1)
      Position   150 degrees E
      Antenna gain toward E2 $g_{r1}(\phi_{12}) = 21$ (dBi)
      Antenna gain toward E1 $g_{r1}(0) = 22$ (dBi)
      Transponder gain   110 (dB)
      Propagation loss from S1 to E1 (37,270 km)   199.4 (dB)
   Earth station 1 (E1)
      Antenna gain toward S1 $G_{R1}(0) = 50$ (dBi)

System 2
   Satellite 2 (S2)
      Position   145 degrees E
   Earth station 2 (E2)
      Position   132.5 degrees E, 43 degrees N
      Antenna gain toward S1 $G_{T2}(\theta_{21}) = 13.4$ (dBi)
      Transmission power $P_{E2} = -33$ (dBW/Hz)
      Propagation loss from E2 to S1 (37,991 km) 199.5 (dB)

By using (3.25), the increases in the equivalent received noise temperature $\Delta T_{S12\Delta}$ of S1 can be calculated as follows:

$$\Delta T_{S12} = \frac{P_{E2} \cdot G_{T2}(\theta_{21}) \cdot g_{r1}(\phi_{12})}{\kappa L_{S1E2}}$$

$$= \frac{10^{-33.4/10} \times 10^{13.4/10} \times 10^{21/10}}{1.38 \times 10^{-23} \times 10^{-199.5/10}} = 1{,}000 \ (\text{K})$$

Therefore, the increase in total equivalent noise is given by (3.27):

$$\Delta T_{12} = r\Delta T_{S12} + \Delta T_{E_{12}}$$

$$= 10^{(110+22-199.4+50)/10} \times 1{,}000$$

$$= 18.2 \ (\text{K})$$

In using (3.27), the second term ΔT_{E12} was not considered because the downlink from S2 to E1 does not cause interference. The ratio of increased noise temperature to system noise free of interference is given by

$$\frac{\Delta T}{T} = \frac{18.2}{80} = 22.8\% \geq 6\%$$

Hence, frequency coordination has to be carried out between two systems.

3.7.3 Methods to Reduce Interference

If interference is evaluated to affect the system, methods to reduce it have to be considered. There are several methods to reduce interference both in uplinks and downlinks, respectively, as follows:

- Uplink (interference from E2 to S1):
  - Reduce EIRP of E2;
  - Increase EIRP of E1;
  - Improve antenna radiation pattern of E2;
  - Improve polarization discrimination;
  - Adopt energy dispersal;
  - Select location of E2;
  - Limit system operations of E2;
  - Limit carrier frequencies of E2;
  - Allocate carrier frequencies from E2;
  - Change satellite position in orbit.

- Downlink (interference from S2 to E1):
  - Reduce EIRP of S1;
  - Increase EIRP of S1;
  - Improve antenna radiation pattern of S1;
  - Improve polarization discrimination;
  - Adopt energy dispersal;
  - Shift antennae beam of S1;
  - Limit carrier frequencies of E2;
  - Allocate carrier frequencies from E2.

References

[1]   Nyquist, H., "Thermal Agitation of Electric Charge in Conductor," *Physical Review*, Vol. 32, July 1928, pp. 110–113.

[2]   Stutzman, W. L. and G. A. Thiele, *Antenna Theory and Design*, New York, NY: John Wiley & Sons, 1981, p. 59.

[3]   Ohmori, S., et al., "Experiments on Aeronautical Satellite Communications Using ETS-V Satellite," *IEEE Trans. on Aerospace and Electronic Systems*, Vol. 28, No. 3, July 1992, pp. 788–796.

[4]   Radio Regulations, Articles 11, 27 and 28.

4

Vehicle Antennas

4.1 Requirements for Antennas [1]

4.1.1 Mechanical Characteristics

4.1.1.1 Compactness and Light Weight

It is self-evident that mobile antennas need to be compact and lightweight. However, a compact antenna has two major disadvantages in electrical characteristics such as low gain and wide beam coverage. The gain is closely related to the beamwidth, and a low-gain antenna should have a wide beamwidth. As the gain of an antenna is theoretically determined by its physical dimensions, reducing the size of an antenna means reducing its gain. Because of low gains and limited electric power supply, it is very difficult for mobile antennas to have enough receiving capability (G/T) and transmission power (EIRP). These disadvantages of mobile terminals can be compensated for by a satellite that has a large antenna and a high-power amplifier with enough electric power. A powerful satellite with high EIRP and G/T performance should permit the fabrication of mobile terminals including compact and lightweight antennas.

The second disadvantage is that a wide-beam antenna is likely to transmit undesired signals to, and receive these from, undesired directions, which will cause interference in and from other systems. The wide beam is also responsible for several fading effects such as that from sea surface reflections in maritime satellite communications and multipath fading in land mobile satellite communications. A compact antenna system is required to prevent fading and interference.

87

4.1.1.2 Installation

Easy installation and appropriate physical shape are very important requirements in addition to compactness and light weight. With shipborne antennas, installation requirements are not as limited compared to aircraft and cars, because even small ships have more space to install an antenna system. However, in the case of cars, especially small private cars, low-profile and lightweight equipment is required. The requirements are the same in aircraft, although more stringent conditions are required to satisfy avionics standards. Low air drag is one of the most important requirements for aircraft antennas [2]. A phased array antenna is considered to be the best candidate for aircraft and small cars because of its low profile and mechanical strength.

4.1.2 Electrical Characteristics

4.1.2.1 Frequency and Frequency Bandwidth

As shown in Table 1.6, typical frequencies allocated to mobile satellite communications are the L (1.6/1.5 GHz) and S (2.6/2.4 GHz) bands, which are being operated in the present and will be operated in future mobile satellite communications systems. The future systems will use Ka (30/20 GHz) and millimeter-wave bands. The required frequency bandwidth to cover transmitting and receiving channels are, in general, about 7% in the L band and 40% in the Ka band. It is a very difficult requirement for antennas that are operated in both the transmitting and receiving frequency bands to achieve frequency bandwidths over 10%.

4.1.2.2 Polarization, Axial Ratio, and Side Lobes

In mobile satellite communication systems, circular-polarized waves are used to avoid polarization tracking. When both satellite and mobile Earth stations use linearly (vertically or horizontally) polarized waves, the mobile Earth stations have to keep the antenna coinciding with the polarization. If the direction of the mobile antenna rotates 90, the antenna cannot receive signals from the satellite. Even if circular-polarization waves are used, polarization mismatch loss caused by axial ratios has to be taken into account, especially in phased array antennas. This will be described in detail in Section 4.2.2.

As mentioned in the previous section, sidelobe levels are required to be under appropriate levels in order to avoid interference to and from other satellite communication systems.

4.1.2.3 Gain and Beam Coverage

Required antenna gains are determined by a link budget, which is calculated by taking into account the satellite capability and the required channel quality.

The channel quality (C/N_0) depends on the *G/T* and EIRP values of the satellite and mobile Earth stations. As mentioned previously, in present systems such as OPTUS, AMSC, and MSAT, a *medium gain* of 8 to 15 dBi is required for voice/high-speed data (about 24 Kbps) channels. In the case of the present INMARSAT-A terminal, a comparatively "high gain" of about 24 dBi is required, due to the difference in satellite capabilities. *Low gain* antennas of about 0 to 4 dBi are used in the INMARSAT-C and other systems to provide low-speed data (nonvoice) services of about 600 to 1,200 bps. The GPS system has adopted low gain antennas because of the extremely low data (50 bps) from the satellites. There are no exact definitions to differentiate between low, medium and high gains. However, in the present and upcoming L-band mobile satellite communication systems, antenna gains will be classified as shown in Table 4.1.

The beams of mobile antennas are required to cover the upper hemisphere independent of mobile motions. Low-gain antennas have advantages in terms of establishing communication channels without tracking the satellite because of their omnidirectional beam patterns. However, high-gain antennas have to track satellites because of their narrow directional beam patterns.

4.1.2.4 Satellite Tracking

It is an essential for high-gain and medium-gain antennas to have a tracking function. Tracking capabilities depend on the beamwidth of the antennas and the speed of mobile motions, and directional antennas with narrow beams have to track the satellite both in elevation and azimuth directions. In general, the required accuracy of tracking is considered to be within 1 dB, which is an angular accuracy within about a half angle of HPBW. However, directional antennas with relatively narrow beams should track the satellite only in the azimuth directions because the elevation angles to the satellite are almost constant, especially in land mobile satellite communications.

There are two types of tracking systems: mechanical and electrical. A mechanical tracking system uses mechanical structures to keep the antenna in the satellite direction through a motor and mechanical drive system. Typical examples are the antenna systems for INMARSAT-A and B. However, an electrical tracking system tracks the satellite by electrical beam scanning. This system has many advantages such as high-speed tracking and being maintenance-free; however, it also has disadvantages such as narrow beam coverage and poor axial ratio with wide scanning angles.

From the standpoint of a tracking algorithm, there are also two types, namely an open-loop method and a closed-loop method. The difference between these two types is whether the signal from the satellite is used or not. The open loop is used to calculate the satellite direction using information on

Table 4.1
Classification of L-band Antennas in Mobile Satellite Communications

Gain Class	Antenna	Typical Gain (dBi)	Typical G/T (dBK)	Typical Antenna (Dimension)	Typical Service
High	Directional	20–24	–4	Dish (1m ϕ)	Voice/high-speed data
		17–20	–8 to –6	Dish (0.8m ϕ)	Ship (INMARSAT-A, B)
Medium	Semidirectional (Only in azimuth)	8–16	–18 to –10	SBF (0.4m ϕ)	Voice/high-speed data
				Phased array (20 elements)	Aircraft (INMARSAT-Aero)
		4–8	–23 to –18	Array (2–4 elements)	Ship (INMARSAT-M)
				Helical, patch	Landmobile
Low	Omnidirectional	0–4	–27 to –23	Quadrifilar	Low-speed data (message)
				Drooping	Ship (INMARSAT-C)
				Dipole	Aircraft
				Patch	Landmobile

Note: SBF = short-backfire antenna.

satellite and mobile positions, without using signals from the satellite. Several kinds of sensors are used in obtaining the mobile position and azimuth angle to the satellite with respect to the heading of the mobile. This method is effective under severe fading conditions such as in land mobile satellite communications, where changing received signals from the satellite are too large and quick to be used as a tracking reference.

The closed-loop method, on the other hand, uses a signal from the satellite to track it. In order to be able to use the closed-loop method, received signals from the satellite must be stable without severe fading, which is expected in aeronautical satellite communications and maritime satellite communications in large vessels. It is very hard to adopt in land mobile satellite communications.

4.2 Basic Knowledge of Antennas [3–5]

4.2.1 Gain

Antenna gain is defined by comparison with an ideal antenna, which has an isotropic radiation pattern without any losses. An isotropic antenna, which does not really exist, radiates power in all directions in uniform intensities. If power P_{in} is input into an isotropic antenna, the power density per unit area P_{ideal} at distance $r(m)$ from the antenna is given by

$$P_{ideal} = \frac{P_{in}}{4\pi r^2} \ (watt/m^2) \qquad (4.1)$$

However, if radiated power density is $F(\theta, \phi)/r^2$ in direction (θ, ϕ) at distance $r(m)$ from the antenna under evaluation, the gain of the antenna can be defined by the following equation:

$$\begin{aligned} Ga(\theta, \phi) &= \frac{F(\theta, \phi)/r^2}{P_{ideal}} \\ &= \frac{F(\theta, \phi)/r^2}{P_{in}/(4\pi r^2)} \\ &= \frac{4\pi \cdot F(\theta, \phi)}{P_{in}} \end{aligned} \qquad (4.2)$$

The gain defined by (4.2) is called an absolute gain or a directive gain, which is determined only by the directivity (radiation pattern) of the antenna without taking account of any losses in the antenna system such as impedance mismatch loss, feeder loss, or spillover loss. If no direction is specified and the gain is not given as function of (θ, ϕ), it is assumed to be maximum gain. The gain is usually expressed in a decibel scale (dB), and is sometimes denoted as dBi meaning that the gain has been defined compared with an isotropic antenna. Further, the gain of circular-polarized antennas is sometimes denoted as dBic.

In actual antennas, it must be noted that radiated power P_{out}, which is really radiated from the antenna is not equal to input power P_{in} which is input to the antenna as defined in (4.2). This is because antennas, in general, have many losses caused by feed lines and impedance mismatches in themselves. The gain defined by (4.2), where P_{out} is substituted for P_{in}, is called effective gain or operating gain. It can easily be understood that absolute (directive) gain is greater than effective gain because $P_{in} \geq P_{out}$. The definition of effective gain does not include losses due to polarization mismatch as a result of the axial ratios of circular-polarized antennas.

There is a general relationship between absolute gain and the physical dimension of the antenna, and this is given by the following equation:

$$Ga = \frac{4\pi}{\lambda^2} \cdot \eta A \qquad (4.3)$$

This is where A and η denote physical aperture and aperture efficiency, respectively. Therefore, $A\eta$ denotes the effective aperture area of the antenna. Using (4.3), it can be seen that compact antennas with small apertures have to have low gains. If an antenna aperture is a dish with a diameter of D, (4.3) can be written as

$$Ga = \left(\frac{\pi D}{\lambda}\right)^2 \eta \qquad (4.4)$$

In decibel expressions, (4.4) can be written as follows:

$$Ga = 10 \cdot \log\left[\left(\frac{\pi D}{\lambda}\right)^2 \eta\right] \text{ (dB)} \qquad (4.5)$$

Figure 4.1 shows the relation between antenna gain and physical aperture size. With the same aperture size, the gain becomes larger when using higher frequencies, because the aperture size becomes larger compared to the wavelength of the radio wave. In Figure 4.1, three frequency bands, the L band

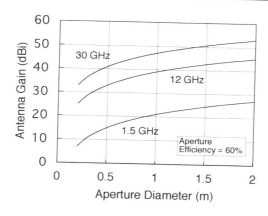

Figure 4.1 Relation between antenna gain and aperture size.

(1.5 GHz), the Ku band (12 GHz), and the Ka band (30 GHz) have been chosen as parameters. A 60% value for aperture efficiency is general in aperture-type antennas such as parabolic antennas. It can be found that the gain in an antenna with a diameter of 1m operated at 1.5 GHz is about 21 dBi, which is a typical value for a shipborne antenna in the INMARSAT system (INMARSAT-A).

4.2.2 Radiation Pattern, Beamwidth, and Side Lobe

Radiation calculation is possible in principle if the electromagnetic field can be described quantitatively at all points of the antenna surface whose boundaries are those of the apertures. In this section, the radiation pattern from a circular aperture is considered as aperture-type antennas have generally been used in mobile satellite communications, especially in maritime satellite communications. This simple problem will provide us with insight into the characteristics of mobile antennas. As shown in Figure 4.2, if aperture S is excited by uniform electric intensity E per unit area of uniform phase and amplitude, the radiation electric field is given by

$$E(\theta, \phi) \propto E \cdot \int_0^{2\pi} \int_0^a \rho^{jkpr\sin\theta \cdot \cos(\phi-\varphi)} d\rho d\varphi$$

$$\propto 2 \cdot \frac{J_1(ka \sin \theta)}{(ka \sin \theta)}$$

$$= 2 \cdot \frac{J_1\left(\pi \frac{D}{\lambda} \sin\theta\right)}{\pi \frac{D}{\lambda} \sin\theta} \tag{4.6}$$

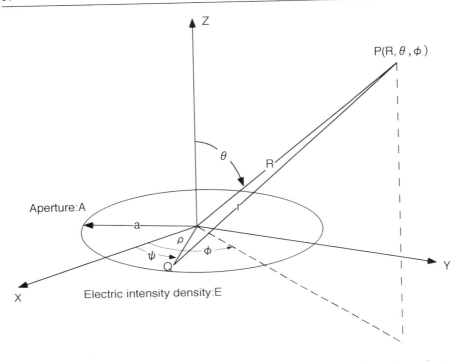

Figure 4.2 Coordinates in calculating the pattern of an aperture excited by uniform phase and amplitude.

where $J_1(x)$ is the first-order Bessel function [6], and $k(= 2\pi/\lambda)$ and a (=D/2) denote the wave number and radius of the aperture, respectively. The other notations denote distances and angles in the coordinates shown in Figure 4.2.

Figure 4.3 shows the radiation power patterns in decibels for circular-aperture antennas calculated by (4.6). Radiation power and physical aperture sizes are normalized by maximum power level and wavelength, respectively. The larger the antennas become, the narrower the beams are, and several side lobes will appear in off-axial angular areas. With uniformly excited apertures, the level of the first side lobe is theoretically predicted to be about −17.6 dB. In actual cases, aperture field distributions are tapered from the center toward the edge in order to lower sidelobe levels and to reduce power spilling outside the dish. In tapered distributions, effective aperture diameter De becomes smaller that physical diameter D, so $De = \eta_D D$, where η_D denotes diameter efficiency. Beamwidth is evaluated by half-power beamwidth (HPBW) $2\theta_{HP}$, where θ_{HP} is the half-power angle when radiated power becomes half the maximum level (−3 dB). The HPBW is given by the following equation by the polynomial approximation of (4.6):

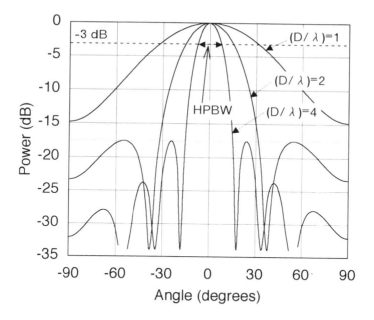

Figure 4.3 Radiation patterns of a circular-aperture antenna.

$$\text{HPBW} = 2\theta_{\text{HP}}$$

$$\approx 56 \times \frac{\lambda}{\eta_D D}$$

$$= 70 \times \frac{\lambda}{D} \text{ (degrees)} \qquad (4.7)$$

where $\eta_D = 0.8$ has been chosen as the general value of actual antennas. By using (4.4) and (4.7), the relation between gain and HPBW is given by

$$\text{HPBW} = 70\pi\sqrt{\frac{\eta}{G_a}} \qquad (4.8)$$

Figure 4.4 shows the relation between half-power beamwidth and the gain of an aperture antenna calculated by (4.8), in which the aperture efficiency is 60% ($\eta_D \approx 0.8$).

4.2.3 Polarization and Axial Ratio

If the electric and magnetic fields are alternating in the plane at all times, this is called a plane-polarized wave, or simply a plane wave. In plane waves, electric

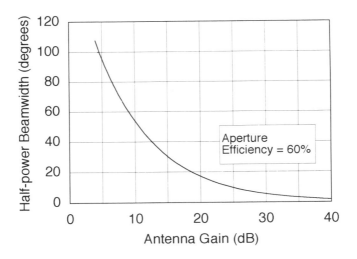

Figure 4.4 Relation between gain and half-power beamwidth of an aperture antenna.

and magnetic fields are orthogonal, and they are also perpendicular to the direction of wave propagation. Electromagnetic waves radiated by an antenna, which are located far from the receiving antenna, are considered to be plane waves near the receiving antenna. If a plane wave is propagating along the z axis, and electric field E is on the x-z plane, E only has an x component as follows:

$$E_x = E_a \cdot e^{j(\omega t - kz + \phi_a)} \tag{4.9}$$

where E_a, ω, k and ϕ_a denote the maximum amplitude of electric field, angular frequency ($=2\pi f$), wave number, and initial phase, respectively. This wave is called a linear (vertical) polarization. In the same manner, a linear (horizontal) polarization wave, whose electric field is on the y-z plane, can be written as

$$E_y = E_b \cdot e^{j(\omega t - kz + \phi_b)} \tag{4.10}$$

where E_b and ϕ_b are the maximum amplitude and initial phase of the wave.

Consider wave **E**, which is composed of E_x and E_y. It can be written as

$$\mathbf{E}(z, t) = \mathbf{x} \cdot E_x + \mathbf{y} \cdot E_y \tag{4.11}$$

where **x** and **y** are unit vectors in the x and y axes, respectively. Equation (4.11) is a general expression of a plane wave propagating along the z direction, and this wave is called an elliptical-polarized wave.

If E_x and E_y satisfy the following conditions that amplitudes are equal and the phase is different by $\pi/2$, then

$$E_a = E_b = E, \ \phi_a = 0, \ \phi_b = \pi/2 \tag{4.12}$$

$$|\mathbf{E}| = E \cdot |e^{j(\omega t - kz)}| \cdot |e^{j\pi/2}|$$
$$= E \tag{4.13}$$

Equation (4.13) indicates that a vector of the electric field rotates around the z axis and its amplitude is constant. This polarization wave is called a circular-polarized wave, which consists of two orthogonal linearly polarized waves with a phase difference of $\pi/2$.

Figure 4.5 shows a circular-polarized wave that is traveling along the z axis. The electric field of the y component advances by $\pi/2$ in phase compared to that of an x component. If an observer at the origin sees the electric wave leaving the observer, the electric vector rotates clockwise while traveling in the z direction. It must be noted, as shown in (4.9), that the electric wave is a function of time (t) and space (z). If the observer stays at the origin and sees the direction of wave propagation (+z direction), the electric vector rotates

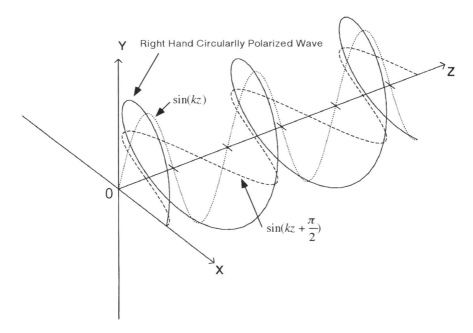

Figure 4.5 Circular-polarized wave propagating along the z axis.

counterclockwise with time proceeding. This wave is called a left-handed circular-polarized wave. It is a very important that the directions of spatial rotation
and time rotation are opposite. This corresponds to the fact that the signs for
time (t) and space (z) are different in (4.9). In electrical engineering including
satellite communications, wave rotation is defined as the wave leaving the
observer, namely that the observer sees the wave from the back of the transmitting antenna. If the wave is traveling towards the observer, rotation becomes
opposite.

In the above equations, if the amplitudes and phase difference between
the two waves are not equal and $\pi/2$, such a wave is called an elliptical-polarized
wave.

Figure 4.6 shows the tip of the instantaneous electric field vector for a
general elliptical wave. The x'-y' axis is chosen to correspond to the major and
minor axes, respectively. In general, the coordinates x-y and x'-y' are not the
same, and the angle between the two coordinates is a function of the amplitudes
and phases of (4.9) and (4.10). The axial ratio (AR) is defined as the ratio of
the major axis electric component to that of the minor axis:

$$|AR| = |E1/E2| \ (1 \leqq |AR| \leq \infty) \tag{4.14}$$

The sign for AR denotes the direction of rotation, however, an absolute
value is usually used to evaluate circular-polarized waves. The AR, in general,
can be expressed in a decibel expression:

$$[AR] = 20 \cdot \log(|E1/E2|) \ (\text{dB}) \ (0 \leqq [AR] \leq \infty) \tag{4.15}$$

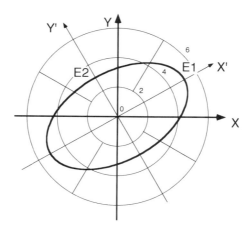

Figure 4.6 General polarization ellipse.

For example, the axial ratio of the wave shown in Figure 4.5 is 1.67 = 5/3 (4.4 dB). The values for $AR = 1$ (0 dB) and $AR = \infty$ (∞ dB) correspond to circular-polarized and linearly polarized waves, respectively. The axial ratio is determined by the performance of the antenna, so the axial ratio is one of the most important parameters of circular-polarized antennas. It can easily be understood that the axial ratio depends on the direction with respect to the axis of the antenna. In general, the AR is the best (smallest) in the boresight direction, and is progressively worse further away from the bore sight.

4.2.4 Polarization Mismatch

In mobile satellite communications, circular-polarized waves are used to avoid polarization tracking. However, in reality, circular-polarized antennas transmit elliptical-polarized waves, not ideal circular-polarized waves with an axial ratio of 0 dB. This characteristic is the same for both transmitting and receiving antennas. Because of degraded axial ratios, polarization mismatch loss has to be taken into consideration. When we assume that axial ratios of transmitting and receiving antennas are AR_a and AR_b, respectively, the minimum and maximum power of loss caused by polarization mismatch are calculated by the following equations:

$$P_{\min} = \frac{(1 + AR_a \cdot AR_b)^2}{(1 + AR_a^2)(1 + AR_b^2)} \tag{4.16}$$

$$P_{\max} = \frac{(AR_a + AR_b)^2}{(1 + AR_a^2)(1 + AR_b^2)}$$
$$(0 \leqq P_{\max}, P_{\min} \leqq 1) \tag{4.17}$$

where the sign for AR is taken into consideration. If the transmitting and receiving antennas have the same rotated polarization, $AR_a \cdot AR_b \geq 0$, and if they have the opposite rotated polarization, $AR_a \cdot AR_b \leq 0$.

Figure 4.7 shows polarization mismatch loss calculated by (4.16) and (4.17). The abscissas and ordinates denote axial ratios of the receiving antenna and polarization mismatch loss, respectively. The parameters are the axial ratios of the transmitting antenna, which have been chosen as 0, 1, 2, 3, 4, and 5 dB. There are maximum and minimum values in received power. This is physically explained by the fact that the received power becomes maximum when the major and minor axes of both waves coincide, and the received power becomes minimum when the major and minor axis of each wave are orthogonal.

In general, the axial ratios of parabolic antennas are small enough to be under 2 dB. The antennas onboard satellites are considered to have efficient

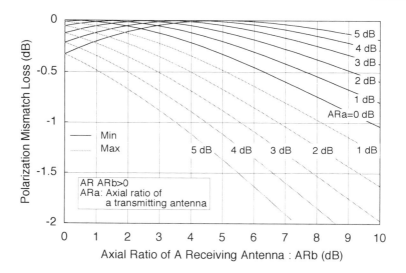

Figure 4.7 Polarization mismatch loss between transmitting and receiving antennas with axial ratios of AR_a and AR_b, respectively.

performance in axial ratios, because they are usually parabolic and their angle to the Earth is sufficiently small at about 17 degrees. However, in mobile satellite communications, a phased array antenna is considered to be the best candidate for mobiles such as aircraft and small cars. As will be mentioned in the following section, phased array antennas have disadvantages in axial ratios, especially when the beam is scanned to larger angles.

Example 4.1

When the axial ratios of antennas onboard a satellite and a mobile are 2 dB and 6 dB, respectively, the polarization mismatch loss can be calculated as follows using (4.16) and (4.17):

$$[P_{\min}] = 10 \log\left\{ \frac{\left(1 + 10^{\frac{2}{20}} \times 10^{\frac{6}{20}}\right)^2}{\left(1 + 10^{\frac{4}{20}}\right) \times \left(1 + 10^{\frac{12}{10}}\right)} \right\} = -0.19 \text{ (dB)}$$

$$[P_{\max}] = 10 \log\left\{ \frac{\left(10^{\frac{2}{20}} + 10^{\frac{6}{20}}\right)^2}{\left(1 + 10^{\frac{4}{20}}\right) \times \left(1 + 10^{\frac{12}{20}}\right)} \right\} = -0.85 \text{ (dB)}$$

The maximum polarization mismatch loss is about 0.9 dB, which can be neglected in actual communications. In order to lower the mismatch loss under 1 dB, the axial ratio of mobile antennas has to be under 7 dB.

4.3 Basic Antennas

4.3.1 Crossed-Dipole Antenna

A dipole antenna with a half-wavelength ($\lambda/2$) is the most widely used, and it is also the most popular having been used in antenna systems such as the parabolic antennas for mobile satellite communications. A half-wavelength dipole is a linear antenna, whose current amplitude varies one-half of a sine wave with a maximum at the center. The radiation pattern of the half-wave dipole #1 shown in Figure 4.8 is calculated in [7] as

$$D1_\phi(\phi) = \left| \frac{\cos[(\pi/2)\ \cos\phi]}{\sin\ \phi} \right| \tag{4.18}$$

Hence, the radiation pattern of dipole #2 is given by

$$D2_\phi(\phi) = \left| \frac{\cos[(\pi/2)\sin\phi]}{\cos\phi} \right| \tag{4.19}$$

The patterns $D1_\phi(\phi)$ and $D2_\phi(\phi)$ are indicated in by the thick and thin lines, respectively, within a coordination system. The half-power beamwidth

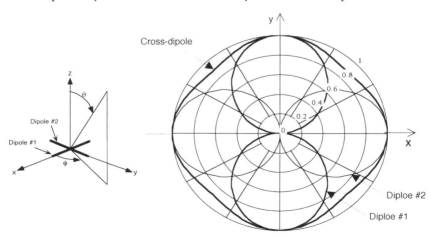

Figure 4.8 Radiation patterns of a dipole, a cross-dipole, and the coordination system.

(HPBW) is about 78. As a dipole antenna radiates linearly polarized waves, two crossed-dipole antennas have been used in order to generate circular-polarized waves. The two dipoles are geometrically orthogonal (x and y axes in Figure 4.8), and equal amplitude signals are fed to them with $\pi/2$ in-phase difference. The radiation pattern of a crossed-dipole antenna is also indicated by the thick line in Figure 4.8, which is nearly omnidirectional in the horizontal plane. A dipole antenna needs a balun to be excited by coaxial cables, which is an unbalanced feed line. Further, a 3-dB hybrid (power divider) is generally used to feed a cross-dipole in order to be able to feed the same power and a phase difference of $\pi/2$ for each dipole element.

Example 4.2

The directivity $D_{\text{dipole}}(\phi)$ of a cross-dipole can be obtained by (4.18) and (4.19) as follows:

$$D_{\text{dipole}}(\phi) = \left| \frac{\cos[(\pi/2)\cos\phi]}{\sin\phi} + j\frac{\cos[(\pi/2)\sin\phi]}{\cos\phi} \right| \qquad (4.20)$$

A crossed-dipole antenna has a maximum gain in the boresight direction (z axis direction in Fig. 4.8). In mobile satellite communications, especially in land mobile communications, elevation angles to the satellite are not 90 except immediately under the satellite. In order to optimize the radiation pattern, a set of dipole antennas are bent toward the ground as shown in Figure 4.9, which is called a drooping dipole antenna. The crossed drooping dipole is one of the most interesting candidates for land mobile satellite communications, where the required angular coverage is narrow and almost constant in elevation. By adjusting the height between the dipole elements and the ground plane and the bending angle of the dipoles, the gain and elevation pattern can be optimized for the coverage region of interest. Figure 4.9 shows the radiation patterns for the antenna designed by Jet Propulsion Laboratory (JPL) which is to be used over the entire continental United States (CONUS). It has a 4-dBi gain [8].

4.3.2 Helical Antenna

A helical antenna can easily generate circular-polarized waves without a balun or a 3-dB power divider, which are required to excite a balanced fed dipole and circular-polarized cross dipoles. Also, it can be operated in a wide frequency bandwidth of up to 200%, because it is a traveling-wave-type antenna. Figure 4.10 is a helical antenna having a ground plane with pitch L, pitch angle A, length L, and diameter D. The diameter of the ground plane is usually

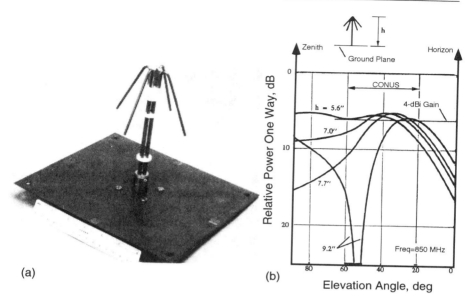

Figure 4.9 (a) Crossed drooping dipole antenna and (b) its radiation pattern [8].

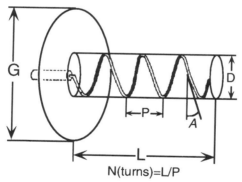

N(turns)=L/P

Figure 4.10 Helical antenna with pitch *L*, pitch angle *A*, length *L*, and diameter *D*. The number of turns is *N* = *L/P*.

selected to be larger than one wavelength. The number of turns is $N = L/P$. It is well known that the parameters for *A* are about 12 to 15 and the circumference of the helix (πD) is about 0.75 to 1.25 wavelengths. Circular-polarized waves with good axial ratios can be transmitted along the *z* axis direction (axial mode) [9].

The gain of a helical antenna depends on the number of *N* turns, and typical gain and half-power beamwidth are about 8 dBi and 50 degrees when $N = 10 \sim 12$.

In mobile antennas, quadrifilar antennas are one of the best candidates. The quadrifilar helical antenna consists of four tape helices equally spaced circumferentially on a cylinder that are fed with equal amplitude signals with relative phases of 0, 90, 180, and 270 degrees [10]. This antenna has two advantages over a conventional unifilar helical antenna. The first is an increase in bandwidth. It can generate axial mode circular-polarized waves in the frequency range from 0.4 to 2.0 wavelengths of the helix circumference. The second is lowered frequency for axial mode operation. The principle disadvantage is an increase in complexity in the feed system [8]. The area of a ground plane is usually about 3 times the diameter of the helix. Figure 4.11 shows a photograph of a quadrifilar antenna designed for the GPS system operated at 1.2 GHz.

4.3.3 Microstrip Patch Antenna

A microstrip patch antenna is very low profile and has mechanical strength, so it is considered to be the best candidate for small cars, and especially in aircraft, which require low air drag. In general, a circular disk antenna element is chosen, and it is printed on a thin dielectric substrate with a ground plane.

Figure 4.12 shows the basic configuration for a circular patch antenna, which has two feed points to generate circular-polarized waves. The resonant frequency excited by basic mode (TM_{11}) is given by [11]

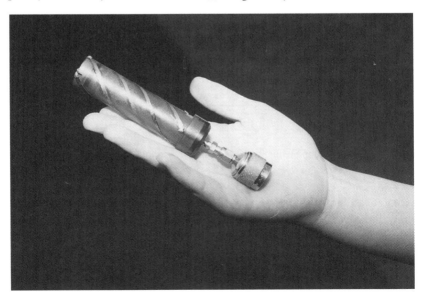

Figure 4.11 Photo of a quadrifilar helical antenna for the GPS.

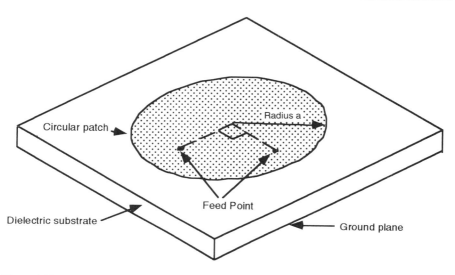

Figure 4.12 Basic configuration for a circular patch antenna with two feed points to generate circular-polarized waves.

$$f = \frac{1.84 \cdot c}{2\pi a\sqrt{\epsilon_r}} \qquad (4.21)$$

where a, c, and ϵ_r are the radius of a circular disk, the velocity of light in free space, and the relative dielectric constant of the substrate, respectively. Figure 4.13 shows the relation between the frequency and diameter of a circular patch antenna.

In land mobile satellite communications, a microstrip antenna with higher order excitation (TM_{21}) is considered better because it can optimize the gain in elevation angles to the satellite the same as a drooping cross-dipole antenna. The resonant frequency of a higher order circular patch antenna is given by (4.19) by changing the coefficient 1.84 by 3.05. Consequently, the area of a higher mode circular patch antenna is about 1.7 times larger in radius.

The far fields of TM_{nm} mode excited circular patch antennas with spherical coordinates can be calculated by the following equations:

$$E_0 \propto \cos n\theta \cdot \lceil J_{n+1}(k_0 a \sin\theta) - J_{n-1}(k_0 a \sin\theta) \rceil$$
$$E_\phi \propto \cos\theta \cdot \sin n\phi \cdot \lceil J_{n+1}(k_0 a \sin\theta) - J_{n-1}(k_0 a \sin\theta) \rceil \qquad (4.22)$$

where k_0 and $J_n(x)$ denote wave number in free space, and the nth order Bessel function [11]. Figure 4.14 shows typical radiation patterns of basic mode (left)

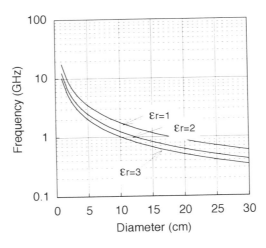

Figure 4.13 Relation between frequency and diameter of a circular patch antenna.

and higher mode (right) circular patch antennas whose gains are about 6 to 8 dBi.

4.4 Phased Array Antennas

Several antennas can be arrayed in space to make a directional pattern or one with a desired radiation pattern. This type of antenna is called an array antenna, which consists of more than two elements. Each element of an array antenna is excited by equal amplitude and phase, and its radiation pattern is fixed. On the other hand, the radiation pattern can be scanned in space by controlling the phase of the exciting current in each element of the array. This type of antenna is called a phased array antenna [13], which has many advantages in terms of mobile satellite communications such as compactness, light weight, high-speed tracking performance, and potentially low cost.

Arrays are found in many geometrical configurations. The most typical type in mobile satellite communications is the planar array, in which elements are arrayed in a plane to scan the beam at both azimuth and elevation angles to track the satellite. Figure 4.15 shows the most simple linear phased array that is composed of four elements, which have the same electrical characteristics, and are arrayed at equal spaces of d along the x axis.

In Figure 4.15, if each element is excited equally in amplitude and phase (the values of phase shifters are the same), the far field of the array antenna is given by

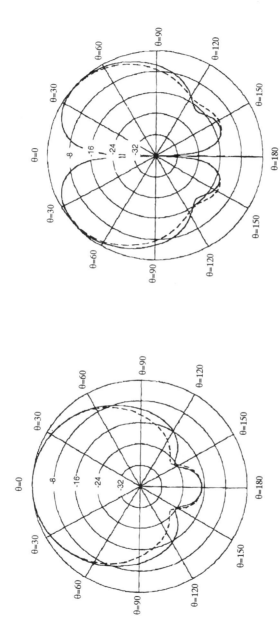

Figure 4.14 Typical radiation patterns of basic mode (left) and higher mode (right) excited circular patch antennas plotted in polar coordinates.

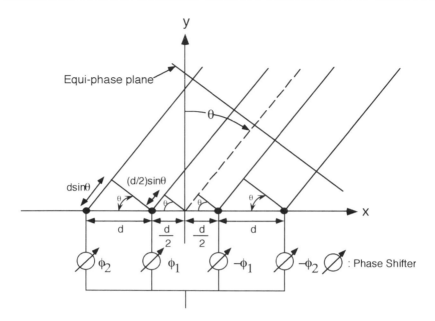

Figure 4.15 A simple linear phased array in which each element is excited by equal amplitude
and different phase. When the phases of excited currents of elements are fixed,
not controlled, the antenna is called an array antenna.

$$E(\theta) \propto \frac{e^{-jkr}}{r} \cdot D(\theta) \cdot \left[e^{-j\left(\frac{3kd}{2}\sin\theta - \phi_2\right)} + e^{-j\left(\frac{kd}{2}\sin\theta - \phi_1\right)} \right.$$

$$\left. + e^{j\left(\frac{kd}{2}\sin\theta - \phi_1\right)} + e^{j\left(\frac{3kd}{2}\sin\theta - \phi_2\right)} \right]$$

$$= \frac{e^{-jkr}}{r} \cdot 2D(\theta) \cdot \left[\cos\left(\frac{kd}{2}\sin\theta - \phi_1\right) + \cos\left(\frac{3kd}{2}\sin\theta - \phi_2\right) \right]$$

$$= \frac{e^{-jkr}}{r} \cdot 2D(\theta) \cdot AF \qquad\qquad (4.23)$$

where the phase center is at the coordinate origin, and $D(\theta)$ is the radiation
pattern of the element. The ϕ_1 and ϕ_2 including their signs are the values of
phase shifters, as shown in Figure 4.15. The coefficient AF is called the array
factor. The radiation pattern for the array antenna is found by multiplying
the radiation pattern of the element antenna and the array factor.

Example 4.3

The array factors AF_2 and AF_4 of linear arrays with two and four elements, excited by equal phase ($\phi_1 = \phi_2 = 0$), whose spacing between elements is half a wavelength ($d = \lambda/2$), are as follows:

$$AF_2 = \cos\left(\frac{\pi}{2}\sin\theta\right)$$

$$AF_4 = \cos\left(\frac{\pi}{2}\sin\theta\right) + \cos\left(\frac{3\pi}{2}\sin\theta\right) \qquad (4.24)$$

Figure 4.16 shows patterns of array factors for (a) two-element and (b) four-element linear arrays. The space between element is half a wavelength. The maximum value was obtained in the boresight direction (y axis).

In (4.21), the array factor will reach maximum in direction θ_0 when the following relations are satisfied. This can physically be explained by the fact that the phases of wave fronts become equal, as shown in Figure 4.15.

$$\frac{kd}{2}\sin\theta_0 - \phi_1 = \frac{3kd}{2}\sin\theta_0 - \phi_2 = n\pi \ (n = 0, \pm1, \pm2,) \qquad (4.25)$$

Therefore, in the case of $n = 0$

$$\phi_1 = \frac{kd}{2}\sin\theta_0 \ \text{and} \ \phi_2 = \frac{3kd}{2}\sin\theta_0 \qquad (4.26)$$

It is found that maximum gain can be obtained in the desired direction, and the beam can be scanned into a desired angle off the boresight direction.

Example 4.4

The radiation pattern of phased array antennas with two and four elements can be calculated by the following equation:

$$D_2(\theta) = \cos\left[\frac{\pi}{2}(\sin\theta - \sin\theta_0)\right]$$

$$D_4(\theta) = \cos\left[\frac{\pi}{2}(\sin\theta - \sin\theta_0)\right] + \cos\left[\frac{3\pi}{2}(\sin\theta - \sin\theta_0)\right] \qquad (4.27)$$

where θ_0 denotes the angle of scanned direction. Each element is assumed to be nondirectional, and element spacing is half a wavelength ($d = \lambda/2$).

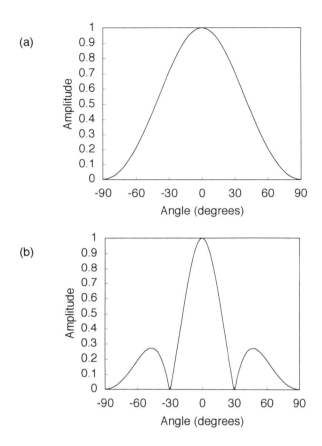

Figure 4.16 Array factors for (a) two-element and (b) four-element linear array antennas with element-spacing of half a wavelength.

Figure 4.17 shows radiation patterns of phased array antennas for (a) two-element and (b) four-element linear arrays. The beam is scanned at an angle of 30 degrees.

By controlling the excitation current of each element, the radiation pattern can be shaped for special applications. In order to suppress side lobes at the required level, excitation with tapered amplitudes is usually adopted. However, the suppression of sidelobe levels cannot be carried out without sacrificing antenna performance such as gain and half-power beamwidth. One way to optimize low side lobes and antenna performance is through Dolph-Tchebysceff distribution [14].

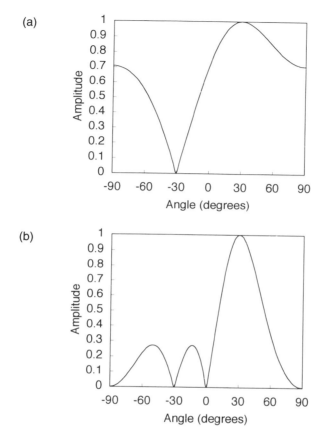

Figure 4.17 Radiation pattern for the phased array antenna with (a) two and (b) four elements, whose main beams are scanned at 30. Element spacing is half a wavelength and each element is assumed to be nondirectional.

Example 4.5

A phased array antenna with four elements, whose elements are excited by nonuniform current amplitudes of 0.7 (−3 dB), 1 (0 dB), 1 (0 dB), and 0.7 (−3 dB), respectively, from left to right along the x axis are shown in Figure 4.16. The directivity can be given by

$$D_4(\theta) = 1 \cdot \cos\left[\frac{\pi}{2}(\sin\theta - \sin\theta_0)\right] + 0.7 \cdot \cos\left[\frac{3\pi}{2}(\sin\theta - \sin\theta_0)\right]$$

Figure 4.18 compares radiation patterns of uniformly and nonuniformly excited phased array antennas. In a nonuniformly excited antenna, the ratios of current amplitudes fed into each element are 0.7 (−3 dB), 1 (0 dB), 1 (0 dB) and 0.7 (−3 dB) from left to right along the x axis, as shown in Figure 4.15. It is found that by tapering the excitation currents of elements, the side lobes can be lowered; however, the beamwidth widens.

From (4.24), the interelement phase difference ϕ required to scan the beam to an angle of θ_0 is given by

$$\phi = \phi_1 - \phi_2 = -\frac{2\pi d}{c} \cdot f \cdot \sin\theta_0 \qquad (4.28)$$

where c and f denote the speed of light and frequency, respectively.

When the frequency is changed by Δf, the beam-scanning and phase deviation angles $\Delta\theta_0$ and $\Delta\phi$ are found by

$$(\phi + \Delta\phi) = -\frac{2\pi}{c} \cdot (f + \Delta f) \cdot \sin(\theta_0 + \Delta\theta_0) \qquad (4.29)$$

When the value of phase is assumed constant or is dependent on frequency, and $\Delta\phi$ equals zero, then the right sides of (4.28) and (4.29) can be equalized as follows:

$$f \cdot \sin\theta_0 = (f + \Delta f) \cdot \sin(\theta_0 + \Delta\theta_0) \qquad (4.30)$$

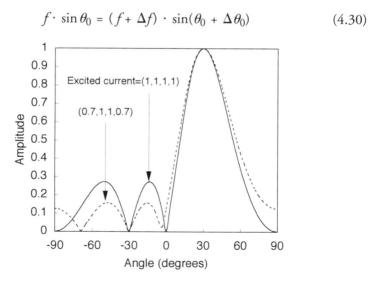

Figure 4.18 Radiation patterns of uniformly excited (solid line) and nonuniformly excited (dotted line) phased array antennas with four elements.

From (4.30), the value of $\Delta\theta_0$ is given by

$$\Delta\theta_0 = -\theta_0 + \sin^{-1}\left[\frac{\sin\theta_0}{1 + \frac{\Delta f}{f}}\right] \qquad (4.31)$$

Figure 4.19 shows the theoretical results calculated by (4.31). Theoretical analysis shows that the absolute value for beam-scanning error increases as beam-scanning angles and frequency differences increase. It is also found that if $\Delta f \geq 0$, then the angle of beam scanning will shift to smaller angles, and if $\Delta f \leq 0$, then it will shift to larger angles compared to the desired scanning angle θ_0 at frequency f. Further, the scanning errors are found to become smaller when $\Delta f \geq 0$ than when $\Delta f \leq 0$. In some cases, the beam-scanning error may become a serious problem in mobile satellite communications because transmitting frequency f_T and receiving frequency f_R are different (for mobile stations, $f_T \geq f_R$), and $|(f_T - f_R)/f_R|$ is about 7% in the L band (1.6/1.5-GHz frequency band), and about 50% in the Ka band (30/20-GHz frequency band). In Fig. 4.19, experimental data measured in the ETS-V experiments [15] is also indicated by the closed circles, where the value of $\Delta f/f \cong 0.066$

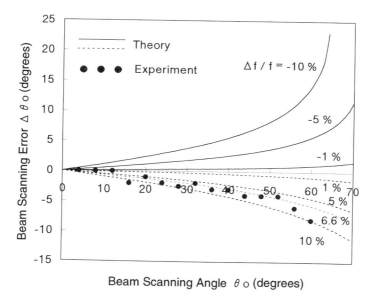

Figure 4.19 Theoretical prediction and experimental results of beam-scanning error caused by frequency differences.

(f_R = 1.545 MHz, f_T = 1.647 MHz). There is good correlation between the theoretical calculations and experimental results.

The reason beam-scanning errors are observed between transmitting and receiving frequencies is that the same phase shifters are used at both frequencies. To eliminate the scanning errors, the following equation must be satisfied, where ϕ_R and ϕ_T denote the values of phase shifts at f_R and f_T, respectively.

$$\frac{\phi_R}{f_R} = \frac{\phi_T}{f_T} \tag{4.32}$$

From (4.32), it is found that the adoption of frequency-dependent phase shifters can potentially reduce the beam-scanning errors between transmitting and receiving frequencies [15].

4.5 Satellite Tracking

4.5.1 Classification of Satellite Tracking

To track a satellite independently of mobile motion is an essential function for directional antenna systems during mobile satellite communications. In general, the required accuracy of tracking needs to be within 1 dB, which corresponds to an angular accuracy within about a half angle of HPBW. The tracking function needs two capabilities, beam steering and tracking control. There are two types of beamsteering methods. The first is mechanical steering, which physically directs the antenna to the satellite. The second is electronic steering, which directs the antenna beam by electronic scanning. A typical example of electronic steering is achieved through a phased array antenna. There are also two methods to control tracking. The first is the closed-loop method, which uses a signal from the satellite to search for and maintain in satellite direction. The step-track method, which is also called the hill-climb method, is a typical example of the closed-loop method. It is commonly used in antenna systems for INMARSAT-A shipborne terminals. Even in a closed-loop method, some sensors such as a gyrocompass and an accelerometer are needed to stabilize the antenna platform; however, the use of satellite signals is an essential factor in the closed-loop method. The second method is the open-loop method, which does not use signals from a satellite. The open-loop method only uses sensors such as a geomagnetic compass and a rate sensor, by which output information such as the looking angle to the satellite is calculated to direct the antenna beam. This method is sometimes called the program-tracking method.

Figure 4.20 classifies satellite-tracking functions. Table 4.2 and 4.3 compare mechanical and electronic tracking methods, and open-loop and closed loop tracking methods, respectively. The appropriate method can be chosen by considering the antenna type, required accuracy, installation, and the environmental conditions of mobiles. There are four to choose from: mechanical/open-loop, mechanical/closed-loop, electronic/open-loop, and electronic/closed-loop. An antenna system for the INMARSAT-A with a gain of about 21 dBi, with a mechanical steering and closed-loop tracking, has generally been adopted. Here, an antenna is mounted on a platform, which is stabilized to obtain a horizontal plane independent of ship motion. In using the step-track method, the condition of received signal levels has to be stable enough, and the beamwidth has to be narrow enough to detect signal variations.

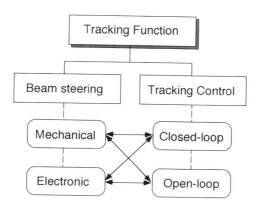

Figure 4.20 Classification of satellite tracking functions.

Table 4.2
Comparison Between Mechanical and Electronic Beam Steering

	Advantages	**Disadvantages**
Mechanical	Technically easy to fabricate Wide beam coverage Good axial ratios in wide beam coverage	Low reliability Low-speed beam scanning Large in volume and heavy
Electronic	Light and low profile High-speed beam scanning High reliability	Technically difficult to fabricate Narrow beam coverage Poor axial ratios in wide scanned coverage Excessive feeder loss

Table 4.3
Comparison Between Open-Loop and Closed-Loop Tracking Methods

Tracking Method	Principle	Advantages	Disadvantages
Open-loop (program)	Calculate the looking angle to the satellite using information from sensors	(1) Not affected by circumstances (2) Stable signal reception	(1) Sensors are required (2) Mobile's position is required
Closed-loop	Seek the maximum level of the signal from the satellite	(1) Mobile's position is not required (2) Sensors are not required	(1) Output of received signal levels are required (2) Inevitable signal variations (3) Affected by circumstances

In land mobile satellite communications, closed-loop tracking is considered to be very difficult. This is because received signal levels from satellites are not as stable as those from maritime satellite communications because of severe propagation environment due to fading, blocking, and shadowing. It is easy to adopt an open-loop method in aircraft and ships because they usually have navigation systems such as the Navy Navigation Satellite System (NNSS) and the Inertial Navigation System (INS). By using these navigation systems, mobiles can obtain information on the looking angles to the satellite. However, land mobiles in general have no navigational function, which is why many studies must be done to obtain a low-cost, open-loop method for land mobile satellite communications.

4.5.2 Antenna Mount Systems

In maritime satellite communications, an antenna is generally mounted on a platform, which has two horizontally stabilized axes (x axis and y axis) achieved by using a gyrostabilizer or sensors such as accelerometers or gyrocompasses. The stabilized platform provides a horizontal plane independent of ship motion such as roll or pitch. As shown in Figure 4.21, ship motion has seven components: roll, pitch, yaw, surge, sway, heave, and turn. Turn means a change in ship heading, which is intentional motion, not caused by wave motion, and the other six components are caused by wave motion. Surge, sway, and heave are caused by acceleration.

An antenna system mounted on a stabilizer is controlled both in elevation and azimuth directions in order to track a satellite. These types of mounts are

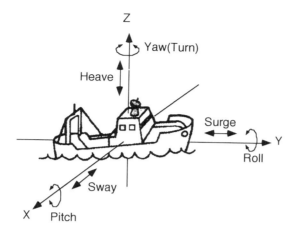

Figure 4.21 The seven components of ship motion.

called four-axis (X/Y/El/Az) mount systems, and are usually adopted for large ships, not for small ones, aircraft, or land mobiles because of their large volume and heavy weight. A two-axis stabilized system is usually adopted as a compact antenna for a small ship because of its compact and lightweight characteristics. There are two types of two-axis stabilized systems: one with an Az-El mount and the other with an X-Y mount. The Az-El mount is more popular because its height is lower than the X-Y mount. However, the Az-El system has an essential disadvantage in tracking accuracy at very high elevation angles near 90. A method to reduce this disadvantage has been proposed by Shiokawa and others [16]. Figure 4.22 shows the basic concepts of four-axis and two-axis (Az-El and X-Y) mount systems.

4.5.3 Closed-Loop Tracking

The most popular closed-loop method is the step track, which is also sometimes called the hill-climb method. Figure 4.23 shows a flow chart for the step-track

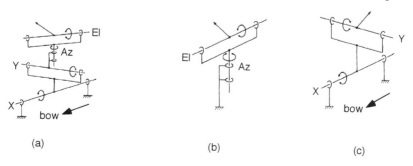

Figure 4.22 (a) Four-axis, (b) two-axis (Az-El), and (c) X-Y mount systems.

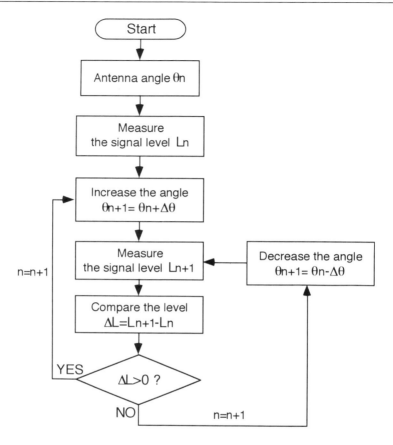

Figure 4.23 Flow chart for step-track algorithm.

algorithm. After the received signal level from the satellite is measured, the angle of the antenna is increased by a step angle of $\Delta\theta$. The levels are compared before and after the antenna angle is increased. If the level is increasing, the antenna angle is increased one more step. If the level is decreasing, the antenna is moved in the opposite direction to decrease the angle by one step. This procedure is carried out alternately between the two axes. As a result, the antenna beam is kept in the satellite direction. The step angle $\Delta\theta$ is chosen depending on system requirements. The performance of step-track tracking depends on the sensitivity to deviation of received the signal level ΔL caused by the increment of antenna angle $\Delta\theta$. For good performance, the antenna beamwidth and the received signal level have to be narrow and stable enough to detect the signal deviation that is caused by the angle increment of the antenna. This method is not applicable to small antennas with wide beamwidths or to land mobile communications, in which severe fading is expected.

4.5.4 Open-Loop Tracking

The open-loop method does not use signals from satellites in the system, so it requires sensors to provide information on the mobile's position. By knowing the position and the heading, mobiles can calculate the looking angle to the satellite, drive motors under the antenna mount, and send information to the antenna beam controller. Aircraft and ships have advantages in adopting the open-loop method because they generally have navigation systems such as a gyrocompass or an Inertial Navigation System (INS). Although land mobiles such as trucks and small cars generally have no navigation systems, car navigation systems using the global positioning satellite system (GPS) or state-of-the-art sensors are becoming popular. The biggest disadvantage of the open-loop method is that sensors such as gyrocompasses and the INSs can only provide the relative angle variations of the mobile (Table 4.4). The antenna has to carry out initial acquisition of the satellite direction by using other methods or sensors. The most attractive sensor to detect absolute direction is a geomagnetic sensor.

4.6 SENSORS

4.6.1 Geomagnetic Sensors

A conventional geomagnetic compass shows that magnetic north has a very simple configuration. However, it is not stable, accurate, or quick. Further, it

Table 4.4
Characteristics of Sensors for Open-Loop Antenna Tracking

Sensor	Accuracy	Response Time	Dynamic range	Characteristics ○ Advantages ● Disadvantages
Fiber-optic gyro	0.01 deg/sec	10 msec	100 deg/sec	○ High accuracy ○ Not affected by traveling conditions ● Cannot measure absolute direction ○ Accumulative errors
Magnetic flux gate	1 deg	Continuous	Depends on inner circuit	○ Can measure absolute direction ○ Not affected by traveling conditions ● Affected by magnetic disturbances

cannot provide information on the direction of electric signals. The most popular geomagnetic sensor for mobile satellite communications is a geomagnetic flux gate compass. The principle behind a flux gate compass is explained in Figure 4.24 (a,b). A flux gate compass consists of a highly permeable core, a toroidal coil, and sensing coils. The toroidal coil excited by dc voltage creates magnetic flux inside the core, as shown in Figure 4.24(a). Two sensing coils with the same performance are wound in series over the toroidal coil at opposite sides of the core, and they pick up current induced by the change in flux intensity inside the core, as shown in (Figure 4.24(b)). If external magnetic force is not applied to the sensor, no current is observed because both induced currents in the sensors are at the same amplitude and in opposite directions. When Earth magnetic flux is applied to the sensor shown in (Figure 4.24(b)), the induced current of the left sensor increases although the current of the right sensor decreases because of the additive and canceling effects of flux intensities, respectively. As a result, current can be measured at the output port of the sensing coil. By using a set of sensing coils, which are wound orthogonally over the core, the direction of real north can be established because north-south and east-west components of Earth magnetic flux can be obtained. Figure 4.25 is a photograph of a geomagnetic sensor.

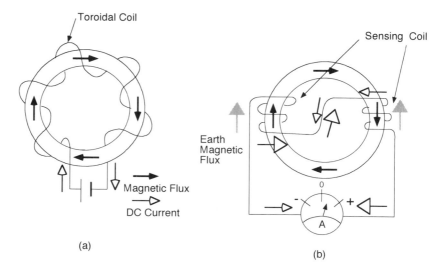

(a)

(b)

Figure 4.24 The principle behind a magnetic flux gate sensor: (a) the toroidal coil excited by dc voltage creating magnetic flux inside the core and (b) two sensing coils with the same performance wound in a series over the toroidal coil at opposite sides of the core picking up current induced by the change in flux intensity inside the core.

Figure 4.25 A photograph of a geomagnetic sensor.

A flux gate compass can give an absolute direction for magnetic north. However, its information has to be calibrated because it uses the magnetic flux of the Earth, which is affected by magnetic circumstances, especially in land mobile satellite communications. The first procedure is to calibrate the directional difference between magnetic north and true north. This calibration is relatively easy because variations depend on the location, and their values are published in marine charts or in scientific tables. The second procedure is to calibrate the magnetic effect of the mobile in which the compass is located. The accuracy of the magnetic flux gate compass is generally within few degrees.

4.6.2 Fiber-Optic Gyro

Gyro sensors are able to detect angular velocities or angles of mobiles in order to control their positions or attitudes without using outer information. The most typical gyro is a "top" spinning at high speed, which uses the law of conservation of momentum. A fiber-optic gyro uses the Sagnac effect, which is the propagation times of two lights propagating mutually in opposite directions in the same closed optical-fiber link. The phase difference $\Delta\phi_S$ between the two lights propagating in a closed optical fiber loop is proportional to the angular velocity Ω of the rotation of the loop. The phase $\Delta\phi_S$ is given by

$$\Delta\phi_S = \left(\frac{8\pi A}{\lambda c}\right) \cdot \Omega$$

$$= \left(\frac{4\pi Rl}{\lambda c}\right) \cdot \Omega \qquad (4.33)$$

where λ, c, A, R, and l are the wavelength of the laser, speed of light in a vacuum, the area of the loop, the radius of the loop, and the total length of the optical-fiber cable [17]. Although the sensitivity of the sensor depends on the area of the loop, a sensor with high sensitivity can be produced by using a multiwound optical-fiber coil.

Compared to mechanical gyros, a fiber-optic gyro has the following advantages:

1. Robustness to acceleration because it is nonmechanical;
2. A quick response time to startup;
3. A wide dynamic range.

Although there are several types of fiber-optic gyros, the most attractive fiber-optic gyro used as an antenna tracking sensor is the phase modulation type. As shown in Figure 4.26, it consists of an optical-fiber loop, a source for solid-state laser light, a phase modulator, and related optoelectronics devices.

Figure 4.27 is a photograph of a fiber-optic gyro. The general performance of a fiber-optic gyro is about 0 degrees ~ ± 100 deg/sec in the dynamic response ranges, and about ±0.01 in angle resolution.

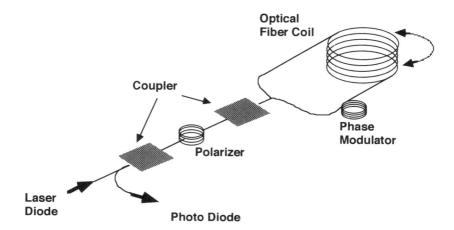

Figure 4.26 Basic configuration for a fiber-optic gyro.

Figure 4.27 A photograph of a fiber-optic gyro.

References

[1]   Ohmori, S., "Vehicle Antennas for Mobile Satellite Communications," *IEICE Trans.*, Vol. E74, No. 10, Oct. 1991, pp. 3210–321.

[2]   "Environmental Conditions and Test Procedures for Airborne Electronics, Electrical Equipment and Instruments," Radio Technical Commission for Aeronautics, RTCA Do-160A.

[3]   Stutzman, W. L., and G. A. Thiele, *Antenna Theory and Design*, New York, NY: John Wiley & Sons, 1981.

[4]   Blake, L. V., *Antennas*, Norwood, MA: Artech House, 1984.

[5]   Fujimoto, K., and J. R. James, *Mobile Antenna Systems Handbook*, Norwood, MA: Artech House, 1994.

[6]   Abramwitz, M., and I. A. Stegun, *Handbook of Mathematical Functions*, New York, NY: Dover Press, p. 370.

[7]   Stuzman, W. L., and G. A. Thiele, *Antenna Theory and Design*, New York, NY: John Wiley & Sons.

[8]   MSAT-X Technical Brochure, "Low Cost Ominidirectional Vehicle Antennas for Mobile Satellite Communications," Jet Propulsion Laboratory.

[9]   Kraus, J., "Helical Beam Antennas for Wideband Applications," *Proc. IRE*, Vol. 36, No. 10, Oct., 1948, p. 1236.

[10]   Adams, A., et al., "The Quadrifilar Helix Antenna," *IEEE Trans. AP*, Vol. AP-22,
        No. 2, March 1974, pp. 173–178.

[11]   Howell, J. Q., "Microstrip Antenna," *IEEE Trans. on Antenna and Propagation*, Vol.
        AP-23, No. 1, Jan. 1975, pp. 90–91.

[12]   Bahl, I. J., and P. Bhartia, *Microstrip Antennas*, Norwood, MA: Artech House, 1980.

[13]   Hansen, R. C., *Microwave Scanning Antenna*, Academic Press, 1966.

[14]   Dolph, C. L., "A Current Distribution for Broadside Arrays Which Optimizes the
        Relationship Between Beam Width and Side-Lobe Level," *Proc. IRE*, No. 34, June 1946,
        pp. 335–348.

[15]   Ohmori, S., S. Taira, and M. W. Austin, "Tracking Error of Phased Array Antenna,"
        IEEE Trans. on Antennas and Propagation, Vol. 39, No. 1, Jan. 1991, pp. 80–82.

[16]   Shiokawa, T., Y. Karasawa and Y. Yuki, "Reduction Method of Pointing Error of
        El/Az Mount for Maritime Satellite Communications," *Electronics and Communication
        in Japan*, Vol. 71, No. 1, Jan. 1988, pp. 106–115.

[17]   Vali, V., and R. W. Shorthill, *Applied Optics*, Vol. 15, No. 15, pp. 1099–1100.

5

Propagation Problems

5.1 Basic Knowledge of Propagation

5.1.1 Background

A mobile satellite communications system should be designed to provide communication services that are as high-quality and reliable as possible while being economical to operate. For digital communications, the service quality is expressed in terms of the bit error rate (BER) performance, which depends on the carrier-to-noise density (C/N_0) ratio. For example, the BER should not be greater than 10^{-4} for voice communication. The service reliability is expressed in terms of service availability, which is defined as the percentage of time that satisfactory communications quality is available or that a link has the specified C/N_0. Although the reliability and availability are related, they actually have different definitions in satellite communications systems.

For mobile satellite communications services, one must take into account outages due to obstruction of the line-of-sight path between a satellite and a mobile terminal as well as signal power fluctuation caused by interference from reflected radio waves. If the statistical characteristics of these propagation effects are not well known, the designed link budget may have insufficient C/N_0 or more C/N_0 than necessary. Mobile satellite communications users would also like to know what percentage of the time or of locations a given good-quality communication service is available in a given service area. For effective mobile satellite communications system design, we must quantitatively know propagation characteristics such as signal fading due to sea surface and ground surface reflections; shadowing from trees, buildings, utility poles, and terrain; Doppler effects due to movement of a mobile terminal; and so forth as well as rainfall attenuation and ionospheric effects. Such propagation characteristics have different statistical properties for land vehicles, vessels, and aircraft. The following is a summary of propagation characteristics for each type of mobile satellite communication channel. Detailed descriptions are given in later sections.

5.1.1.1 Land Mobile Satellite Communications Channels

When the direct signal transmitted from a satellite is obstructed by roadside trees, utility poles, buildings, or terrain, communication outages will occur. If the outage continues for a sufficiently long period of time, the communication channel will be disconnected. We have to know statistical characteristics of signal attenuation and its frequency as well as the duration of outages in given service areas such as urban, suburban, and rural areas.

Radio waves reflected from mountains, hills, and artificial structures such as buildings and bridges cause interference to the direct signal from a satellite, and cause amplitude and phase fluctuations in the received signal. Such multipath fading may degrade the BER performance for digital modulation.

5.1.1.2 Maritime Satellite Communications Channels

Multipath fading caused by reflections from the sea surface can significantly impair maritime satellite communications channels, especially at low elevation angles. Fading characteristics depend on the antenna gain of the ship's earth station, elevation angle, sea surface conditions (rough or smooth), and so forth. A number of multipath fading reduction techniques such as polarization shaping, antenna pattern shaping, multielement antenna systems, and error-correction coding have been proposed, and some of them are discussed in Section 5.3.

5.1.1.3 Aeronautical Satellite Communications Channels

Propagation conditions in aeronautical satellite channels are superior to those in land mobile and maritime channels, because there are no obstacles between a satellite and an aircraft earth station. However, at low elevation angles, and when a low-gain antenna is used, multipath fading caused by reflection from the sea or ground surface occurs, although it is less than that in the maritime case. In an aeronautical satellite communication system, an aircraft demodulator must track the received signal and remove the Doppler effect due to the high speed of flight using digital signal processing or using a pilot signal.

5.1.2 Free-Space Loss

To calculate the C/N_0 for the satellite link budget, it is necessary to know the following relationship between transmitted power and received power. The power flux density P_0, which is the signal power per unit area of a spherical surface at a sufficiently large distance R away from a transmitting antenna, can be expressed by

$$P_0 = \frac{P_t G_t}{4\pi R^2} \tag{5.1}$$

where P_t is the transmitted power and G_t is the transmitting antenna gain. The received power P_r at a receiving antenna is expressed by

$$P_r = P_0 A\eta = \frac{P_t G_t A\eta}{4\pi R^2} \tag{5.2}$$

where A is the aperture area of the receiving antenna and η is the antenna efficiency. The receiving antenna gain of a parabolic antenna can be expressed by

$$G_r = \frac{4\pi A\eta}{\lambda^2} \tag{5.3}$$

where λ is the wavelength. Then, (5.2) can be rewritten as

$$P_r = P_t G_t G_r \left(\frac{\lambda}{4\pi R}\right)^2 = \frac{P_t G_t G_r}{L_f} \tag{5.4}$$

where L_f is the free-space transmission loss, which can be expressed in decibels as

$$L_f = 20 \log_{10}\left(\frac{4\pi R}{\lambda}\right) = 92.448 + 20 \log_{10} f + 20 \log_{10} R \text{ (dB)} \tag{5.5}$$

where f is the frequency in gigahertz and R is the distance in kilometers. Figure 5.1 shows the free-space loss as a function of frequency for a distance of 36,000 km.

5.1.3 Rainfall Attenuation

Rainfall attenuation, which is caused by radio absorption and scattering from raindrops, can impair a satellite communication link at frequencies above 10 GHz, as well as increase the noise temperature and impairment of cross-

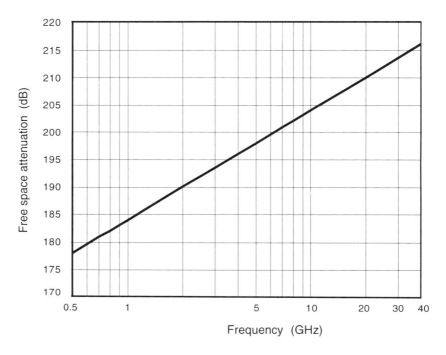

Figure 5.1 Free-space loss.

polarization discrimination. A number of methods have been proposed for predicting the rainfall attenuation. In this section, we introduce one recommended by the International Telecommunication Union (ITU) [1].

5.1.3.1 ITU Method of Predicting Rainfall Attenuation

The prediction method consists of the following steps:

Step 1

Calculate the effective rain height h_R:

$$h_R(\text{km}) = \begin{cases} 3.0 + 0.028\ \phi & \text{for } 0 \le \phi < 36 \text{ degrees} \\ 4.0 - 0.075(\phi - 36) & \text{for } 36 \text{ degrees} \le \phi \end{cases} \quad (5.6)$$

where ϕ is the latitude of the Earth station.

Step 2

Calculate the slant path length L_s (km) below the rain height:

$$L_S(\text{km}) = \frac{h_R - h_S}{\sin\ \theta} \quad \text{for } \theta \ge 5° \quad (5.7)$$

where θ is the elevation angle and h_S (km) is the Earth station's height above sea level. For $\theta < 5$ degrees, the path length is given by

$$L_S(\text{km}) = \frac{2(h_R - h_S)}{\left(\sin^2\theta + \dfrac{2(h_R - h_S)}{R_e}\right)^{1/2} + \sin\theta} \quad \text{for } \theta < 5° \quad (5.8)$$

where R_e is the equivalent radius of the Earth (8,500 km for $h_S \leq 1$ km).

Step 3

Calculate the horizontal projection L_G from the slant-path length.

$$L_G \text{ (km)} = L_S \cos\theta \quad (5.9)$$

Step 4

Evaluate the rain rate $R_{0.01}$ exceeded for 0.01% of an average year. If local statistical data for rain rate is not available, the ITU recommendation provides the estimated rain rate for various climatic zones.

Step 5

Calculate the reduction factor $r_{0.01}$ for 0.01% of the time:

$$r_{0.01} = \frac{1}{1 + L_G/L_0} \quad (5.10)$$

where

$$L_0(\text{km}) = 35 \exp(-0.015 \, R_{0.01}) \quad \text{for } R_{0.01} \leq 100 \text{ (mm/h)}$$

For $R_{00.1} > 100$ (mm/h), $R_{0.01} = 100$ (mm/h); that is, L_0 (km) $= 35 \exp(-1.5)$, is used.

Step 6

Calculate the attenuation coefficient γ_R (dB/km) using two parameters k and α:

$$\gamma_R \text{ (dB/km)} = k \, (R_{0.01})^\alpha \quad (5.11)$$

For linear and circular polarizations, these parameters can be obtained from Table 5.1 [2] and the following equations:

Table 5.1
Regression Coefficients for Estimating Rain Attenuation [2]

Frequency (GHz)	k_H	k_V	α_H	α_V
1	0.0000387	0.0000352	0.912	0.880
2	0.000154	0.000138	0.963	0.923
4	0.000650	0.000591	1.121	1.075
6	0.00175	0.00155	1.308	1.265
8	0.00454	0.00395	1.327	1.310
10	0.0101	0.00887	1.276	1.264
12	0.0188	0.0168	1.217	1.200
20	0.0751	0.0691	1.099	1.065
30	0.187	0.167	1.021	1.000
40	0.350	0.310	0.939	0.929

$$k = [k_H + k_V + (k_H - k_V) \cos^2 \theta \cos 2\tau]/2 \qquad (5.12)$$

$$\alpha = [k_H\alpha_H + k_V\alpha_V + (k_H\alpha_H - k_V\alpha_V) \cos^2 \theta \cos 2\tau]/2k \qquad (5.13)$$

where θ is the path elevation angle and τ is the polarization tilt angle relative to the horizon (τ = 45 degrees for circular polarization).

Step 7

The rainfall attenuation (dB) exceeded for 0.01% of an average year can be expressed by

$$A_{0.01} \text{ (dB)} = \gamma_R \cdot L_S \cdot r_{0.01} \qquad (5.14)$$

Step 8

We can estimate the rainfall attenuation A_P for other percentages P of an average year, in the range 0.001% to 1%, as follows:

$$A_P = A_{0.01} \cdot 0.12 \cdot P^{-(0.546 + 0.043 \log P)} \qquad (5.15)$$

Step 9

The worst-month percentage of time P_w when a given attenuation is exceeded is derived from the average annual percentage of time P, using the following relationship:

$$P = 0.3 \, P_w^{1.15} \qquad (5.16)$$

5.1.3.2 Rainfall Attenuation in a Link Budget

For a linearly operated satellite transponder, the total $(C/N_0)_t$ can be given, in absolute values, by

$$\left(\frac{C}{N_0}\right)_t^{-1} = \left[\frac{1}{A_u}\left(\frac{C}{N_0}\right)_u\right]^{-1} + \left[\frac{1}{A_u A_d}\left(\frac{C}{N_0}\right)_d\right]^{-1} \qquad (5.17)$$

where $(C/N_0)_u$ is the uplink C/N_0 and $(C/N_0)_d$ is the downlink C/N_0 in the clear sky condition, A_u is uplink rainfall attenuation, and A_d is downlink rainfall attenuation. Since most mobile satellite communication systems use the L-band or S-band frequencies for the satellite-mobile link, rainfall attenuation does not have any severe effects. However, the satellite-hub link uses frequencies above 10 GHz, for example, the Ku band or Ka band, and the total $(C/N_0)_t$ is affected by rainfall attenuation through A_u or A_d. The service availability is defined as the percentage of the time during which $(C/N_0)_t$ is larger than a specified value C/N_0, which can be calculated using (5.14) or (5.15). The required service availability is obtained by assigning a margin against rainfall attenuation to $(C/N_0)_t$ in the link budget.

Future mobile satellite communication systems will use the Ka band or millimeter waves for satellite-mobile links. In that case, rainfall attenuation may impair both the uplink and the downlink. Several methods [3] of compensating for rainfall attenuation, such as uplink power control [4,5], site diversity [6], and adaptive control of data rate [7,8] or forward error correction [9], have been proposed, and some of them have already been introduced into fixed satellite communication services.

5.1.4 Ionospheric Scintillation

Ionospheric effects are important at frequencies below 1 GHz. They may even be important at frequencies above 1 GHz. Ionospheric scintillation occurs as short-term rapid signal fluctuations, and is mainly caused by irregularities in the ionosphere ranging from altitudes of 200 to 600 km. The frequency dependence depends on the ionospheric conditions, but the attenuation varies approximately as the square of the wavelength. The effect is greater for lower frequencies and at lower latitudes. In the L band and S band, this effect can be ignored at medium latitudes except during periods of solar activity. When the Sun was very active, L-band enhancement and fading of 6 dB and −36 dB, respectively, were observed even at 37 degrees latitude [10].

5.1.5 Faraday Rotation

The polarization rotation of a linearly polarized wave, which is caused by free electrons in the path between a satellite and an Earth station, is referred to as Faraday rotation. For a total electron content of 10^{18} electrons/m^2, the polarization rotation angles are approximately 70 degrees at 850 MHz and 20 degrees at 1.6 GHz. The rotation angles vary approximately with $1/f^2$. When a linearly polarized wave is used for an Earth-satellite link, there is significant signal loss of 9.3 dB at 850 MHz and 0.5 dB at 1.6 GHz. This effect can be avoided by using circular-polarized signals or higher frequency bands.

5.2 Propagation in Land Mobile Satellite Communications

Recently, in the United States, Canada, Australia, Europe, and Japan, domestic mobile satellite communication services have started. The main purpose of these systems is to extend mobile telephone services to rural and remote areas where terrestrial cellular services are not provided. In this section, we present propagation characteristics for land vehicles. Since these recent mobile satellite communication systems use the L band (1.5 to 1.6 GHz), we mainly describe L-band propagation characteristics for high-elevation satellite channels. For low-elevation channels, several studies have been also carried out in Europe, Canada [11], and the United States[12], but due to the limited total pages of this book, we will confine ourselves to the high-elevation channels.

Propagation characteristics for terrestrial cellular systems have been studied by many researchers [13–15]. For cellular systems, in a typical urban environment, line-of-sight communication is not usually available due to blockage by buildings and other structures, but a mobile terminal receives many waves reflected from these structures and can conduct communication using these signals. On the other hand, mobile satellite communication can be expected to use the line-of-sight signal from a satellite because of the high elevation angles of satellites. Since the distance between a satellite and a mobile station is much greater than that between a moving vehicle and a fixed base station in a terrestrial cellular system, the received signal power at the mobile station is too weak, in most cases, for reflected signals to be used. Therefore, when the line-of-sight is blocked by roadside trees, buildings, overpasses, or utility poles, satellite communication will not be available. To design land mobile satellite communication systems, one needs information about the propagation statistics of multipath fading and shadowing.

Recently, the wideband properties of the mobile-satellite channel have been studied for broadband applications using geostationary or nongeostationary satellites. The wideband measurements have been carried out using a signal spread by a pseudonoise to obtain the power delay profiles. Lutz and others [16,17] found that the wideband properties are characterized by few echoes attenuated by 10 to 15 dB, with small delays, usually less than 600 nsec, and the delay spread tends to decrease for higher elevation angles.

5.2.1 Propagation Environment

Figure 5.2 shows a typical propagation environment for scattering, multipath fading, and shadowing. A vehicle runs at a distance of 5 to 20m from roadside trees and has an omnidirectional antenna, which has azimuthally uniform gain but elevational directivity, or a medium- or high-gain antenna with automatic tracking capability. When the direct signal transmitted from a satellite is obstructed by building, trees, utility poles, or other structures, the communication quality is degraded or communication may be completely blocked since a significant fraction of the total energy arrives at the mobile antenna through the direct component.

Other components are a specular wave reflected from buildings or flat terrain with some propagation delay and attenuation and diffuse waves randomly scattered by surrounding objects. The former component causes frequency-selective fading with deep fades [18]. However, for most service conditions with elevation angles greater than 20 degrees, the specular compo-

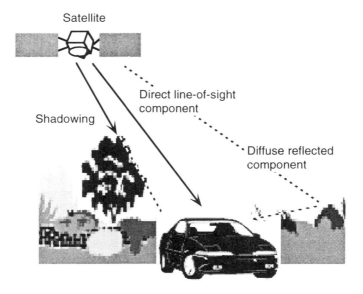

Figure 5.2 Typical propagation environment for land mobile satellite communications.

nents reflected from buildings are rarely received at the mobile antenna, and the ground reflected components can be reduced by the mobile antenna directivity, which decreases its gain rapidly at lower elevation angles. The diffusely reflected component causes multipath fading, which is described in the following section.

5.2.2 Shadowing and Multipath Measurements

Figures 5.3 and 5.4 show received power of an L-band continuous wave (CW) signal transmitted from the Japanese Engineering Test Satellite Five (ETS-V)

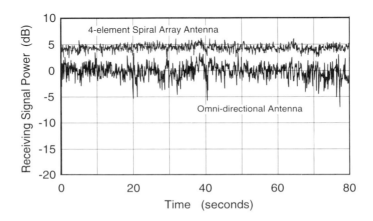

Figure 5.3 Received signal powers measured with an omnidirectional antenna and a medium-gain antenna under line-of-sight conditions.

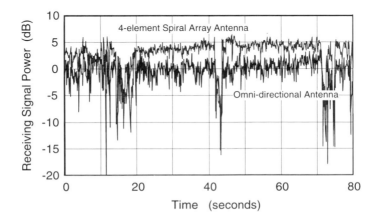

Figure 5.4 Received signal powers measured with the same antennas as in Figure 5.3 under shadowing by utility poles and overpasses.

and received by two antennas on the roof of the test van shown in Figure 5.5 under two conditions: 1) line-of-sight in Figure 5.3 and 2) shadowing in Figure 5.4. One antenna was a helical antenna, which had azimuthally omnidirectional gain of about 4 dBi and the other was a medium-gain four-element spiral array antenna with gain of about 12 dBi. Both antennas measured the signal power simultaneously. In Figures 5.3 and 5.4, the ordinate shows relative received powers with respect to each line-of-sight level. For clear comparison, the line-of-sight level of the medium-gain antenna is shown to be about 5 dB in these figures.

Figure 5.3 shows received power fluctuations of the line-of-sight signal. Both antennas receive not only the direct component but also diffuse components reflected from surrounding objects near the vehicle. The higher the directivity of the receiving antenna, the lower the fluctuation in received power because the antenna directivity can reduce the received power of scattered components. Such fading caused by various reflections from surrounding objects is referred to as multipath fading.

Figure 5.4 shows rapid signal attenuation, *shadowing,* caused by obstruction of the satellite path by roadside trees, utility poles, and overpasses. During shadowing, the received power decreases below the noise level of the receiver. If shadowing continues for too long, the satellite communication link will be disconnected.

Figure 5.5 Test van for land mobile satellite propagation measurements.

Based on a given threshold of the signal level, one can determine whether a propagation channel is in the fading state (below the threshold) or nonfading state (above the threshold). Statistical characteristics of fading, the fading and nonfading durations, depend on the service area, such as urban, suburban, or rural areas. To design land mobile satellite communication systems, one must study the statistical properties of propagation channels in these service areas.

5.2.3 Theoretical Model for Fading

Based on the above-mentioned scenario of land mobile satellite propagation, this section presents theoretical models for a fading channel. We identify three components of the received signal at a mobile antenna: the direct line-of-sight component, the specularly reflected component, and the diffuse multipath component. By combining these three components, we can model the propagation in a fading channel.

5.2.3.1 Rician Fading—The Direct and Multipath Components

When a mobile antenna receives the direct component and N reflected waves from surrounding objects, the received signal can be expressed by

$$r(t) = A \cos(\omega_0 t + \theta_0) + \sum_{i=1}^{N} c_i \cos(\omega_0 t + \omega_i t + \theta_0 + \theta_i) \qquad (5.18)$$

Here, the first term is the direct component, where A is the amplitude, ω_0 is the angular frequency, and θ_0 is the phase of the direct wave; and the second term is the N reflected waves, where c_i, $\omega_0 + \omega_i$, and $\theta_0 + \theta_i$ are the amplitude, angular frequency, and phase of the ith reflected signal, respectively. This equation can be rewritten as

$$r(t) = A \cos(\omega_0 t + \theta_0) + x(t) \cos(\omega_0 t + \theta_0) - y(t) \sin(\omega_0 t + \theta_0)$$
$$(5.19)$$

where

$$x(t) = \sum_{i=1}^{N} c_i \cos(\omega_i t + \theta_i)$$
$$y(t) = \sum_{i=1}^{N} c_i \sin(\omega_i t + \theta_i) \qquad (5.20)$$

Then, $x(t)$ and $y(t)$ are each the sum of statistically independent scattered waves, so they can be approximated by Gaussian random processes with an average of 0 and variance of σ^2 due to the central limit theorem. The amplitude $\rho(t)$ and phase $\phi(t)$ of signal $r(t)$ can be expressed by

$$r(t) = \rho(t)\cos(\omega_0 t + \theta_0 + \phi(t)) \tag{5.21}$$

where

$$\rho(t) = \sqrt{(A + x(t))^2 + y^2(t)} \tag{5.22}$$

and

$$\phi(t) = \arctan\left(\frac{y(t)}{A + x(t)}\right) \tag{5.23}$$

Since $x(t)$ and $y(t)$ are Gaussian, the joint probability density function (pdf) of Gaussian variables $x'(t)$ $(=A + x(t))$ and $y(t)$ can be expressed by

$$p(x', y) = \frac{1}{2\pi\sigma^2}\exp\left(-\frac{(x' - A)^2 + y^2}{2\sigma^2}\right) \tag{5.24}$$

Transformed by

$$x' = \rho\cos\phi$$
$$y = \rho\sin\phi \tag{5.25}$$

the joint pdf of amplitude ρ and phase ϕ is given by

$$p(\rho, \phi) = \frac{\rho}{2\pi\sigma^2}\exp\left(-\frac{\rho^2 + A^2 - 2\rho A\cos\phi}{2\sigma^2}\right) \tag{5.26}$$

We can get $p(\rho)$ by integrating (5.26) over ϕ from 0 to 2π:

$$p(\rho) = \int_0^{2\pi} p(\rho, \phi)\, d\phi$$

$$= \frac{\rho}{\sigma^2}\exp\left(-\frac{\rho^2 + A^2}{2\sigma^2}\right) I_0\left(\frac{\rho A}{\sigma^2}\right) \quad \text{for } \rho \geq 0 \tag{5.27}$$

where $I_0(z)$ is the zeroth-order modified Bessel function of the first kind:

$$I_0(z) = \frac{1}{2\pi} \int_0^{2\pi} \exp(z \cos \vartheta) d\vartheta \qquad (5.28)$$

Then, the envelope of the combined signal of the direct signal and diffuse components can be described by the Nakagami-Rice distribution or by the Rician distribution. The fading channel, which can be described by the Rician distribution, is referred to as the Rician fading channel. The power ratio of the direct component to the diffuse multipath components is referred to as the Rician factor, the K factor, or the Rician K factor, as defined below in absolute value:

$$K = \frac{A^2}{2\sigma^2} \qquad (5.29)$$

Usually, this factor is presented in dB:

$$K(\text{dB}) = 10 \log_{10}\left(\frac{A^2}{2\sigma^2}\right) \qquad (5.30)$$

The pdf of (5.27) is expressed by two parameters, A and σ^2. Using the normalized amplitude ρ', the pdf can be written with only one parameter, K. If the K factor is known, the fading distribution is described perfectly. When the normalized envelope is defined by

$$\rho' = \frac{\rho}{\sqrt{A^2 + 2\sigma^2}} \qquad (5.31)$$

the pdf of ρ' is rewritten as

$$p(\rho') = 2(1 + K)\rho' \exp[-(1 + K)\rho'^2 - K] I_0(2\sqrt{K(K + 1)}\rho') \qquad (5.32)$$

It can be shown that the pdf of ϕ can be written as

$$p(\phi) = \frac{1}{2\pi}\exp(-K)[1 + \sqrt{\pi K}\exp(K \cos^2 \phi)\cos \phi\{1 + \text{erf}(\sqrt{K}\cos \phi)\}] \qquad (5.33)$$

where the error function of $\text{erf}(z)$ is defined by

$$\text{erf}(z) = \frac{2}{\sqrt{\pi}} \int_0^z e^{-t^2} dt \tag{5.34}$$

Values of the Rician K factor depend on the propagation environment, satellite elevation angle, antenna radiation pattern, and so forth. For rural areas, the Rician K factors are about 10 to 20 dB for omnidirectional antennas. For a medium-gain antenna of 12 dBi, propagation measurements carried out in Japan using a satellite signal found K factors of 20 to 25 dB on expressways, as shown in Section 5.2.4, and 15 to 20 dB along ordinary roads in urban and suburban areas.

5.2.3.2 Rayleigh Fading—No Direct Signal

If the direct signal and specularly reflected components are totally blocked, the diffuse components dominate, and the envelope of the received signal can be described by the Rayleigh distribution. Such a channel is referred to as a Rayleigh fading channel. The Rayleigh distribution is a special case of the Rician distribution with no direct signal; that is $A = 0$. According to (5.26), the pdf of the Rayleigh distribution can be written as

$$p(\rho, \phi) = p(\rho)p(\phi) \tag{5.35}$$

where

$$p(\rho) = \frac{\rho}{\sigma^2} \exp\left(-\frac{\rho^2}{2\sigma^2}\right) \qquad \text{for } \rho \geq 0 \tag{5.36}$$

and

$$p(\phi) = \frac{1}{2\pi} \qquad \text{for } 0 \leq \phi < 2\pi \tag{5.37}$$

The Rayleigh distribution has only a single parameter of the scattered signal power, σ^2. Figure 5.6 compares the Rician and Rayleigh pdfs. The Rician pdf approaches the Gaussian one as the K factor tends to infinity.

5.2.3.3 Log-Normal Fading—Slow Shadowing

Slow shadowing is likely to be expressed by a log-normal distribution. When a random variable z has a log-normal distribution, its logarithm $\ln z$ has a

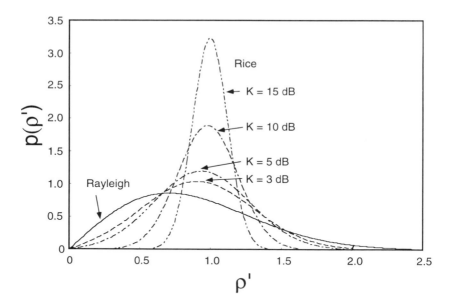

Figure 5.6 Rician probability density function and Rayleigh probability density function.

Gaussian (normal) distribution. Transforming the variables, we can get the pdf $p(z)$ and the cumulative density $P(z)$,

$$p(z) = \frac{1}{z} \frac{1}{\sqrt{2\pi\sigma^2}} \exp\left[-\frac{(\ln z - m)^2}{2\sigma^2}\right] \quad \text{for } z \geq 0 \tag{5.38}$$

$$P(z) = \frac{1}{2}\left[1 + \text{erf}\left(\frac{\ln z - m}{\sqrt{2\sigma^2}}\right)\right] \quad \text{for } z \geq 0 \tag{5.39}$$

where m and σ^2 are the mean and variance of $\ln z$, respectively.

When the power w in dB has a Gaussian distribution, the power s in watts has a log-normal distribution as follows:

$$w = \ln z = 10 \log s \tag{5.40}$$

$$p(s) = \frac{1}{s} \frac{(10/\ln 10)}{\sqrt{2\pi\sigma^2}} \exp\left[-\frac{(10 \log s - m)^2}{2\sigma^2}\right] \quad \text{for } s \geq 0 \tag{5.41}$$

where m and σ^2 are the mean and variance of $10 \log s$ or w, respectively. Figure 5.7 shows the log-normal distribution.

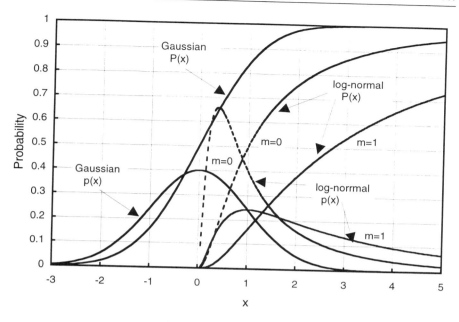

Figure 5.7 Normal (Gaussian) and log-normal distributions.

In Figure 5.7, Gaussian $p(x)$ shows the Gaussian pdf with mean of 0 and variance of 1, and Gaussian $P(x)$ is its cumulative density function. The log-normal pdfs with $m = 0$ and $m = 1$ and $\sigma^2 = 1$ are shown, and their cumulative density functions are also shown.

5.2.3.4 Power Spectrum—Doppler Effects

The above sections described the pdfs of the amplitude and phase of received signals. This section discusses the Doppler effect due to movement of mobile terminals. As shown in Figure 5.8, a mobile vehicle running at velocity of v, receives a sinusoidal wave coming from direction θ at transmitted frequency f_0. Due to the Doppler frequency shift, the received frequency of this signal is expressed by

$$f = f_0 + f_0 \frac{v}{c} \cos\,\theta = f_0 + f_m \cos\,\theta \qquad (5.42)$$

where c is the velocity of electromagnetic waves, $f_m = f_0(v/c) = v/\lambda$, and λ is wavelength. This f_m is referred to as the maximum Doppler shift, which characterizes fading speed. We can get the power spectral density $S(f)$ of the received diffuse components, which consists of reflected waves coming from

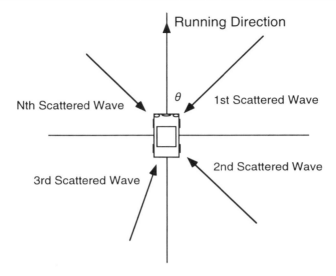

Figure 5.8 Scattered waves received at a moving vehicle.

many different directions, using the scenario shown in Figure 5.8. The power spectral density within the range of frequencies from f to $f + df$ is equal to that within the angular direction range from θ to $\theta + d\theta$ and that from $-\theta$ to $-\theta - d\theta$. Therefore, the power spectral density can be written as

$$S(f)df = -[\sigma^2 G(\theta)p(\theta) + \sigma^2 G(-\theta)p(-\theta)]d\theta \qquad (5.43)$$

where $p(\theta)$ is the pdf of waves coming inside the angular direction range from θ to $\theta + d\theta$, and $G(\theta)$ is the antenna gain for the direction θ. For an isotropic antenna, $G(\theta) = 1$, and for a half-wavelength dipole antenna, $G(\theta) = 1.6$. Then, σ^2 is defined as the signal power received by an isotropic antenna.

From (5.42), we can get

$$df = -f_m \sin \theta \, d\theta = -\sqrt{f_m^2 - (f - f_0)^2}\, d\theta$$

Then, the power spectral density is given by

$$S(f) = \frac{\sigma^2[G(\theta)p(\theta) + G(-\theta)p(-\theta)]}{\sqrt{f_m^2 - (f - f_0)^2}} \qquad (5.44)$$

Assuming that incidental angles of the diffuse components are uniformly distributed and the received antenna is omnidirectional, the power spectral density is expressed by

$$S(f) = \frac{1.6\sigma^2}{\pi\sqrt{f_m^2 - (f - f_0)^2}} \quad \text{for } |f - f_0| \le f_m$$

$$= 0 \quad \text{for } |f - f_0| > f_m \quad (5.45)$$

Figure 5.9 shows this power spectral density. The effective bandwidth of the power spectrum corresponds to twice the maximum Doppler shift f_m. The larger the bandwidth of the power spectrum, the faster the fading speed. Many interesting statistical properties, such as level crossing rates and fading duration, can be obtained from the power spectrum.

5.2.3.5 Power Delay Profile

When the satellite transmits a narrowband pulse signal and a moving vehicle receives it, the vehicle may also receive many echoes coming through many different paths. The power delay profile is defined as the received power as a function of delay time and a received frequency, and provides information on the propagation characteristics of multipath channels. Several parameters for characterizing the multipath channels are described as follows. The mean excess delay d is the mean of the normalized power delay profile $E(\tau)$, and is expressed as

$$d = \int_0^\infty \tau E(\tau) d\tau \quad (5.46)$$

The delay spread S is defined as the standard variation of τ.

$$S = \sqrt{\int_0^\infty \tau^2 E(\tau) d\tau - d^2} \quad (5.47)$$

The upper bound of transmission rates in a given propagation channel without serious frequency-selective fading depends on the coherent bandwidth,

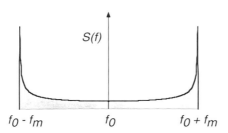

Figure 5.9 Power spectral density of reflected diffuse waves.

which is defined as the frequency separation at which the correlation of two signals is 0.5. The power delay profile is related to amplitude frequency characteristics, which is an amplitude variation as a function of frequency, by the Fourier transform. When the power delay profile $E(\tau)$ can be approximated by an exponential function $\exp(-\tau/\sigma)$, the coherent bandwidth B_c is expressed as $B_c = 1/(2\pi S)$.

Instead of transmitting a pulse signal, a spread spectrum signal can be used to measure the impulse response of the propagation channel. Several studies have been carried out to investigate the broadband characteristics of satellite-to-mobile channels. Ikegami and others [19] measured the power delay profile using the spread spectrum method via the ETS-V at 1.5 GHz. In land mobile satellite channels at moderate elevation angles of about 45 degrees, they did not observe multipath components with an excess delay of more than 1 ms, which is a typical value to terrestrial mobile channels. They concluded that wideband systems more than 1 MHz wide are available in most urban areas. Jahn and others [17] measured wideband channel characteristics at 1.8 GHz using a spread spectrum signal transmitted from an aircraft. They found that the wideband properties of the land mobile satellite channel are characterized by few echoes with small delays, usually less than 600 nsec, the echoes are strongly attenuated by 10 to 15 dB, and the delay spread decreases as the elevation angles increase. Some wideband trials carried out for low Earth orbit (LEO) systems suggest that the delay spread is in the order of less than 0.25 ms, and indicate no serious problems associated with excess delays.

5.2.4 Empirical and Statistical Models for Fading

This section presents empirical models and statistical models for shadowing and multipath fading based on propagation measurements using satellites, helicopters, and towers.

5.2.4.1 Empirical Roadside Shadowing Model

The statistical characteristics of fading in land mobile satellite channels, which were derived from helicopter-mobile and satellite-mobile measurements in central Maryland, performed by Vogel and Goldhirsh [20,21], are shown in Figure 5.10. The abscissa shows the signal fade with respect to the line-of-sight level, and the ordinate shows the percentage of the distance for which the fade was larger than the value on the abscissa. The empirical roadside shadowing (ERS) model, which is obtained by the best fit to the measured cumulative distributions of the fade at 1.5 GHz, is expressed by

$$A(P, \theta) = -M(\theta) \ln P + N(\theta) \tag{5.48}$$

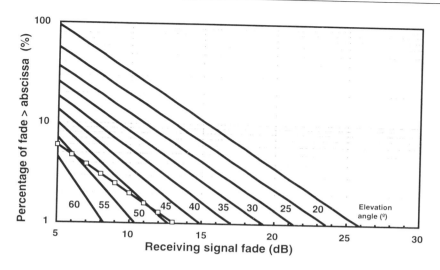

Figure 5.10 Cumulative fade distribution at 1.5 GHz derived from the empirical roadside shadowing model for different elevation angles. Squares show a fade distribution obtained from Australia runs at elevation angle of 51 degrees (*After:* [21]).

where A is the signal fade in dB, P is the percentage of the distance over which the fade is exceeded, and θ is the satellite elevation angle. The elevation angle dependence in the coefficients of M and N can be expressed by second- and first-order polynomials, respectively, as follows:

$$M(\theta) = -0.002\theta^2 + 0.0975\theta + 3.44$$

and

$$N(\theta) = -0.443\theta + 34.76 \qquad (5.49)$$

As shown in Figure 5.10, for example, the percentage of the distance over which the fade exceeds 5 dB is 4.6% at an elevation angle of 60 degrees, 9.9% at 50 degrees, 18.3% at 40 degrees, and 36.9% at 30 degrees. The lower the satellite elevation angle, the worse the signal fade. At an elevation angle of 20 degrees, the percentage of the distance reaches 95.0%. This shows that the propagation condition at low elevation angles becomes very severe for land mobile satellite communication services. Conversely, high-elevation satellites have definite advantages for mobile satellite communications.

5.2.4.2 Fade Statistics Measured in Japan

This section presents fade statistics measured along 3,513 km of expressways, which connect almost all main cities and play a very important role in land transportation in Japan [22].

A 1.5-GHz left-hand circular-polarized continuous wave transmitted from the Japanese ETS-V satellite was received by an electronically steerable 19-element phased array antenna, which was attached to the top of a test van, with an antenna gain of 12 dBi at an elevation angle of 45 degrees and a 3-dB beamwidth of 28 degrees. When in line of sight to the satellite, the signal-to-noise ratio was approximately 25 dB. The antenna beam tracked the satellite under the control of two vehicle motion sensors: a geomagnetic sensor and a fiber-optic gyroscope. The propagation data were acquired along ten main expressways in Japan (abbreviated as TOHO, TOME, HOKU, KANZ, CHUO, MEIS, CHGO, KYUS, NAGA, and MIYA in Figure 5.11). Table 5.2 shows the distances and satellite elevation angles, ranging from 40 to 48 degrees, along these roads.

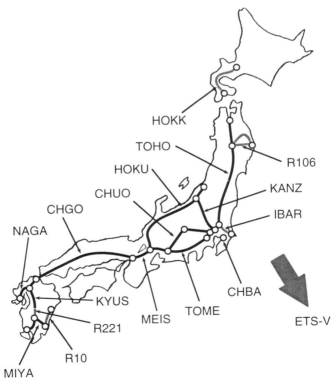

Figure 5.11 Expressways used for propagation measurements in Japan.

Table 5.2

Routes, K-Factors, Fade Levels at 1% and 10% Probabilities, Cumulative Probabilities at 5 and 10 dB Fades, Elevation Angles, and Total Distances for Propagation Measurements

Route	K factor (dB)	1% (dB)	10% (dB)	5 dB (%)	10 dB (%)	El. (deg.)	Distance (km)
TOHO	22.2	26.0	1.0	2.9	2.8	42–47	681
TOME	20.7	28.4	1.4	3.6	3.3	47	401
HOKU	23.6	31.4	23.4	12.6	12.5	45–47	495
KANZ	24.7	29.6	1.7	8.9	8.7	45–47	254
CHUO	19.8	27.7	2.9	8.0	7.5	46–47	292
MEIS	22.3	25.3	1.4	3.9	3.0	47	180
CHGO	19.7	29.0	3.7	9.0	7.8	46–47	566
KYUS	20.6	29.2	7.8	10.3	9.9	46–47	331
NAGA	21.4	28.9	3.2	9.0	8.8	46	229
MIYA	22.0	27.7	1.0	3.8	3.6	47–48	84

Figure 5.12 shows cumulative fade distributions of received signals with respect to the line-of-sight level. The ordinate shows the percentage of the distance over which the signal fade exceeded the abscissa value. Since the ordinate has a Gaussian scale, a straight line in this figure indicates that the pdf of the received signal envelope is a normal distribution, or a Rician distribution with a large Rician K factor. Since every curve is approximately straight

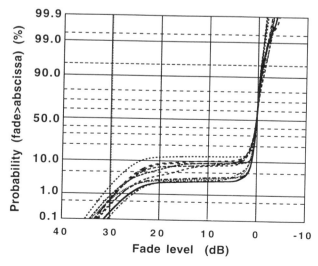

Figure 5.12 Cumulative fade distribution measured on expressways in Japan.

between the fades of −2 dB and +2 dB, the pdf of the received signal envelope can be expressed by (5.32).

The second column in Table 5.2 shows that the Rician K factors, which were obtained from the best fit of (5.32) to the measured fade data between the fades of −2 dB and +2 dB, ranging from 20 to 25 dB. The narrower the antenna beam, the fewer the diffuse components scattered from surrounding objects; hence, the larger the Rician K factor. Low-gain or azimuthally omnidirectional antennas receive more multipath diffuse components, so their fade distributions have smaller Rician factors.

Table 5.2 also shows the fade levels at 1% and 10% probabilities, and probabilities at the fades of 5 and 10 dB. The fade level at 10% probability, or the 10% probability level, is defined as the fade level for which the signal attenuation exceeded 10% of the total length of the roads. Among expressways, the 10% probability levels range from 1 to 4 dB, except for the HOKU and KYUS roads, which have mountainous areas and many tunnels. The 1% probability levels reached a noise level of 25 dB in our measuring system. Various investigations into fade statistics have been reported; for example, at similar elevation angle measurements, fade statistics for Australia show that the 10% probability levels are 6.1 dB [21] and 8.8 dB [23]. The 10% probability level was 5.6 dB [24] and 5.0 dB [25] in England and 6.1 dB in the United States [21]. Therefore, the 10% probability levels measured along Japan's expressways are less than those measured in those countries, owing to less shadowing and a narrower antenna beamwidth.

As shown in the fifth and sixth columns of Table 5.2, the percentages of distance over which the fade exceeded 5 dB, or the 5-dB fade probability, were approximately the same as those for the 10-dB fade probability. Therefore, a fade margin of more than 5 dB assigned in a satellite link design to combat shadowing and blockage cannot extend the availability of communication services. However, with fade margins of about 5 dB, land mobile satellite communication services will be available over at least 90% of the total length of expressways, even taking into account tunnels.

In Figure 5.12, every curve of the cumulative fade distributions inclines moderately in the fades between 5 and 20 dB. This is because the probability of fades occurring between 5 and 20 dB is lower. This indicates that land mobile propagation channels along expressways can be classified into two states: 1) line-of-sight and 2) shadowing or blockage by man-made structures such as tunnels, signposts, and overpasses.

5.2.4.3 Fade Duration Statistics

To design a digital modem, an error-correction scheme, and a data format for such propagation channels, it is important to know the time duration or the distance over which the channel is available and unavailable for sending and receiving data.

Vogel and others [26,27] measured statistical characteristics of fade, fade duration, and nonfade duration for L-band land mobile satellite propagation using the ETS-V in southeastern Australia. The receiving antenna was a crossed drooping dipole antenna having a 4-dBi gain, an azimuthally omnidirectional radiation pattern, and a relatively flat elevation pattern over 15 degrees to 75 degrees.

The cumulative distribution of the fade duration was modeled by the following log-normal distribution [27]:

$$P(d_f > d \mid A > A_q) = \frac{1}{2}\left[1 - \text{erf}\left\{\frac{(\ln d - \ln \alpha)}{\sqrt{2\sigma^2}}\right\}\right] \qquad (5.50)$$

where $P(d_f > d|A > A_q)$ represents the probability that the fade duration distance d_f exceeds duration d under the condition that fade A exceeds threshold A_q (in the fade state), erf( ) is the error function (cf. (5.34)), $\ln \alpha$ is the mean of $\ln d$, and σ^2 is the variance of $\ln d$.

The joint probability that the channel is in the fade state $(A > A_q)$ and that d_f exceeds d is given by

$$P(d_f > d, A > A_q) = P(d_f > d|A > A_q)P(A > A_q) \qquad (5.51)$$

where $P(A > A_q)$ is the cumulative probability that fade A exceeds threshold A_q and is given by their ERS model, (5.48), as shown in Figure 5.13 for *extreme* shadowing and *moderate* shadowing.

Figure 5.14 shows an example of the best fit cumulative distributions (5.50) of the fade duration for threshold A_q of 5 dB, which exhibits extreme, moderate, and light shadowing. This figure shows a 10% probability of a fade duration exceeding 1m and only about 1.5% of exceeding 3m. Measured fade statistics show that $P(A > 5 \text{ dB})$ is 6% for moderate shadowing and 35% for extreme shadowing. Therefore, the probability that the channel is in the fade state and that the fade duration exceeds 1m is 0.6% for moderate shadowing and 3.5% for extreme shadowing.

Table 5.3 shows fade duration regression values α and σ for the threshold A_q of 5 dB for light, moderate, and extreme shadowing, measured in Australia.

5.2.4.4 Nonfade Duration Statistics

Using the same propagation data set [26,27], the cumulative distribution of the nonfade duration can be modeled by the following power distribution:

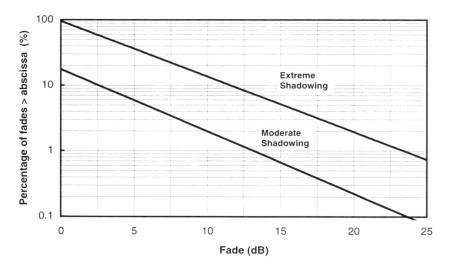

Figure 5.13 ERS model for best fit distribution from measurements in Australia at elevation
angle of 51 degrees (*After:* [21]).

$$P(d_{nf} > d | A < A_q) = \beta(d)^{-\gamma} \qquad (5.52)$$

where $P(d_{nf} > d | A < A_q)$ is the probability that nonfade duration d_{nf} exceeds
nonfade duration d under the condition that fade A is smaller than the threshold
A_q (in the nonfade state). The values of β and γ are shown in Table 5.4.
Figure 5.15 shows the cumulative distributions of nonfade duration for light
and extreme shadowing. This figure shows a 6.2% probability of a nonfade
duration exceeding 10m for light shadowing and 1.8% of exceeding 10m for
extreme shadowing. The joint probability that the channel is in the nonfade
state and that the nonfade distance duration exceeds d can be calculated by
an equation similar to the one used for fade duration probability.

5.2.5 Attenuation Caused by Trees, Buildings, and Utility Poles

5.2.5.1 Vegetative Attenuation—Frequency and Seasonal Dependence

The attenuation of a signal passing through a canopy of red pine foliage was
measured by Ulaby and others [28] at an elevation angle of 50 degrees. The
path length through the canopy was approximately 5.2m and the average
attenuation coefficient was approximately 1.8 dB/m in the L band (1.6 GHz).
The results at 1.6 GHz (f_L) and 870 MHz (f_{UHF}) can be expressed by

$$A(f_L) = A(f_{UHF}) \sqrt{\frac{f_L}{f_{UHF}}} \qquad (5.53)$$

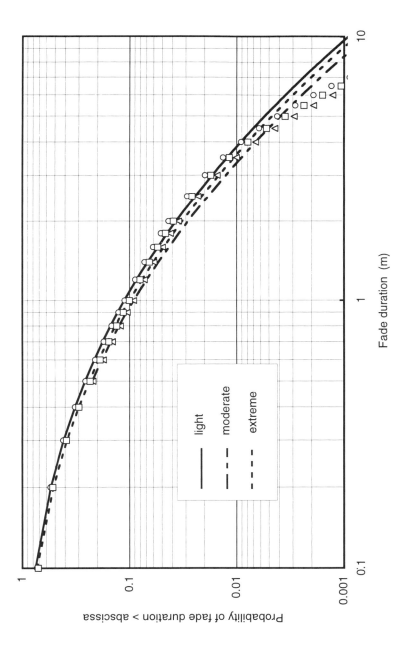

Figure 5.14 Log-normal cumulative distributions of fade duration for the threshold of 5 dB for light, moderate, and extreme shadowing, which are shown by a solid, a dot-and-dash, and a broken line, respectively. Open circles, triangles, and squares are fade duration simulated by a Markov transition model, as shown in Section 5.2.6.2.

Table 5.3
Values of Parameters for Fade-Duration Model [27]

Run	Shadowing	α	σ
383	Light	0.23	1.21
342	Moderate	0.20	1.21
343	Extreme	0.21	1.22
409	Extreme	0.47	1.38

Table 5.4
Values of Parameters for Nonfade Duration Model [27]

Run	Shadowing	β	γ
383	Light	0.215	0.55
342	Moderate	0.196	0.61
343	Extreme	0.117	0.84
409	Extreme	0.234	0.74

where $A(f_L)$ and $A(f_{UHF})$ are the attenuation at 1.6 GHz and 870 MHz, respectively. The results shows that the higher the frequency, the greater the vegetative attenuation.

The vegetative attenuation depends on the type of foliage, density of branches, tree height, and so forth. The attenuation caused by Callery pear trees was measured in October 1985 (full foliage) and March 1986 (bare branches) by Goldhirsh and Vogel [29] at 870 MHz. The results at elevation angles from 15 degrees to 40 degrees show

$$A(\text{Leaf}) \approx 1.35A(\text{Bare}) \tag{5.54}$$

which shows that, in the Callery pear tree, the attenuation contribution from trees leaves is only 35% greater in dB than the attenuation from trees without leaves. Therefore, the majority of the attenuation is caused by scattering and absorption from the tree branches.

5.2.5.2 Vegetative Attenuation—Theoretical Model

Since the shape of a tree is complicated, it is not easy to calculate the attenuation effect due to vegetative shadowing. A simple model [30,31] has been proposed to calculate vegetative attenuation from the cross-section between the first

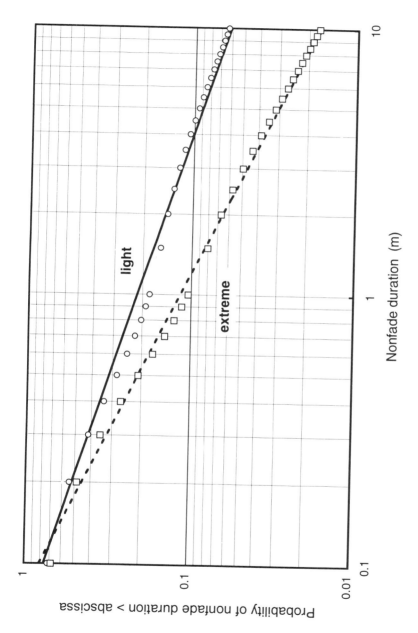

Figure 5.15 Cumulative distribution of nonfade duration for the threshold of 5 dB for light and extreme shadowing, which are shown by a solid and a broken line, respectively. Open circles and squares are nonfade duration simulated by a Markov transition model, as shown in Section 5.2.6.2.

Fresnel zone and a triangle that is a model of a tree, as shown in Figure 5.16. The area S_0 of the first Fresnel ellipsoid can be expressed by

$$S_0 = \frac{\lambda \pi (d \cos \theta + \lambda/4)}{\cos^3 \theta} \qquad (5.55)$$

where λ is the wavelength, d is the distance between a tree and an antenna, and θ is the elevation angle. Then, using the area S_1 occupied by a triangle-shaped tree inside the first Fresnel ellipsoid S_0, the attenuation A due to vegetative shadowing is given by

$$A = -20 \log\left(\frac{S_0 - S_1}{S_0}\right) \qquad (5.56)$$

Yoshikawa and others [30,31] reported that the theory correctly explains the average decrease of signal attenuation with increasing distance d between the tree and antenna in the fade range <10 dB, although, with decreasing distance d, the calculated attenuation increases monotonically whereas the measured attenuation increases vibrationally. For fades >10 dB, in particular, large vibrational changes in the fade are observed, as the distance decreases. This fade variation is caused by diffraction and interference of radio waves passing through leaves and the spaces between tree branches.

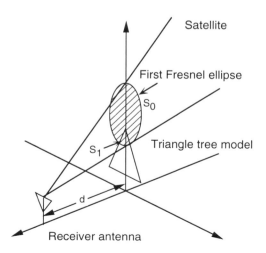

Figure 5.16 Vegetative shadowing model using the first Fresnel zone.

5.2.5.3 Attenuation by Roadside Utility Poles

Modeling a roadside utility pole as an infinitely long cylinder with perfect conductivity, we can calculate the diffraction pattern using the electromagnetic theory. The received signal power at a mobile antenna is expressed as the sum of the direct wave and scattered waves. Figure 5.17 shows the theoretical calculation and results measured using the ETS-V satellite at an elevation angle of 47 degrees. This good coincidence shows that the fluctuations in the received signal power due to roadside poles can be explained by a diffraction model with a perfectly conducting cylinder of infinite length. This model calculation can be used to evaluate amplitude and phase behaviors at a mobile modem under different propagation conditions: frequency, antenna pattern, elevation angle, distance from a pole, and so forth.

5.2.5.4 Attenuation by Buildings—Knife-Edge Model

Attenuation of the line-of-sight signal caused by trees, buildings, and irregular terrain can be estimated by modeling the form of these obstacles as a diffracting knife-edge of negligible thickness or as a thick smooth object. The exact solution to the knife-edge diffraction problem is well known.

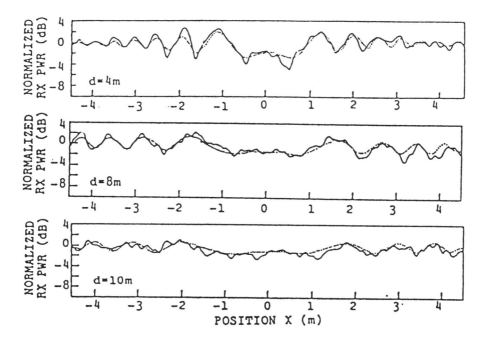

Figure 5.17 Measured and calculated signal amplitude and phase fluctuation as a function of perpendicular location x toward a satellite, in the L band (1,544 MHz).

For a single knife-edge obstacle, as shown in Figure 5.18, the attenuation can be expressed by

$$A(\text{dB}) = 10 \log 8 - 10 \log \left[(1 - 2C(v))^2 + (1 - 2S(v))^2 \right] \quad (\text{dB})$$

$$(5.57)$$

where $C(v)$ and $S(v)$ are the Fresnel integrals,

$$C(v) = \int_0^v \cos\left(\frac{\pi}{2} t^2 \right) dt$$

$$S(v) = \int_0^v \sin\left(\frac{\pi}{2} t^2 \right) dt$$

$$v = h\sqrt{\frac{2}{\lambda}\left(\frac{1}{d_1} + \frac{1}{d_2} \right)}$$

$$(5.58)$$

where h is the height of the obstacle above the straight line joining the two ends of the path (if the height is below this line, h is negative), d_1 and d_2 are the respective distances between the two ends and O, and λ is the wavelength.

For $v > -0.7$, the attenuation can be expressed approximately by [32]

$$A \ (\text{dB}) = -6.9 - 20 \log(\sqrt{(v - 0.1)^2 + 1} + v - 0.1) \quad (5.59)$$

Figure 5.19 shows knife-edge diffraction attenuation calculated from (5.59).

Yoshikawa and others [30,31] used this knife-edge diffraction model to calculate attenuation due to a building using the following parameter exchanges. As shown in Figure 5.20, because $d_1 \gg d_2$, we have

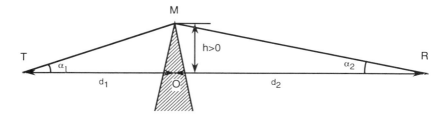

Figure 5.18 Knife-edge obstacle.

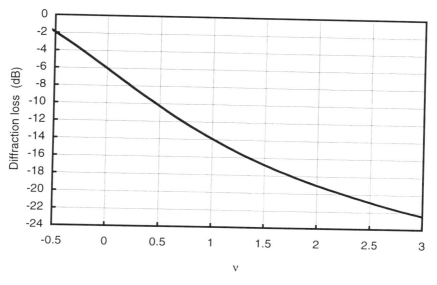

Figure 5.19 Knife-edge diffraction attenuation.

$$\nu = h\sqrt{\frac{2}{\lambda}\frac{1}{d_2}} \tag{5.60}$$

and

$$d_2 = r\cos\delta$$
$$h = r\sin\delta \tag{5.61}$$

where

$$r = \sqrt{d_3^2 + (H - h_a)^2} \tag{5.62}$$

and

$$d_3 = d/\sin\theta$$
$$\delta = \theta_3 - El \tag{5.63}$$
$$\theta_3 = \arctan[(H - h_a)/d_3]$$

This calculated loss agreed well with the loss measured using ETS-V and INTELSAT satellites. If we know the elevation angle of the satellite, building height, antenna height, distance of antenna from the building, and frequency,

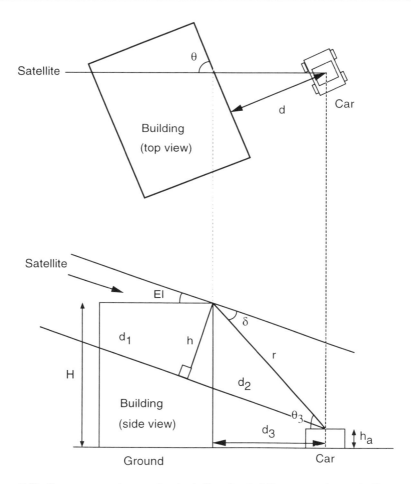

Figure 5.20 Parameter exchanges for the knife-edge building model. (*After:* [30]).

then we can estimate the diffraction loss near a building and estimate available or unavailable areas for satellite communications.

5.2.6 Propagation Model

We have discussed probability distribution models based on several pdfs, empirical regression models, and theoretical models based on the electromagnetic theory. In this section, we present propagation models that can be used to evaluate the performance of modems, voice codecs, and forward error-correction codecs.

5.2.6.1 Loo's Propagation Model

A statistical propagation model based on pdfs for shadowing and multipath fading has been proposed by Loo [33]: 1) multipath fading under the line-of-sight condition is described by Rician fading, and 2) shadowing of the line-of-sight signal is described by log-normal distribution. Combining 1) and 2), multipath fading is expressed by a Rician process with log-normally distributed Rician K factor. The pdf of the signal amplitude r is given by

$$p(r) = \frac{r}{\sqrt{2\pi}s\sigma^2} \int_0^\infty I_0\left(\frac{rz}{\sigma^2}\right)\frac{1}{z} \cdot \exp\left[-\frac{(\ln(z) - m)^2}{2s^2} - \frac{(r^2 + z^2)}{2\sigma^2}\right] dz \qquad (5.64)$$

where σ^2 is the mean of multipath scattered power, and m and s^2 are the mean and variance of the line-of-sight envelope, respectively. The first term of the exponential function in the integral shows that the amplitude of the line-of-sight signal z has a log-normal distribution. The second term in the exponential function and the modified Bessel function I_0 show that the signal envelope, which consists of the line-of-sight signal and the multipath scattered waves, has the Rician distribution.

Best-fit values of parameters obtained from measured propagation data using the INMARSAT satellite at 1542 MHz are

$$10 \log_{10} s = 0.7$$
$$10 \log_{10} m = -0.5 \qquad (5.65)$$
$$10 \log_{10} \sigma^2 = -9.0$$

Using this model, Loo estimated that the coherent minimum shift keying (MSK) requires fade margins of 11 and 12 dB for bit error rates of 10^{-3} and 10^{-4}, respectively. This result includes log-normal shadowing effects as well as multipath fading.

5.2.6.2 Markov Transition Model

This propagation model proposed by Wakana [34] includes three distinct concepts: 1) multipath fading, which is represented by the Rician statistics; 2) a Markov model, which presents Markov transitions between fade states and nonfade states; and 3) an attenuation model for fade states. Figure 5.21 shows the block diagram of a channel simulator based on the concept of this propagation model.

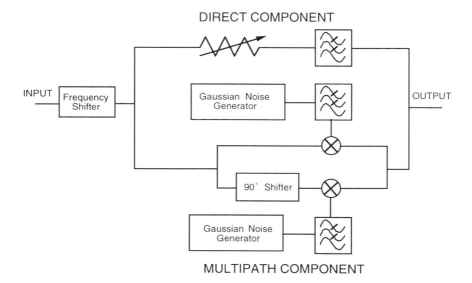

Figure 5.21 Block diagram of a Markov transition model for multipath fading and shadowing.

Multipath Fading Model

Multipath diffuse components can be produced by passing white Gaussian noise through filters that have the following transfer function $H(f)$, in both in-phase and quadrature channels:

$$|H(f)|^2 = \frac{\sigma^2}{\pi\sqrt{f_m^2 - f^2}} \qquad \text{for } |f| \leq f_m$$

$$= 0 \qquad \text{for } |f| > f_m \qquad (5.66)$$

To approximate this transfer function, several numerical low-pass filters can be used; for example, a third-order Butterworth digital filter. As mentioned in Section 5.2.3.4, the cutoff frequency f_m of the low-pass filter corresponds to the maximum Doppler shift $f_m = f_0(v/c)$. When the direct component is not attenuated at the variable attenuator in Figure 5.21, the ratio of the direct signal power to the multipath power is the Rician K factor. By controlling output powers from two Gaussian noise generators, the Rician K factor can be adjusted to simulate various propagation conditions.

Markov Model for State Transition

Based on a given threshold of a signal level, we can determine whether the propagation channel is in a fade or nonfade state. To simulate shadowing of the line-of-sight signal by trees, buildings, and utility poles, the transition between fade and nonfade states is described by a four-state or five-state Markov model. Two states, called fade #1 and #2 (or short and long fade states), are used for fade states; and either two or three states, called nonfade #1, #2, and #3, are used for nonfade states. Transition probabilities are determined for this model to have the same cumulative distributions of fade and nonfade durations as those of measured data produced by using a parameter fitting procedure.

Figure 5.22 shows the transition scheme of the five-state Markov model. It assumes no transition between different fade states or between different nonfade states. Therefore, the five-state Markov model has only eight parameters.

Attenuation Model for Fade States

When a propagation state is the fade state, the direct component is attenuated to simulate shadowing using one of the following two attenuation algorithms.

#1 State Dependent Attenuation Model (SDA model)
    Direct Signal Level in dB

$$
\begin{aligned}
&= 0 && \text{for nonfade states} \\
&= \text{Att1} && \text{for fade \#1} \\
&= \text{Att2} && \text{for fade \#2}
\end{aligned}
\qquad (5.67)
$$

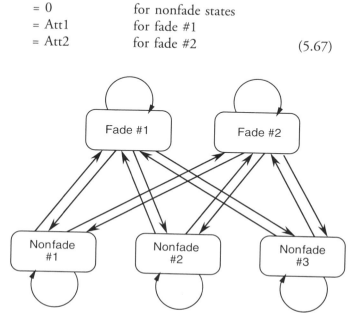

Figure 5.22 Transition scheme of five-state Markov model.

#2 Duration Dependent Attenuation Model (DDA model)
    Direct Signal Level in dB

$$= 0 \qquad\qquad\qquad \text{for nonfade states}$$
$$= a^*(\text{fade duration}) + b \qquad \text{for fade states} \qquad\qquad (5.68)$$

In the SDA model, each fade state is characterized by its attenuation of the direct signal component. Since average fade durations are different between fade #1 and fade #2, the result is a relationship between fade duration and attenuation of the direct component.

In the DDA model, the attenuation of the direct component is assumed to be a function of fade duration. For example, the longer a fade state continues, the more the direct component is attenuated. This assumption is based on the observation that a larger obstacle produces more attenuation.

Moreover, for smooth changes in signal power at the Markov transition between fade and nonfade states, a low-pass filter is inserted at the output of the above-mentioned attenuator. Since this filter is artificial, parameters for characterizing it (e.g., the cutoff frequency) are determined by a best-fit procedure.

Transition Probability

Using the proposed Markov model, the probability $P_{\text{FADE}}(m)$ that fade duration is m and the probability $P_{\text{NONFADE}}(m)$ that nonfade duration is m can be written by

$$P_{\text{FADE}}(m) = p_1 q_1^{m-1}(1 - q_1) + (1 - p_1)q_2^{m-1}(1 - q_2)$$

$$P_{\text{NONFADE}}(m) = p_1' q_1'^{m-1}(1 - q_1') + p_2' q_2'^{m-1}(1 - q_2')$$
$$+ (1 - p_1' - p_2')q_3'^{m-1}(1 - q_3') \qquad (5.69)$$

where p_1 is the probability that the fade state is fade #1; therefore, $1 - p_1$ is the probability that the fade state is fade #2. The probability of the fade state staying the same after the next transition is q_1 when the fading state is fade #1 and q_2 when it is fade #2. The transition probabilities for nonfade transition are defined in the same way. The total transition matrix is defined as follows:

$$\begin{pmatrix} q_1 & 0 & (1-q_1)p_1' & (1-q_1)p_2' & (1-q_1)(1-p_1'-p_2') \\ 0 & q_2 & (1-q_2)p_1' & (1-q_2)p_2' & (1-q_2)(1-p_1'-p_2') \\ (1-q_1')p_1 & (1-q_1')(1-p_1) & q_1' & 0 & 0 \\ (1-q_2')p_1 & (1-q_2')(1-p_1) & 0 & q_2' & 0 \\ (1-q_3')p_1 & (1-q_3')(1-p_1) & 0 & 0 & q_3' \end{pmatrix}$$

$$(5.70)$$

The values of the transition probabilities are determined by using a fitting optimization algorithm (or by using a software program like the solver in Microsoft Excel). Table 5.5 shows best-fit values of eight parameters, shown in (5.70), which were obtained by fitting fade and nonfade duration statistics of this model to those of Vogel's empirical model (cf. (5.50) and (5.52) and Tables 5.3 and 5.4). Both Vogel's fade duration model and nonfade duration model can be represented by a single Markov transition model. Calculated cumulative distributions of fade and nonfade durations are shown by circles, triangles, and squares in Figures 5.14 and 5.15, respectively.

Comparison between Measured and Simulated Data

Figures 5.23 and 5.24 show simulated fading and measured fading. These field measurements were carried out in Australia using a helicopter. Figure 5.25 shows the cumulative distribution of the fade level, and Figure 5.26 shows cumulative distributions of fade and nonfade durations from measured and simulated propagation data. The comparison shows that simulated signal fluctuation has the same statistical characteristics of the fade depth, fade duration, and nonfade duration.

5.3 Propagation in Maritime Satellite Communications

Maritime satellite communication services were started in 1976 by the COMSAT Marisat system and are being expanded by the INMARSAT, which was

Table 5.5(a)
Markov Transition Parameters for Fade Duration

Run #	Shadowing	p_1	q_1	q_2
383	Light	0.20340	0.92535	0.68624
342	Moderate	0.18208	0.92058	0.65641
343	Extreme	0.19266	0.92314	0.66568
409	Extreme	0.34369	0.95745	0.79719

Table 5.5(b)
Markov Transition Parameters for Nonfade Duration

Run #	Shadowing	p_1'	p_2'	q_1'	q_2'	q_3'
383	Light	0.13744	0.79565	0.97216	0.65581	0.99798
342	Moderate	0.06390	0.42516	0.99706	0.64732	0.90916
343	Extreme	0.86020	0.10579	0.64214	0.94909	0.99286
409	Extreme	0.78505	0.04077	0.78452	0.99763	0.96996

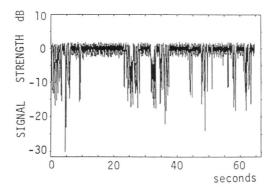

Figure 5.23 Simulated fading using the Markov transition model.

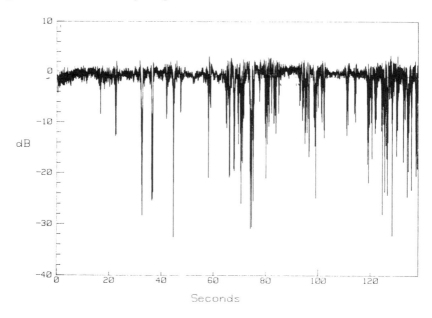

Figure 5.24 Measured multipath fading in Australia.

established in 1979 as an international organization mandated with providing worldwide maritime satellite communication services. By September 1994, INMARSAT had provided satellite communication services to more than 31,628 ships via four second-generation satellites. The original analog services via INMARSAT A terminals provide high-quality telephone, fax, telex, and data communications. Of the 25,216 INMARSAT A terminals in operation, more than two-thirds are installed on large oceangoing vessels. For small vessels,

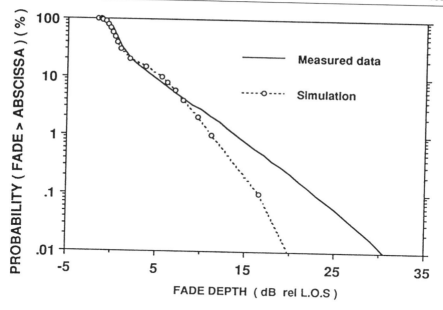

Figure 5.25 Cumulative distribution of fade using the Markov model.

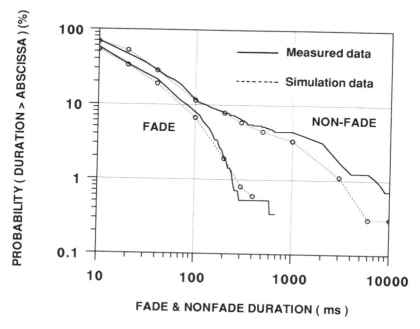

Figure 5.26 Cumulative distribution of fade and nonfade duration using the Markov model.

ship terminals should be designed to be lightweight and compact with a low-gain or a medium-gain antenna, which receives waves reflected from the sea surface as well as the direct wave, because the antennas have a broad beamwidth. Interference between the direct and reflected radio waves causes multipath fading and significantly impairs maritime satellite communications channels, especially at low elevation angles. This fading effect is a function of wave height and elevation angle. This section presents propagation characteristics of maritime satellite communication channels.

The Communication Research Laboratory (CRL) has started to develop a small Earth station for small vessels of about 30 tons based on digital modulation and coding using the multipath fading reduction technique. Field tests have been carried out in several coastal areas in Japan using a transmitting tower. Since 1987, onboard field experiments have been carried out using the Japanese ETS-V satellite, which was launched in 1987, to evaluate the performance of digital modems and codecs, multiple access techniques, antenna tracking, the fading reduction system, and a communication and radiodetermination system, as well as to study propagation characteristics. This section also includes results of field experiments using ETS-V.

5.3.1 Multipath Fading—Theoretical Model

Multipath waves reflected from the sea surface consist of a specular coherent component and a diffuse incoherent component. The signal received at a ship antenna is the sum of the direct wave and these two components. For a calm sea surface, the coherent component dominates, but it diminishes rapidly as the sea surface becomes rough. In this section, we present a theoretical model of sea reflection.

As shown in Figure 5.27, the phase difference between two waves reflected from the sea surface is given by

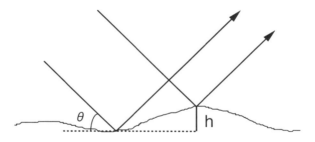

Figure 5.27 Waves reflected from the sea surface.

$$u = 2hk \sin \phi$$

$$= \frac{4\pi}{\lambda} h \sin \phi \tag{5.71}$$

where u is referred to as the roughness factor [35] of the sea surface, λ is the wavelength of the radio wave, ϕ is the elevation angle, and h is the rms wave height. According to the Rayleigh criterion, the sea surface is rough for $u > \pi/2$ and smooth for $u < \pi/2$.

Assuming that the sea surface height has a Gaussian distribution, the significant wave height H, which is defined as the average wave height of the highest one-third of all waves, is given by

$$H = 4h \tag{5.72}$$

The sea state is also characterized by the significant wave height $H(m)$; the sea state is calm for $H < 0.15$, smooth for $0.15 \leq H \leq 2.0$, moderate for $2.0 < H \leq 4.0$, rough for $4.0 < H \leq 9.5$, and high for $H > 9.5$. For $u > 2$, the specular coherent component becomes fairly small: for the L band (1.5 GHz), $H > 1.4$ at an elevation angle of $5°$, and $H > 0.7$ at an elevation angle of 10 degrees.

5.3.1.1 Specular Coherent Component

The received power of the specular coherent component is given by

$$\frac{P_r}{P_t} = \frac{\lambda^2}{(4\pi)^2} \left(\frac{1}{R_1 + R_2} \right)^2 G_t G_r R_f^2 |\chi|^2 \tag{5.73}$$

where P_r and P_t are received and transmitted powers, respectively, G_r and G_t are antenna gains of a receiver and transmitter, respectively, R_1 is the distance between the receiver and scattering sea surface, and R_2 is the distance between the transmitter and sea surface. The factor $[\lambda/4\pi(R_1 + R_2)]^2$ is the free-space loss, as shown by (5.5) in Section 5.1.2., R_f^2 is the power reflection coefficient of the smooth surface, and $|\chi|^2$ is the power loss due to scattering, which depends on the sea surface condition.

For $u > 1$, the amplitude of the coherent component E_c can be approximated by [36]

$$E_c = E_0 \exp(-u^2/2) I_0(u^2/2) \tag{5.74}$$

where u is the roughness factor, E_0 is the amplitude of the coherently reflected wave for $u = 0$, and I_0 is the zeroth-order modified Bessel function. The coherent component decreases rapidly with increasing wave height, elevation angle, and frequency. For $u = 2$, for example, the power of the coherent component decreases by one-tenth; that is, $(E_c/E_0)^2 = 1/10$. Then, the incoherent component becomes dominant.

5.3.1.2 Diffuse Incoherent Components

The diffuse incoherent power can be expressed by the following radar equation:

$$\frac{P_r}{P_t} = \frac{\lambda^2}{(4\pi)^3} \int \frac{G_t(\mathbf{i})\,G_r(\mathbf{o})}{R_1^2 R_2^2} \sigma(\mathbf{i},\,\mathbf{o})\,dS \qquad (5.75)$$

where $\mathbf{i}$ and $\mathbf{o}$ are incident and outgoing vectors, respectively, on a scattering surface dS and $\sigma(\mathbf{i},\mathbf{o})$ is the scattering cross-section. To calculate the scattering cross-section for the incoherent component, several approximation methods have been proposed. Under the condition that the sea surface is perfectly conductive without shadowing or multiple scattering, the Kirchhoff approximation gives the following equation [37]:

$$\sigma = \frac{u^2}{\beta^2}\sec^4\gamma \sum_{m=1}^{\infty} \frac{u^{2m}}{m!m}\, e^{-u^2\left(1 + \frac{\tan^2\gamma}{m\beta^2}\right)} \qquad (5.76)$$

where

$$\gamma = \tan^{-1}\left\{\frac{(\sin^2\theta_i - z\sin\theta_i\sin\theta_s\cos\phi_s + \sin^2\theta_s)^{1/2}}{\cos\theta_i + \cos\theta_s}\right\} \qquad (5.77)$$

The values θ_i and θ_s are the incidental and outgoing angles, respectively, ϕ_s is the change in horizontal scattering angle, and β is the variance of the sea surface slope, $\beta = 2h/\lambda_0$, where λ_0 is the correction length for sea waves. For a calm sea ($u^2 \ll 1$) and a rough sea ($u^2 \gg 1$), (5.76) can be approximated by

$$\sigma = \begin{cases} \dfrac{u^4}{\beta^2}\sec^4\gamma\; e^{-u^2\left(1 + \frac{\tan^2\gamma}{\beta^2}\right)} & \text{for } u^2 \ll 1 \\[4mm] \dfrac{1}{\beta^2}\sec^4\gamma\; e^{-\frac{\tan^2\gamma}{\beta^2}} & \text{for } u^2 \gg 1 \end{cases} \qquad (5.78)$$

Experiments carried out by Beard [38] showed that the reflection coefficient of incoherent components has a maximum value at $u = 1.2$ for frequencies ranging from 5.7 GHz to 34.9 GHz, and decreases gradually with increasing u. Ohmori and others [39] found that the reflection coefficient (actually, they measured the fade depth) had a maximum value at $h = 2\lambda$ ($u \approx 2.2$) in L-band measurements at a 5-degree elevation angle. Karasawa and others [40] reported that the reflection coefficient was nearly constant for $u > 2$ at elevation angles above 7 degrees, while it became smaller with increasing u or decreasing elevation angle when the sea waves were composed of wind waves and swell waves. However, the coefficient kept constant with increasing u at elevation angles less than 5 degrees for swell-dominated sea. They considered that radio waves reflected from the sea surface are obstructed a second time by sea waves, which causes multiple scattering. At low elevation angles, power absorbed by multiple scattering increases with increasing steepness of the sea surface, and the increase in multiple scattering causes the decrease in the reflection coefficient of incoherent components.

5.3.1.3 Received Signal Fluctuation

The signal received at a ship earth station consists of the direct component, the coherent component, and the incoherent component. When the incoherent component is approximated to be a Gaussian process, as mentioned in Section 5.2.3, the statistical characteristics of the received signal can be described by a Rician distribution with a direct power-to-incoherent multipath power ratio, or a Rician K factor.

Using (5.74) and (5.76) for the coherent and incoherent components, we can express the probability distribution function (pdf) of the received signal by

$$p(E, \phi) = \left(\frac{2E}{\sigma^2}\right) \exp\left[-(E^2 + 1 + 2E_c\cos \phi + E_c^2)/\sigma^2\right] \qquad (5.79)$$
$$\cdot I_0(2E\sqrt{1 + 2E_c\cos \phi + E_c^2}/\sigma^2)$$

where E is the amplitude of the received signal, ϕ is the phase difference between the direct and coherent reflected components, E_c is the amplitude of the coherent component, and σ^2 is the average power of the incoherent component as seen in Figure 5.28. While a ship earth station is receiving the signal wave, the phase difference ϕ may vary because the ship is moving. In practice, the pdf should be used by calculating the integral of (5.79) with ϕ, which can be assumed to have a uniform distribution over the range 0 to 2π. For $u > 2$, since the coherent component diminishes, (5.79) becomes a simple Rician distribution.

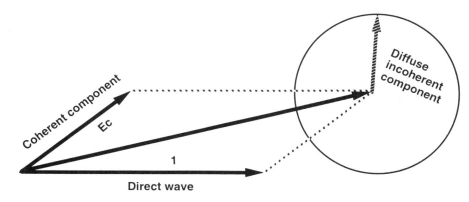

Figure 5.28 Direct, coherent, and incoherent components.

For a given Rician factor, the fade depth exceeded for p (%) of the time can be calculated using (5.79). However, it is not easy to determine values of the Rician K factor for different wave heights, elevation angles, antenna beamwidths, and so forth. Statistical models for predicting multipath power or fade depth have been proposed, and these are presented in the next section.

5.3.2   Statistical Model for Multipath Fading

Sandrin and others [41] proposed a generalized model for the maritime multipath fading, based on experimental data from a number of sources. They obtained the Rician factors as a function of elevation angles for different ship antennas. In general, nearly all of the experimental data fit the Rician distribution well, at least in the range of percent-of-time values from 90 to 99%, which are of greatest interest to maritime communications engineers. This model is not a worst-case model, but represents typical fading characteristics, shown in Figure 5.29, which can be described by:

1. For antenna gains from 0 to 16 dBi, the Rician K factor in dB is given by

$$K = E\ell + 4 \text{ for } 2° \leq E\ell \leq 4° \tag{5.80}$$

   where $E\ell$ is the elevation angle in degrees;

2. For other antenna gains and elevation angles, the Rician factor is given by a straight line, as shown in Figure 5.29.

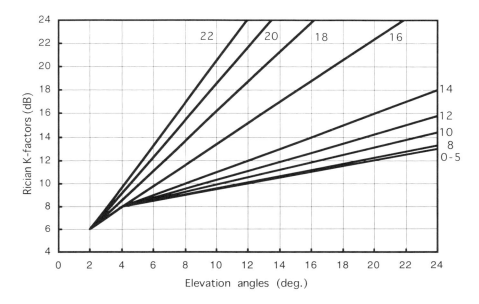

Figure 5.29 Rician *K* factors as a function of elevation angles. (*After:* [41]).

5.3.3 Multipath Fading Reduction Technique

A number of fading reduction techniques, such as polarization compensation, multielement antenna systems, and forward error-correction code interleaving, have been proposed [41]. Some typical examples are presented in this section.

5.3.3.1 Polarization Compensation

When a circular-polarized wave is reflected from a smooth sea surface, the reflected component having the same polarization (referred to as the copolarized component) diminishes with increasing elevation angle while the oppositely polarized component (the cross-polarized component) increases. A horizontally polarized wave is reflected with a small amount of attenuation and phase reversal, while a vertically polarized wave is reflected with a large amount of attenuation and no phase reversal.

Using such electromagnetic characteristics, several methods to reduce multipath fading have been proposed. Here, we present one developed by the CRL [42,43]. This fading reduction method uses both the electromagnetic characteristics of a circular-polarized wave and an electrical function of a hybrid combiner of a ship-borne antenna.

A circular-polarized wave reflected from a smooth sea surface is described in terms of horizontally and vertically polarized waves as follows. Complex reflection coefficients R_H and R_V are given by

$$R_H = \frac{\cos\theta - \sqrt{n^2 - \sin^2\theta}}{\cos\theta + \sqrt{n^2 - \sin^2\theta}}$$

$$R_V = \frac{n^2 \cos\theta - \sqrt{n^2 - \sin^2\theta}}{n^2 \cos\theta + \sqrt{n^2 - \sin^2\theta}} \tag{5.81}$$

and

$$n^2 = \epsilon_r + j\,\frac{\sigma}{\omega\epsilon_0} \tag{5.82}$$

where θ, ϵ_o, ω, ϵ_r, and σ are the incident angle, dielectric constant in vacuum, angular frequency, relative dielectric constant, and conductivity of the sea surface, respectively. The complex reflection coefficients, R_R and R_L, of the right and left circular-polarized waves are rewritten as

$$R_R = \frac{1}{2}(R_H - R_V)$$

$$R_L = \frac{1}{2}(R_H + R_V) \tag{5.83}$$

which correspond to cross-polarized and copolarized components, respectively.

Figure 5.30 shows the reflection coefficients calculated using (5.81) with $\epsilon_r = 70$, $\sigma = 5.5$ s/m, and $f = 1{,}540$ MHz. For elevation angles less than 5 degrees, which corresponds to an incidental angle of more than 85 degrees, the amplitude of the cross-polarized component is larger than that of the copolarized component, while the phase difference between the two components is relatively insensitive to the incidental angle.

Figure 5.31 shows the configuration of the proposed fading reduction antenna system. Symbols T1, T2, T3, and T4 denote four terminals of a hybrid. Terminal T3 has both the direct copolarized component E_0 and the reflected copolarized component $E_L e^{j\phi_L}$, and terminal T4 has the reflected cross-polarized component $E_R e^{j(\phi_R - \pi/2)}$:

$$\text{T3: } E_0 + E_L e^{j\phi_L}$$

$$\text{T4: } E_R e^{j(\phi_R - \pi/2)} \tag{5.84}$$

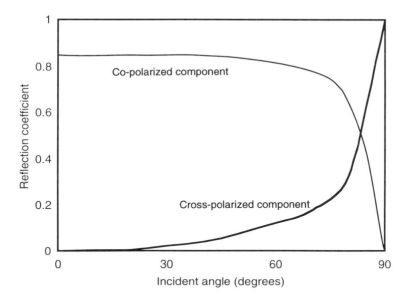

Figure 5.30 Reflection coefficients R_R and R_L vs. elevation angle.

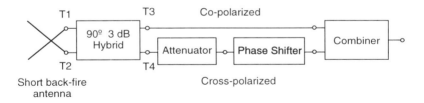

Figure 5.31 Configuration of CRL's fading reduction system.

Therefore, two components from T3 and T4 are combined at the combiner after the component from T4 has been attenuated by A and shifted in phase by ϕ as follows:

$$E_L = E_R/A$$
$$\phi_L - (\phi_R - \pi/2) + \phi = \pm\pi \quad (5.85)$$

Then, the reflected copolarized component is eliminated at the output of the combiner, which reduces the multipath fading. Field tests for evaluating the performance of the fading reduction technique have been carried out using a shipborne short-backfire antenna. The experimental results are presented in the next section.

5.3.3.2 Diversity Techniques

To reduce multipath fading, several diversity techniques, such as antenna spatial diversity, polarization diversity, angle diversity, frequency diversity, and time diversity, have been proposed. The simplest implementation is switched spatial diversity, in which antennas are installed at different places on a ship. An array of closely spaced antenna elements is controlled to block the reflected waves from the sea surface and to provide maximum signal strength by maximum ratio prediction diversity combination. Another proposed fading reduction approach uses shaping of the antenna pattern, which has a sharp cutoff below the horizon. The success of diversity techniques depends on the degree to which the signals received on different diversity branches have independent fading statistics.

5.3.3.3 Interleaved FEC Codes

Interleaved FEC codes have been studied to randomize and spread burst errors caused by deep fading due to multipath fading. However, the performance of interleaved FEC codes alone is limited in the maritime satellite channel because fade durations are relatively long compared to transmission rates and burst rates. Similarly, the single control element used in the polarization shaping scheme limits performance in a diffuse multipath environment. When the two schemes are used together, however, the fading reduction provided by polarization shaping and by a simplified interleaving scheme has performance equivalent to that of a more complex interleaving technique.

5.3.4 Experimental Results Using the ETS-V Satellite

5.3.4.1 Experimental System Configuration

CRL [43] has been carrying out field experiments on a ship earth station based on digital modulation and coding techniques by using the Japanese ETS-V satellite. In order to obtain a better understanding of multipath fading phenomena under various sea conditions and elevation angles, the experiments have been conducted in the East China Sea, South China Sea, Indian Ocean, and North and South Pacific Oceans.

The experimental link consists of a coastal Earth station at Kashima City located at 36 degrees north latitude and 141 degrees east longitude in Japan, the ETS-V satellite in a geostationary orbit of 150 degrees east longitude, and a ship Earth station (SES) installed on a ship of the Hokkaido University, named "Oshoro Maru" (gross tonnage: 1,779 tons). The SES antenna is an improved short-backfire antenna 40 cm in diameter and has a two-axis mount (Az/El) with a program tracking function slaved to the shipborne navigation system. Using a motion detector installed at the center of gravity of the ship, it can compensate for ship motion and keep the antenna pointing toward the satellite. The addition of a second smaller reflector in front of the main subreflector of the short-backfire antenna improves the antenna gain by 1.4 dB and the axial ratio by 0.4 dB, compared with the conventional short-backfire antenna. The SES has two receivers: one is for a left-hand circular-polarized signal (copolarized component) and the other for a right-hand circular-polarized signal (cross-polarized component). The latter is used to reduce the multipath fading caused by sea surface reflection, as mentioned above.

5.3.4.2 Experimental Results for Fading Statistics

Figure 5.32 shows typical cumulative distributions of received signal power at several elevation angles. The measured distributions had a good fit to the Rician distribution with Rician factors of 5 to 9 dB, 6 to 12 dB, and 15 dB at elevation angles of 3 degrees, 6 degrees, and 10 degrees, respectively. Moreover, our measured data can be fit to the generalized model for fading statistics by $K = El + 4$ in (5.80).

Figure 5.33 shows fade depth as a function of the standard deviation of wave height at various elevation angles. At elevation angles less than 2 degrees,

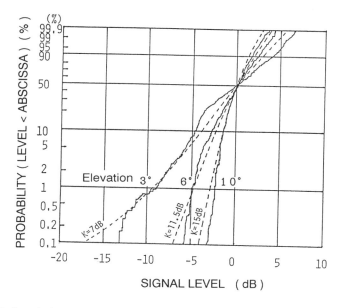

Figure 5.32 Cumulative distributions of received signal power with respect to the medium values at several elevation angles. Dashed curves are the Rician distributions with several Rician K factors. Antenna gain is 15.4 dBi.

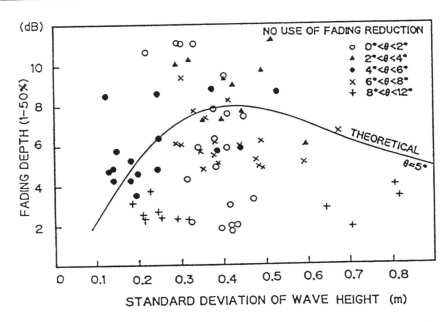

Figure 5.33 Fade depth vs. standard deviation of wave height at various elevation angles.

there is not a strict relationship between the fade depth and wave height. As the standard deviation of wave height increases to 0.5m, the fade depth also increases. Theoretical calculation of multipath fading at low elevation angles has suggested that fade depth decreases with wave height when the wave height is above 0.4m ($h = 2\lambda$) [39]. Not enough data was obtained during these experiments to confirm this, but for 6 degrees $< \theta <$ 8 degrees and 8 degrees $< \theta <$ 12 degrees, the fade depth data had such a tendency.

5.3.4.3 Experimental Results for Multipath Fading Reduction Technique

Using the multipath fading reduction system shown in Figure 5.31 and mentioned in Section 5.3.3, field experiments were carried out.

Figure 5.34 shows received signal powers without and with the fading reduction technique, the cross-polarized component power, and the waveheight, which were measured simultaneously at an elevation angle of 5.9 degrees when the standard deviation of wave height was 0.4m. As shown in Figure 5.31, the cross-polarized component was added to the copolarized component with attenuation of 0 dB and a fixed phase shift of 225 degrees, which are theoretical values for a specular sea. Without the fading reduction technique, both copolarized and cross-polarized components had peak-to-peak power fluctuations of more than 10 dB. The former fluctuation is caused by interference between

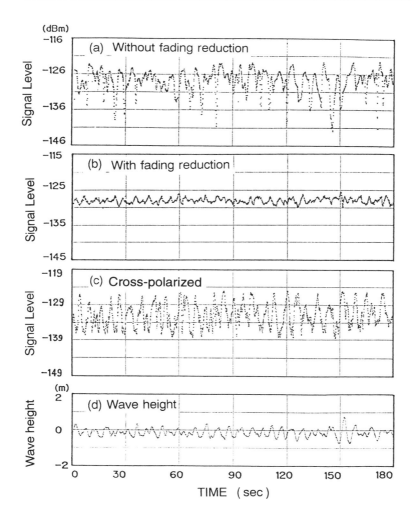

Figure 5.34 Received signal powers (a) without and (b) with the fading reduction technique, (c) cross-polarized component power, and (d) waveheight, measured at elevation angle of 5.9 degrees when the standard deviation of wave height was 0.4m.

the direct wave and the reflected copolarized component. As shown in this figure, this technique can reduce such power fluctuation by canceling the reflected copolarized component using the cross-polarized component. Values of both attenuation and phase shift agreed with their theoretical values.

Figure 5.35 shows a cumulative distribution of C/N_0 (the ratio of carrier power to noise power density) with respect to the medium of C/N_0 measured without the fading reduction technique. The fading depth (1 to 50%), which

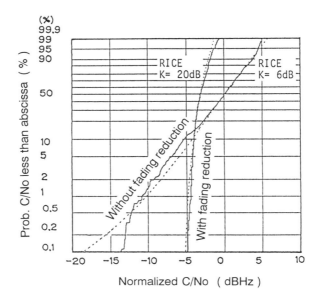

Figure 5.35 Cumulative distribution of C/N_0. Dashed curves are Rician distributions with Rician K factors of 20 and 6 dB.

is defined as the difference between 1% and 50% values of C/N_0 in the cumulative distribution, was improved from 10.9 to 1.4 dB by the fading reduction technique. Both the cumulative statistics with and without the fading reduction technique follow the Rician distribution with Rician K factors of 20 and 6 dB, respectively. An increase in the Rician K factor from 6 to 20 dB means an effective reduction of reflected copolarized components.

Improvement in fading depth (1 to 50%) is plotted as a function of elevation angle in Figure 5.36. In the proposed fading reduction system shown in Figure 5.31, thermal noise is also added to the received signal through the cross-polarized component route, and this degrades the total C/N_0. When the attenuation of the cross-polarized component in this reduction system is 0 dB, the degradation of the total C/N_0 is at least 3 dB. Therefore, when this improvement is more than 3 dB, it can be considered that this technique actually improves communication channel quality. Figure 5.36 indicates that this technique has a definite advantage at elevation angles less than 10 degrees.

These results indicate that this simple fading reduction technique, which consists of an additional low-noise amplifier for the cross-polarized component, a phase shifter, and a power combiner, is effective in improving the quality of maritime communication channels at low elevation angles.

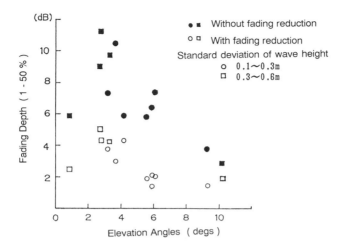

Elevation angle dependence of fading reduction

Figure 5.36 Fading depth (1 to 50%) vs. elevation angles with and without the fading reduction techniques.

5.4 Propagation in Aeronautical Satellite Communications

A typical aeronautical satellite communication situation is illustrated in Figure 5.37. In 1987, the International Telecommunication Union (ITU) assigned 1.5 GHz (1,545 to 1,555 MHz) and 1.6 GHz (1646.5 to 1656.5 MHz) to aeronautical satellite communication services. Since 1991, INMARSAT has been providing a digital voice and data service between aircraft and ground stations, which are connected to the public switched telephone network. In 1994, United Airlines, British Airways, Air Canada, Lufthansa, and other airlines installed INMARSAT aeroterminals, and the number of aeroterminal installations grew from 285 to 493 in 1994.

When an aircraft antenna with gain of about 10 dBi for telephony or 0 dBi for low-speed data is used, multipath fading due to the sea and ground surface reflections can impair communication quality, as in maritime satellite communications. The aircraft body affects antenna radiation patterns and may cause multipath fading by reflections from it. The high speed of an aircraft causes a large Doppler shift and spectrum spread. Since aircraft require high integrity of communications, even short-term propagation effects are important considerations when designing aeronautical satellite communications systems.

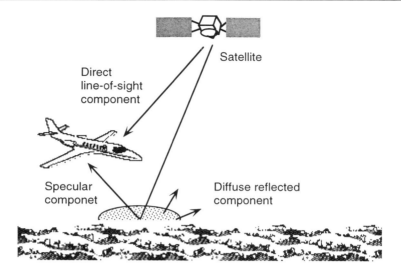

Figure 5.37 Aeronautical satellite communication situation.

One difference between aircraft and ships in terms of the multipath fading caused by reflection from the sea or ground surface is that, considering the high speed and altitude of aircraft, it may be necessary to consider the sphericity of the Earth. Yasunaga and others [44] modified the sea surface reflection theory for maritime communications into one for the aeronautical case considering the Earth's sphericity. In this section, we present the results of their theoretical modification and of in-flight experiments carried out by several organizations [44–47].

5.4.1 Field Measurement Using a Helicopter With a Low-Gain Antenna

A field test was carried out by Yasunaga and others [44] using an INMARSAT V satellite (63 degrees east) at 1,538 MHz. Two antennas, a nine-element microstrip array (MSA) antenna having gain of 15 dBi and a circular patch antenna (CPA) with 7 dBi gain, were installed one on each side of a helicopter. At a 5.5-degree elevation angle and with a rough sea condition, 99% fade depths, which are defined as the fade not exceeding 99% of the time, were 7 to 11 dB for CPA and 7 to 9 dB for MSA at aircraft altitudes ranging 100 to 10,000m. Figure 5.38 shows the 99% fade depth as a function of the antenna altitude for CPAs and MSAs. The hatched region corresponds to the theoretical characteristics. The fade depth for an aeronautical satellite communication channel at an altitude of 10,000m was only about 2 dB less than that for a maritime channel at a low elevation angle of 5.5 degrees. The fade depth

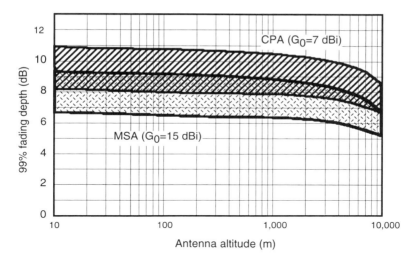

Figure 5.38 Fade depth not exceeded for 99% of the time vs. antenna altitude for circular-polarization at 1.54 GHz for wave heights from 1.5 to 3m.

decreased as the elevation angle increased. For example, the 99% fade depth for CPA and MSA were about 6 and 3 dB, respectively, at a 10-degree elevation angle and a 5,000m altitude.

5.4.2 Multipath Fading Measurements Using ATS-5 and ATS-6

Experiments using the NASA ATS-5 satellite and a KC-135 jet aircraft were conducted in the 1.5 to 1.6 GHz band [45,48]. The direct signal was received with a 15-dBi quad-helix antenna and the sea-reflected signal was received with a 13-dBi crossed-dipole array antenna. Figure 5.39 shows the spectrum bandwidth of the reflected waves versus elevation angle at 1.6 GHz.

Another experiment using the NASA ATS-6 satellite and a KC-135 jet aircraft was conducted at 1.6 GHz [48]. The propagation characteristics were measured with a two-element waveguide array in the aircraft nose radome, with a 1-dB antenna beamwidth of 20 degrees in azimuth and 50 degrees in elevation. Table 5.6 summarizes multipath parameters obtained from ATS-6 ocean measurements. The mean-square scattering coefficient is the ratio of the direct power to the multipath power. The delay spreads are the widths of power delay profile of incoherent waves. Correlation bandwidth is the 3-dB bandwidth of the frequency autocorrelation function, and the decorrelation time is the 3-dB width of the time autocorrelation function.

In this ATS experiment, multipath parameters for land-reflected multipath effects were also measured. However, land multipath signals were found to be

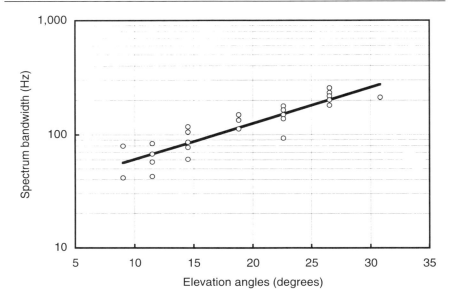

Figure 5.39 Spectrum bandwidth of the reflected wave vs. elevation angle. (*Source:* [48].)

Table 5.6
Parameters for Sea Surface-Reflected Multipath Effects

Parameters	Typical Values for Elevation Angles		
	8-Degree	**15-Degree**	**30-Degree**
Mean square scatter coefficients for horizontal polarization	−2.5 dB	−1 dB	−1 dB
Mean square scatter coefficients for vertical polarization	−14 dB	−9 dB	−3.5 dB
Delay spread (3 dB)	0.6 msec	0.8 msec	0.8 msec
Correlation bandwidth (3 dB)	160 kHz	200 kHz	200 kHz
Doppler spread (3 dB)	5 Hz	70 Hz	140 Hz
Decorrection time (3 dB)	7.5 msec	3.2 msec	2.2 msec

Source: [48].

highly variable, and no consistent dependence on elevation angle was established, perhaps because the ground terrain was highly variable.

5.4.3 Field Measurement Using the ETS-V Satellite

The CRL carried out in-flight experiments using a transoceanic B-747F freighter and the ETS-V satellite. Most experiments were conducted on the route between

Narita and Anchorage. The airborne antenna, which is a set of two 16-element phased array antennas, was installed on the top of the fuselage within a fairing. Figure 5.40 shows an example of C/N_0 measured over a period of 3 minutes. As shown in Figure 5.40(a), most data show constant C/N_0 and no fading. Figure 5.40(b) shows fading when the antenna beam direction coincided with that of the main wing and when the aircraft was rolling slowly.

Figure 5.41 shows the standard deviation of the received signal level versus the beam direction. The standard deviation was relatively small, but the deviation between 120 and 150 degrees in azimuth was slightly larger than that in any other direction. Analyzing the frequency spectrum of the received signal, we found that frequency components under 0.1 Hz were dominant. Since components of several hertz are dominant in the fading spectrum for sea reflection, this fading is not caused by sea reflection, but by waves reflected from the main wings of the aircraft.

During in-flight experiments, no fading caused by the sea surface reflection was observed except at very low elevation angles less than 0 degrees. We found that sea surface-reflected waves were blocked out by the fuselage and the wings, so the antenna on the top of the fuselage did not receive the sea surface-reflected waves. Conversely, the antenna received fading of about 2 dB due to reflection from the main wings when the antenna beam coincided with their direction.

In an aeronautical satellite channel, propagation conditions are superior to those of land mobile and maritime satellite channels because there are no obstacles in the link between a satellite and an aircraft earth station. Fading due to sea surface reflections is small because reflected waves are blocked by aircraft structures when antennas are installed on top of the fuselage, compared with when antennas are installed on the side of the aircraft.

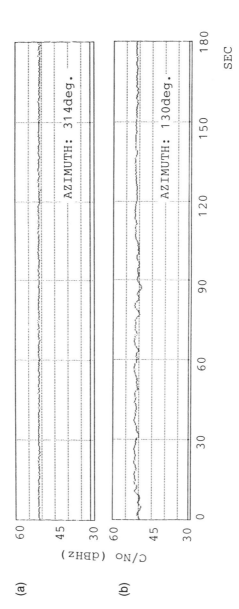

Figure 5.40 *C/N₀* measured during in-flight experiments on the route between Narita and Anchorage.

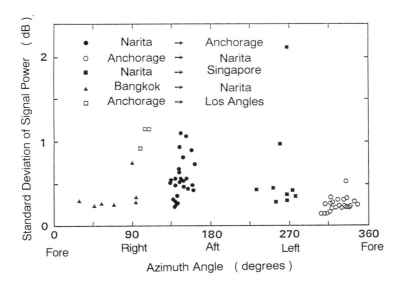

Figure 5.41 Standard deviation of the received signal vs. the antenna beam direction measured in in-flight experiments.

References

[1]   ITU-R Recommendations, "PN618-3: Propagation data and prediction methods required for the design of earth-space telecommunications systems," ITU, 1994.

[2]   ITU-R Recommendations, "PN838: Specific attenuation model for rain for use in prediction methods," ITU, 1994.

[3]   Arnbak, J., "Adaptive control of satellite resources," *Proc. Symposium on Advanced Satellite Communications Systems*, 1978, pp. 43–49.

[4]   Nishiyama, I., R. Miura, and H. Wakana, "Closed-loop uplink power control experiments in K-band using CS-2," *Proc. ISTS*, Tokyo, Japan, 1986.

[5]   Hentinen, V. O., "Error performance for adaptive transmission on fading control," *IEEE Trans. Communications*, Vol. COM-22, No. 9, Sept. 1974, pp. 1331–1337.

[6]   CCIR Rep. 564–4, "Propagation data and prediction methods required for earth-space telecommunication systems," *Recommendations and Reports of the CCIR*, Vol. 5, ITU, Geneva, 1991.

[7]   Levitt, B. K., "Rain compensation algorithm for ACTS mobile terminal," *IEEE J. on Selected Areas in Communications*, Vol. 10, No. 2, 1992, pp. 358–363.

[8]   Sato, M., et al., "Adaptive data rate control TDMA system as a rain attenuation compensation technique," *Proc. 3rd Int. Mobile Satellite Conf.*, Pasadena, CA, June, 1993, pp. 505–510.

[9]   Hagenauer, J., "Rate compatible punctured convolutional code," *Proc. ICC'87*, Seattle, WA, 1987, pp. 29.1.1–29.1.5.

[10]   Wakana, H., "Propagation research in Japan," *Proc. 15th NASA Propagation Experimenters Meeting*, London, Canada, July 1991, pp. 88–94.

[11]   Butterworth, J. S., "Propagation measurements for land mobile satellite services in the 800 MHz band," Communications Research Centre Technical Note, No. 724, Aug. 1984.

[12]   Vogel, W. J., and J. Goldhirsh, "Multipath fading at L band for low elevation angle, land mobile satellite scenarios," *IEEE J. on Selected Areas in Communications*, Vol. 13, No. 2, Feb. 1995, pp. 197–204.

[13]   Lee, W.Y.C., *Mobile Communications Engineering*, New York, NY: McGraw Hill, 1982.

[14]   Steele, R., (ed.), *Mobile Radio Communications*, NJ: IEEE Press, 1992.

[15]   Jakes, W. C., (ed.), *Microwave Mobile Communications*, NJ: IEEE Press, 1993.

[16]   Lutz, E., et al., "DLR activities in the field of personal satellite communications systems," *Space Communications*, Vol. 14, 1996, pp. 111–121.

[17]   Jahn, A., et al., "Narrow- and wide-band channel characterization for land mobile satellite systems: experimental results at L-band," *Proc. Int. Mobile Satellite Conf., IMSC'95*, Ottawa, Canada, June 1995, pp. 115–121.

[18]   Kennedy, R. S., *Fading Dispersive Communication Channels*, New York, NY: Wiley Interscience, 1969.

[19]   Ikegami, T., et al., "Measurement of multipath delay profile in land mobile satellite channels," *Proc. Int. Mobile Satellite Conf., IMSC'93*, Pasadena, CA, June 1993, pp. 331–336.

[20]   Vogel, W. J., and J. Goldhirsh, "Mobile satellite system propagation measurements at L-band using MARECS-B2," *IEEE Trans. on Antennas and Propagation*, Vol. AP-38, No. 2, Feb. 1990, pp. 259–264.

[21]   Goldhirsh, J., and W. J. Vogel, "Propagation effects for land mobile satellite systems: overview of experimental and modeling results," NASA Reference Publication 1274, 1992.

[22]   Wakana, H., et al., "Fade statistics measured by ETS-V in Japan for L-band land-mobile satellite communication systems," *Electronics Letters*, Vol. 32, No. 6, March 1996, pp. 518–520.

[23]   Bundrock, A. and R. Harvey, "Propagation measurements for an Australia land mobile-satellite system," *Proc. Int. Mobile Satellite Conf.*, May 1988, pp. 119–124.

[24]   Renduchintala, V.S.M., et al., "Communications service provision to land mobiles in northern Europe by satellites in high elevation orbits—propagation aspects," *40th Int. Conf. on Vehicular Technologies*, 1990.

[25]   Butt, G., B. G. Evans, and M. Richharia, "Multiband propagation experiment for narrowband characterisation of high elevation angle land mobile-satellite channels," *Electronics Letters*, Vol. 28, No. 15, July 1992, pp. 1449–1450.

[26]   Vogel, W. J., J. Goldhirsh, and Y. Hase, "Land-mobile-satellite fade measurements in Australia," *AIAA Journal of Spacecraft and Rockets*, July-Aug., 1991.

[27]   Hase, Y., W. J. Vogel, and J. Goldhirsh, "Fade-duration derived from land-mobile satellite measurements in Australia," *IEEE Trans. Communications*, Vol. 39, No. 5, May 1991, pp. 664–668.

[28] Ulaby, F. T., M. W. Whitt, and M. C. Dobson, "Measuring the propagation properties of a forest canopy using a polarimetric scatterometer," *IEEE Trans. on Antennas and Propagation*, Vol. AP-38, No. 2, Feb. 1990, pp. 251–258.

[29] Goldhirsh, J., and W. J. Vogel, "Roadside tree attenuation measurements at UHF for land-mobile satellite systems," *IEEE Trans. on Antennas and Propagation*, Vol. AP-35, No. 5, May 1987, pp. 589–596.

[30] Yoshikawa, M., and M. Kagohara, "Propagation characteristics in land mobile satellite systems," *39th IEEE Vehicular Technology Conf.*, May 1989, pp. 550–556.

[31] Yoshikawa, M., and M. Kagohara, "A report on mobile satellite propagation in ETS-V/EMSS experiments," *IEICE Technical Report*, SAT89-21, 1989, pp. 17–22 (in Japanese).

[32] Boithias, L., *Propagation des ondes radioelectriques dans l'environnement terrestre*, 2nd edition, Paris, France: Dunod, 1984.

[33] Loo, C., "Measurements and models for a land mobile satellite channel and their applications to MSK signals," *IEEE Trans. on Vehicular Technology*, Vol. VT-35, No. 3, Aug., 1987, pp. 114–121.

[34] Wakana, H., "A propagation model for land-mobile-satellite communication," *Proc. IEEE Antennas and Propagation Society Symposium*, Canada, June 1991, pp. 1526–1529.

[35] Beckman, P., and A. Spizzichino, *The Scattering of Electromagnetic Waves From Rough Surfaces*, New York, NY: Pergamon, 1963.

[36] Miller, A. R., R. M. Brown, and E. Vegh, "New derivation for the rough surface reflection coefficient and for the distribution of sea-wave elevations," *IEEE Proc.*, Vol. 131, Pt. H., April 1984, pp. 114–116.

[37] Karasawa, Y., and T. Shiokawa, "Characteristics of L-band multipath fading due to sea surface reflection," *IEEE Trans. on Antennas and Propagation*, Vol. AP-32, No. 6, June 1984, pp. 618–623.

[38] Beard, C. I., "Coherent and incoherent scattering of microwaves from the ocean, " *IRE Trans. on Antennas and Propagation*, Vol. AP-9, No. 5, 1961, pp. 740–783.

[39] Ohmori, S., et al., "Characteristics of sea reflection fading in maritime satellite communications, " *IEEE Trans. on Antennas and Propagation*, Vol. AP-33, No. 8, Aug. 1985, pp. 838–845.

[40] Karasawa, Y., and T. Shiokawa, "A simple prediction method for L-band multiple fading in rough sea conditions," *IEEE Trans. Communications*, Vol. 36, No. 10, Oct. 1988, pp. 1098–1104.

[41] Sandrin, W. A., and D. J. Fang, "Multipath fading characterization of L-band maritime mobile satellite links," *COMSAT Technical Review*, Vol. 16, No. 2, Fall 1986, pp. 319–338.

[42] Ohmori, S., and S. Miura, "A fading reduction method for maritime satellite communications," *IEEE Trans. on Antennas and Propagation*, Vol. AP-31, No. 1, Jan. 1983, pp. 184–187.

[43] Wakana, H., et al., "Experiments on maritime satellite communications using the ETS-V satellite," *J. Commun. Res. Lab.*, Vol. 38, No. 2, 1991, pp. 223–237.

[44] Yasunaga, M., et al., "Characteristics of multipath fading due to sea surface reflection in aeronautical satellite communications," *IEICE Trans. B-II*, Vol. J72-B-II, No. 7, July 1989, pp. 297–303 (in Japanese).

[45]   Sutton, R. W., et al., "Satellite-aircraft multipath and ranging experiment results at L band," *IEEE Trans. Communications*, Vol. COM-21, No. 5, May 1973, pp. 639–647.

[46]   Hagenauer, J., et al., "The aeronautical channel characterization," *DFVLR Report*, No. NE-NT-T-87-17, 1986.

[47]   Jedrey, T. C., K. I. Dessouky, and N. E. Lay, "An aeronautical-mobile satellite experiment," *IEEE Trans. Vehicular Technology*, Vol. 40, No. 4, Nov. 1991, pp. 741–749.

[48]   Reports of the CCIR, Annex to Volume V, 1990.

6

Digital Communications Technologies

This chapter surveys digital communication technologies that are used in current mobile satellite communication systems and ones in the planning or future stages. Section 6.1 presents requirements for digital modulation and demodulation techniques used in mobile satellite communications systems. Several modulation schemes, band-limited signal design techniques, power spectral characteristics, bit error performance, and carrier recovery techniques are described. Especially, phase shift keying (PSK) modulation schemes, which are mostly used for satellite communications, are focused on. Section 6.2 presents error-control techniques such as automatic repeat request and forward error-correction techniques. Section 6.3 considers coded modulation, which is a combination of modulation and error-correction coding, including trellis-coded modulation and block-coded modulation. Section 6.4 describes digital speech coding such as adaptive differential pulse code modulation (ADPCM), and several linear predictive coding methods to provide high-quality speech signals at low bit rates. Section 6.5 presents multiple access used for satellite communication systems: frequency division multiple access (FDMA), time division multiple access (TDMA), and code division multiple access (CDMA).

6.1 Modulation and Demodulation

The first generation of worldwide mobile satellite services (MSS), which the International Maritime Satellite Organization (INMARSAT) provided by Standard-A mobile terminals (recently renamed INMARSAT-A mobile terminals), used analog FM for the voice circuits and PSK for the telex and signaling traffic [1]. In the experimental stage of the next-generation MSS systems, Japan's

Communications Research Laboratory (CRL) [2], Canada's Communications Research Centre (CRC), and other research organizations have developed analog amplitude-companded single-sideband (ACSSB) [3,4] modems, which offer high-quality voice service with graceful degradation against multipath fading and vegetative shadowing and with narrow bandwidth of about 3 to 4 kHz. However, the progress in digital signal processing technologies and the development of low-bit-rate digital voice codecs has promoted a dramatic shift from analog to digital techniques. The mobile satellite communication systems currently in use and in the planning stage in Australia, Canada, the United States, Japan, and Europe use digital modulation schemes, such as $\pi/4$-QPSK, OQPSK, or MSK [5]; low-bit-rate digital voice codecs of about 4.8 to 6.7 Kbps, such as vector sum excited linear prediction (VSELP), low-delay code-exited linear prediction (LD-CELP), adaptive differential pulse code modulation (ADPCM), regular pulse excited linear prediction code with long-term prediction (RPE-LTP); and powerful forward error-correction (FEC) techniques.

In this section, we briefly survey digital modulation schemes, their power and bandwidth efficiencies, and bit error performances for designing mobile satellite communications systems.

6.1.1 Requirements for Modulation and Demodulation

The MSS modems may suffer from the following propagation impairments as well as rainfall attenuation, ionospheric scintillation, and atmospheric attenuation: 1) varying Doppler frequency shift due to motion of mobile terminals; 2) varying channel conditions due to multipath fading and shadowing (this may sometimes result in rapid signal loss); and 3) low signal-to-noise ratio relative to terrestrial mobile communication systems.

For good performance of mobile satellite communication services, MSS modems have to overcome the above propagation channel characteristics, and we need to determine design choices of MSS terminals to best match them. Moreover, for suitable MSS modems, the following requirements for good power and bandwidth efficiencies, immunity to nonlinearity, robust synchronization under multipath fading and shadowing, and simple implementation should be considered.

6.1.1.1 Power and Bandwidth Efficiencies

One of the most important performance parameters for MSS modulation schemes is efficiency, which includes both power efficiency and bandwidth efficiency, since mobile satellite communication systems usually have limited availability of both power and bandwidth. These two efficiencies are defined as follows. The power efficiency is defined as the ratio of required signal energy per bit to noise density (E_b/N_0) needed to achieve a given bit error rate (BER) over an additive white Gaussian noise (AWGN) channel. For mobile satellite applications, the BER performance over a Rician fading channel is also important, but in this section we deal with only the performance over AWGN. The bandwidth efficiency is defined as the ratio R/W, where R is the information rate in bits per second and W is the required channel bandwidth in hertz. There is no universal definition of the bandwidth W, but there are various definitions that depend on a variety of system requirements. In the next section, we describe several definitions of W [6]. Figure 6.1 illustrates examples of bandwidths based on different definitions for binary phase shift keying (BPSK), and Table 6.1 shows bandwidths, E_b/N_0 for bit error probability of 10^{-5}, nonlinear immunity and implementation simplicity for BPSK, quadrature phase shift keying (QPSK), offset QPSK (OQPSK), $\pi/4$-PSK, and minimum shift keying (MSK) modulation schemes.

6.1.1.2 Immunity to Nonlinearity

Nonlinear amplification of a satellite transponder or a high-power amplifier of a mobile terminal, which may be operated near saturation, regenerates side lobes of the modulated signal that were removed at transmitting mobile terminals to suppress interference to adjacent channels. Moreover, envelope fluctua-

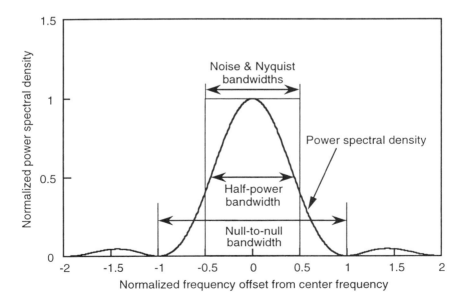

Figure 6.1 Definitions of several bandwidths of a modulated signal (BPSK).

Table 6.1
Bandwidths, E_b/N_0 for Bit Error Probability of 10^{-5}, Nonlinear Immunity, and Implementation Simplicity for BPSK, QPSK, OQPSK, $\pi/4$-PSK, and MSK

Modulation Schemes	99%-Power BW	Null-to-Null BW	35-dB BW	Half-Power BW	Noise BW	E_b/N_0 for BER = 10^{-5}	Nonlinearity Immunity	Implementation Simplicity
BPSK	20.56	2.00	35.12	0.88	1.00	9.6	D	A
QPSK	10.28	1.00	17.56	0.44	0.50	9.6	C	B
OQPSK and $\pi/4$-PSK	10.28	1.00	17.56	0.44	0.50	9.6	B	C
MSK	1.18	1.50	3.24	0.59	0.62	9.6	A	D

Source: [6,10].

tion will be translated into an unwanted phase modulation. Therefore, constant-envelope modulation schemes after filtering are preferable for MSS modems.

6.1.1.3 Synchronization

Three types of synchronization are needed for demodulation at MSS modems. First, carrier recovery or carrier synchronization is required for coherent demodulation. Second, symbol clock recovery or symbol synchronization is required to estimate the timing of state transitions in modulation schemes. Third, word synchronization is required to estimate the end of one word and the beginning of the next word. For mobile satellite communication channels, multipath fading, shadowing, Doppler shifts, and phase noise may degrade the modem performance more severely than additive white Gaussian noise. Therefore, fast synchronization and resynchronization are important performances for MSS modems.

6.1.1.4 Implementation Simplicity

Many types of modulation schemes have been proposed. For simple and small MSS modems, implementation simplicity of the modulation and demodulation algorithm is also an important parameter for choosing a suitable modulation scheme. Recent advances in digital signal processing, in both software and hardware, are also making complicated modulation schemes possible candidates for MSS terminals. Compatibility with terrestrial and satellite mobile communication systems will also be another important aspect for future MSS modem design.

6.1.2 Definitions of Bandwidth

This section describes several definitions of bandwidth and comparisons of bandwidth efficiencies of BPSK, QPSK, OQPSK, $\pi/4$-PSK and MSK, as shown in Table 6.1.

6.1.2.1 Fractional Power Containment Bandwidth (99% Bandwidth)

The 99% (energy containment) bandwidth is defined as the bandwidth inside which 99% of the signal power exists. The value of 99% is the one most commonly used, but other percentages, such as 90 or 95%, are also used sometimes. For example, the 99% bandwidth efficiencies of BPSK, QPSK, and MSK are 20.56, 10.28, and 1.18 Hz/bits/sec, respectively.

6.1.2.2 Null-to-Null Bandwidth

The null-to-null bandwidth is defined as the bandwidth of the main spectral lobe. This definition is clear for a modulation scheme whose power spectrum has a main lobe bounded by well-defined spectral nulls, but for several modulation schemes, which do not have frequencies where the power spectra are nil, the null-to-null bandwidth cannot be defined. For example, the null-to-null bandwidths of BPSK, QPSK, and MSK are 2.0, 1.0, and 1.5 Hz/bits/sec, respectively. MSK has a larger null-to-null bandwidth than QPSK, but a much less 99%-power bandwidth than QPSK.

6.1.2.3   35- or 50-dB Bandwidths

The 35- or 50-dB bandwidths are defined as the bandwidths outside of which the power spectral density is less than the power density of the center frequency by 35 or 50 dB, respectively. For example, the 35-dB bandwidths of BPSK, QPSK, and MSK are 35.1, 17.6, and 3.2 Hz/bits/sec, respectively. This definition of bandwidth is useful in specifications concerned with adjacent channel interference.

6.1.2.4   Noise Bandwidth

The noise bandwidth is defined as the ratio P/P_c, where P is the total signal power and P_c is the power spectral density at the center frequency. This definition of noise bandwidth can be applied to evaluate data transmission performance against intentional interference or jamming. The noise bandwidths of BPSK, QPSK, and MSK are 1.0, 0.5, and 0.6 Hz/bits/sec, respectively.

6.1.2.5   Half-Power Bandwidth (3-dB Bandwidth)

The half-power bandwidth is defined as the bandwidth between frequencies at which the power spectral density has dropped to half its level, or 3 dB below the peak power at the center frequency. The half-bandwidths of BPSK, QPSK, and MSK are 0.9, 0.4, and 0.6 Hz/bits/sec, respectively.

6.1.2.6   Nyquist Bandwidth

According to the Nyquist criterion, the minimum bandwidth W for M-ary PSK at baseband with no intersymbol interference is half the symbol rate, $R_s/2$; thus, the bandwidth W at the carrier frequency is twice that (i.e., $W = R_s$). Since $R_s = R/\log_2 M$ for M-ary modulation in terms of the bit rate R, the Nyquist bandwidth is given by $R/\log_2 M$. Hence, the bandwidth efficiency for MPSK is $\log_2 M$.

On the other hand, for orthogonal M-ary frequency shift keying (MFSK), the Nyquist minimum bandwidth at the carrier frequency is $W = MR_s$. Therefore, the bandwidth efficiency $R/W = R_s\log_2 M/(MR_s) = \log_2 M/M$ in bits/sec/Hz. As M increases, the bandwidth efficiency R/W increases for MPSK, while it decreases for MFSK. Conversely, as M increases, E_b/N_0, which is required to achieve a given bit error probability, decreases for MFSK. This means that the bit error performance of MFSK is improved.

6.1.3 Band-Limited Signal Design

In this section, we consider a signal pulse design for efficiently utilizing a band-limited MSS channel.

A real-valued bandpass signal $s(t)$, which has a common form for several types of modulation schemes, is expressed by [7]

$$s(t) = A(t) \cos\{2\pi f_c t + \theta(t)\} \tag{6.1}$$

where $A(t)$ is the amplitude, $\theta(t)$ is the phase, and f_c is the carrier frequency of $s(t)$. Using the complex-valued equivalent low-pass signal $u(t)$, we have

$$s(t) = Re[u(t) e^{j2\pi f_c t}] \tag{6.2}$$

where

$$u(t) = A(t) \exp\{j\theta(t)\} \tag{6.3}$$

and $Re[\ ]$ is the real part of the complex value in the bracket. In linear modulation methods such as pulse amplitude modulation (PAM) and phase shift keying (PSK), $u(t)$ can be expressed by

$$u(t) = \sum_{n=0}^{\infty} a_n g(t - nT) \tag{6.4}$$

where $\{a_n\}$ is the information-bearing sequence of symbols, $g(t)$ is a signal pulse, and T is the time interval of a symbol. For example, the BPSK signal can be given by $a_n = \exp(j\theta_n)$, where $\theta_n = 0$ or π. Assuming that the propagation channel has an ideal frequency response, the equivalent low-pass output $v(t)$ of a matched filter at a mobile receiver is given by

$$v(t) = \sum_{n=0}^{\infty} b_n y(t - nT - t_d) + n(t) \tag{6.5}$$

where $\{b_n\}$ is the received sequence of symbols, $y(t)$ is the received pulse, t_d is the time delay, and $n(t)$ is the response of the matched filter to the received noise. The transmitted signal $u(t)$ is actually passed through numerous bandpass filters through a satellite link to avoid overloads, spurious interference, and adjacent channel interference. Here, we consider only transmission and reception filtering at Earth terminals by neglecting the filter distortion effects in a satellite communication link.

When $v(t)$ is sampled at time $t_m = mT + t_d$, we have

$$v(t_m) = \sum_{n=0}^{\infty} b_n y(mT - nT) + n(t_m)$$

$$= b_m y(0) + \sum_{\substack{n=0 \\ n \neq m}}^{\infty} b_n y(mT - nT) + n(t_m) \qquad (6.6)$$

Since $y(0)$ can be regarded as a scaling factor, the first term is the desired information symbol, the second term represents the intersymbol interference, and $n(t_m)$ is the additive Gaussian noise.

If the Fourier transform $Y(f)$ of $y(t)$ is band-limited, that is, $Y(f) = 0$ for $|f| > B$, according to the sampling theorem, $y(t)$ can be expressed by

$$y(t) = \sum_{n=-\infty}^{\infty} y\left(\frac{n}{2B}\right) \frac{\sin 2\pi B(t - n/2B)}{2\pi B(t - n/2B)} \qquad (6.7)$$

When the sampling rate $(1/T)$ is the Nyquist rate (i.e., $1/T = 2B$), the above equation can be rewritten as

$$y(t) = \sum_{n=-\infty}^{\infty} y(nT) \frac{\sin \pi(t - nT)/T}{\pi(t - nT)/T} \qquad (6.8)$$

According to (6.6), the condition of no intersymbol interference gives

$$y(nT) = \begin{cases} 1 & n = 0 \\ 0 & n \neq 0 \end{cases} \qquad (6.9)$$

Then, from (6.8), the pulse $y(t)$ is given by

$$y(t) = y(0) \frac{\sin \pi t/T}{\pi t/T} \qquad (6.10)$$

and its power spectral density $S_y(f)$ is shown by the following rectangular function:

$$S_y(f) = \begin{cases} \dfrac{y(0)^2}{2\,T^2} & |f| \le B(=1/2\,T) \\ 0 & |f| > B(=1/2\,T) \end{cases} \tag{6.11}$$

Figure 6.2 shows the pulse $y(t)$ and its rectangular power spectral density $S_y(f)$. However, this ideal case has a problem: the ideal rectangular spectrum $S_y(f)$ is not physically attainable and slow decay of the pulse $y(t)$ causes an infinite series of intersymbol interference due to error in sample timing.

6.1.3.1 Spectral Shaping

The ideal rectangular power spectrum $S_y(f)$ causes long tails in the pulse $y(t)$ decay, but a strictly time-limited pulse conversely results in wider bandwidth of its power spectrum, which causes adjacent channel interference as well. Next, we consider designing a pulse using a spectral shaping technique with no intersymbol interference. Selecting the bandwidth $2B$ of the pulse spectrum between $1/T$ (the Nyquist bandwidth) and $2/T$, we will search for a suitable pulse shape that has fast decay of its tail and no intersymbol interference.

For no intersymbol interference, $y(t)$ is expressed by

$$y(t) = g_\beta(t)\frac{\sin(\pi t/T)}{\pi t/T} \tag{6.12}$$

where $g_\beta(t)$ is an arbitrary function and its Fourier transform $G_\beta(f)$ is band-limited within $-\beta/2\,T \le f \le \beta/2\,T$ for $0 \le \beta \le 1$.

Then, the Fourier transform $Y(f)$ of $y(t)$ is given by the convolution of $G_\beta(f)$ and the rectangular spectrum of (6.11), which is band-limited within $-1/2\,T \le f \le 1/2\,T$. Therefore, $Y(f)$ is band-limited within $-(1 + \beta)/2\,T \le f \le (1 + \beta)/2\,T$ for $0 \le \beta \le 1$.

6.1.3.2 Spectrally Raised Cosine (SRC) Pulse

A popular example of $G_\beta(f)$ is the following raised cosine function:

$$G_\beta(f) = \frac{\pi T}{2\beta}\cos\left(\frac{\pi Tf}{\beta}\right) \qquad -\beta/2\,T \le f \le \beta/2\,T$$

$$= 0 \qquad\qquad\qquad \text{otherwise} \tag{6.13}$$

and $g_\beta(t)$ is calculated by

$$g_\beta(t) = \frac{\cos\beta\pi t/T}{1 - 4\beta^2 t^2/T^2} \tag{6.14}$$

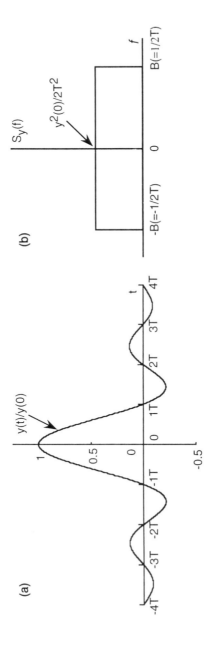

Figure 6.2 (a) A pulse without intersymbol interference and (b) its power spectral density.

Calculating the convolution, we have

$$
Y(f) = \begin{cases} T & 0 \leq |f| < (1 - \beta)/2T \\ \dfrac{T}{2}\left[1 - \sin\dfrac{\pi T}{\beta}\left(|f| - \dfrac{1}{2T}\right)\right] & (1 - \beta)/2T \leq |f| \leq (1 + \beta)/2T \\ 0 & (1 + \beta)/2T < |f| \end{cases}
$$

$$(6.15)$$

where β is called the rolloff factor. The spectrally raised cosine pulse $y(t)$ is expressed as

$$
y(t) = \frac{\cos\beta\pi t/T}{1 - 4\beta^2 t^2/T^2}\,\frac{\sin\pi t/T}{\pi t/T} \tag{6.16}
$$

The tail of the pulse decays as $1/t^3$. For $\beta = 0$, the bandwidth of $Y(f)$ is the Nyquist bandwidth, and for $\beta = 1$, the bandwidth is twice the Nyquist bandwidth. In other words, by reducing the symbol rate to less than $2B$ symbols/sec, we can have physically attainable pulses without intersymbol interference at the sampling instants. Figure 6.3 shows $G_\beta(f)$, $Y(f)$, and $y(t)$ for several values of β.

When we remove the condition of no intersymbol interference, we have a class of partial-response signals. For example, the duobinary signal pulse [7] is given by

$$
y(nT) = \begin{cases} 1 & n = 0,\ 1 \\ 0 & \text{otherwise} \end{cases} \tag{6.17}
$$

In this case, having two nonzero coefficients at the sampling instants results in deterministic intersymbol interference, which can be controlled at the receiving terminal. The merit of the duobinary pulse is that the frequency characteristics can be more easily approximated by attainable filters and narrower bandwidth can be achieved by using controlled intersymbol interference.

6.1.3.3 *L* Rectangular (LREC) and *L* Raised Cosine (LRC) Pulses

Below, we present another example of popular pulse shapes. Here, L denotes the length of symbol intervals.

For an L rectangular pulse $y(t)$, where $y(t) = 1/(2LT)$ for $-LT/2 \leq t \leq LT/2$ and $y(t) = 0$ otherwise, its Fourier transform $Y(f)$ is given by

$$
Y(f) = \frac{1}{2}\frac{\sin\pi fLT}{\pi fLT} \tag{6.18}
$$

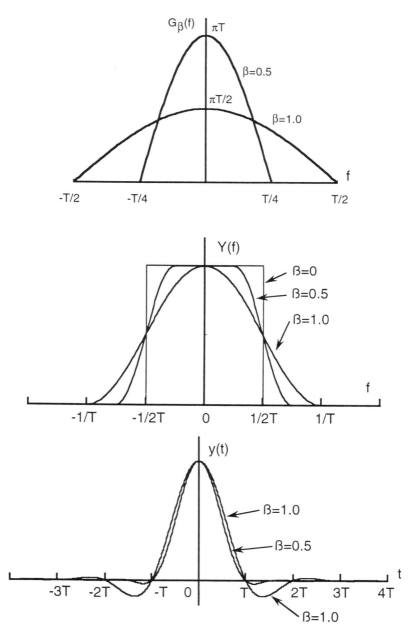

Figure 6.3 $G_\beta(f)$, $Y(f)$, and spectrally raised cosine pulse $y(t)$ for several values of the rolloff factor β. $G_\beta(f)$, $Y(f)$, and $y(t)$ are given by (6.13), (6.15), and (6.16), respectively.

On the other hand, the raised cosine pulse $y(t)$ is defined as

$$y(t) = \frac{1}{2LT}\left[1 - \cos\frac{2\pi}{LT}\left(t - \frac{LT}{2}\right)\right] \quad \text{for } -LT/2 \leq t \leq LT/2$$
$$= 0 \qquad \text{otherwise} \qquad (6.19)$$

Therefore, its Fourier transform is given by

$$Y(f) = \frac{LT}{2}\frac{\sin \pi fLT}{\pi fLT(1 - f^2L^2T^2)}e^{-j\pi fLT} \qquad (6.20)$$

Figure 6.4 shows the raised cosine pulse and its power spectrum $|Y(f)|^2$, which has nulls at $f = n/(LT)$ for $n = \pm2, \pm3, \ldots$. Compared to the power spectrum of the rectangular pulse, the spectrum of the raised cosine pulse has

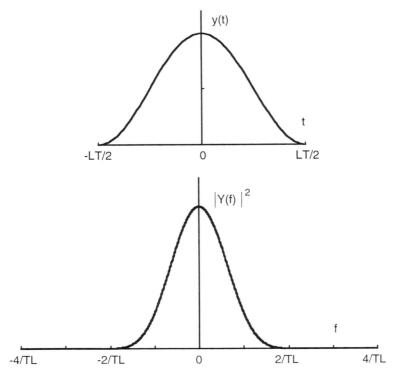

Figure 6.4 *L* raised cosine pulse $y(t)$ and its power spectrum $|Y(f)|^2$, which are given by (6.19) and (6.20), respectively.

a broader main lobe, but the tails decay faster as $1/f^6$. Choosing a pulse having $L > 1$ causes overlapping between succeeding pulses, and this is called the partial-response technique. The pulse with $L \leq 1$ is called the full-response pulse. Table 6.2 shows several pulses and their Fourier transforms.

6.1.4 Modulation Techniques—BPSK, QPSK, OQPSK, and MSK

This section presents a family of PSK, which is most popular for satellite communications [8–10]. In particular, BPSK, QPSK, OQPSK, $\pi/4$-PSK, and MSK have been used for MSS modems. Section 6.1.5 describes $\pi/4$-QPSK and Gaussian-filtered MSK (GMSK) and Section 6.1.6 describes continuous phase modulation (CPM).

6.1.4.1 Representation of Signals

As shown in the previous section, a bandpass signal is expressed by

$$s(t) = Re[u(t)e^{j2\pi f_c t}] \tag{6.21}$$

where f_c is the carrier frequency and $u(t)$ is the equivalent low-pass signal.

$$u(t) = \sum_{n=-\infty}^{\infty} a_n g(t - nT) \tag{6.22}$$

where $g(t)$ represents a pulse and $\{a_n\}$ are information carrying symbols.

6.1.4.2 BPSK

For binary PSK, binary data is expressed by $a_n = \exp(j\phi_n)$ for $\phi_n = 0$ or π, and the phase changes every data bit of duration T_b. When $g(t)$ is a rectangular pulse over symbol duration $T (= T_b)$, the BPSK signal is expressed by

$$\begin{aligned} s(t) &= Aa_n \cos2\pi f_c t \\ &= A\cos\{2\pi f_c t + \phi_n\} \end{aligned} \qquad nT \leq t < (n+1)T \tag{6.23}$$

6.1.4.3 QPSK

For M-ary PSK, a_n is selected from M signals $\{\exp[j2\pi(m-1)/M] \; ; \; m = 1, 2, \ldots, M\}$, and the resulting signal $s(t)$ is written as

$$s(t) = A\cos\left\{2\pi f_c t + \frac{2\pi}{M}(m-1)\right\} \qquad m = 1, 2, \ldots, M \tag{6.24}$$

Table 6.2
Pulse Shapes and Their Fourier Transforms

Pulse	Pulse Shape, $y(t)$	Fourier Transform $Y(f)$
SRC	$y(t) = \dfrac{\cos\beta\pi t/T}{1 - 4\beta^2 t^2/T^2}\,\dfrac{\sin\pi t/T}{\pi t/T}$	$Y(f) = \begin{cases} T & 0 \le \lvert f\rvert < (1-\beta)/2T \\ \dfrac{T}{2}\left[1 - \sin\dfrac{\pi T}{\beta}\left(\lvert f\rvert - \dfrac{1}{2T}\right)\right] & (1-\beta)/2T \le \lvert f\rvert \le (1+\beta)/2T \\ 0 & (1+\beta)/2T < \lvert f\rvert \end{cases}$
LREC	$y(t) = 1/(2LT)$ for $-LT/2 \le t \le LT/2$	$Y(f) = \dfrac{1}{2}\dfrac{\sin\pi fLT}{\pi fLT}$
LRC	$y(t) = \dfrac{1}{2LT}\left[1 - \cos\dfrac{2\pi}{LT}\left(t - \dfrac{LT}{2}\right)\right]$ for $-LT/2 \le t \le LT/2$	$Y(f) = \dfrac{LT}{2}\dfrac{\sin\pi fLT}{\pi fLT(1 - f^2 L^2 T^2)}e^{-j\pi fLT}$
GMSK	$y(t) = \dfrac{1}{2T}\left[Q\left(2\pi B\dfrac{t - T/2}{\sqrt{\ln 2}}\right) - Q\left(2\pi B\dfrac{t + T/2}{\sqrt{\ln 2}}\right)\right]$ $Q(t) = \dfrac{1}{\sqrt{2\pi}}\displaystyle\int_t^\infty e^{-\tau^2/2}d\tau$	$Y(f) = \dfrac{1}{2}\sqrt{\dfrac{\ln 2}{\pi B^2}}\exp\left[-\dfrac{f^2}{(B^2/\ln 2)}\right]\dfrac{\sin\pi fT}{\pi fT}$

Note: SRC: spectrally raised cosine, LREC: L-rectangular, LRC: L-raised cosine, GMSK: Gaussian-filtered MSK.

For QPSK, $\{\phi_m\}$ is a set of $\{0, \pi/2, \pi, 3/2\pi\}$. Then, the signal $s(t)$ is given by

$$s(t) = \frac{A}{\sqrt{2}} a_n^I \cos\left(2\pi f_c t + \frac{\pi}{4}\right) + \frac{A}{\sqrt{2}} a_n^Q \sin\left(2\pi f_c t + \frac{\pi}{4}\right) \qquad (6.25)$$

where $\{a_n^I\}$ and $\{a_n^Q\}$ are the ±1-valued data, which are converted from input data sequence $\{a_n\}$ into the in-phase channel (I channel) and quadrature channel (Q channel), respectively. Then, (a_n^I, a_n^Q) is $(1, 1)$ for $\phi_n = 0$, $(1, -1)$ for $\phi_n = \pi/2$, $(-1, -1)$ for $\phi_n = \pi$, and $(-1, 1)$ for $\phi_n = 3\pi/2$. Figure 6.5 shows a typical quadrature modulator and demodulator. The durations of both symbols $\{a_n^I\}$ and $\{a_n^Q\}$ are doubled to become $T\ (= 2T_b)$, where T_b is the

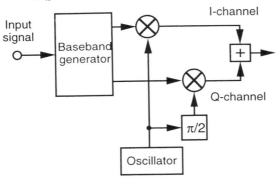

(a)

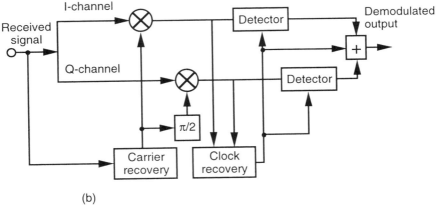

(b)

Figure 6.5 (a) Quadrature modulator and (b) demodulator.

information bit duration and T is the symbol duration. Therefore, the symbol rate on channels is $1/(2T_b)$ symbols/sec, although the information data rate is $1/T_b$ bits/sec. Data sequences for I and Q channels are multiplied by cosinusoidal and sinusoidal waves, respectively.

6.1.4.4 OQPSK

OQPSK delays the quadrature bit stream by T sec relative to the in-phase bit stream to restrict the phase transitions to phase changes of 0 or $\pi/2$ every T sec. Using $\{a_n^I\}$ and $\{a_n^Q\}$, the equivalent low-pass signal is expressed by

$$u(t) = \sum_{n=-\infty}^{\infty} \left[a_n^I g(t - 2nT) - j a_n^Q g(t - 2nT - T) \right] e^{j\pi/4} \qquad (6.26)$$

The signal is written by

$$s(t) = \left[\sum_n a_n^I g(t - 2nT) \right] \cos(2\pi f_c t + \pi/4)$$
$$+ \left[\sum_n a_n^Q g(t - 2nT - T) \right] \sin(2\pi f_c t + \pi/4) \qquad (6.27)$$

The conventional QPSK, where the data transition occurs at the same time in both the I and Q channels, has larger phase changes than OQPSK, which has phase changes of at most $\pm\pi/2$, because either the I or Q channel changes its phase at each bit transition and large envelope fluctuations do not occur as they do with π-phase changes in QPSK. Figure 6.6 shows the comparison of QPSK and OQPSK modulated signals.

INMARSAT aeronautical satellite communication systems uses aviation-QPSK (A-QPSK) and aviation-BPSK (A-BPSK) schemes. The A-QPSK uses the same concept as OQPSK. The 99% bandwidth of the A-QPSK signal using a filter for the SRC pulse given by (6.15) with a rolloff factor β of 1.0 is designed to be 1.64. On the other hand, A-BPSK uses I channel and Q channel signals of QPSK alternatively at each bit interval. In this case, the phase transition is also restricted to $\pm\pi/2$. The 99% bandwidth of the A-BPSK signal passed through the filter given by (6.15) with $\beta = 0.4$ is 1.2.

6.1.4.5 MSK

Minimum shift keying (MSK) is a special case of OQPSK with a sinusoidal pulse defined as

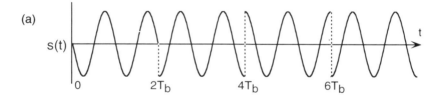

(a)

(b)

Tb: information bit interval

Figure 6.6 QPSK and OQPSK modulation signals (a) QPSK and (b) OQPSK.

$$g(t) = \begin{cases} \dfrac{A}{\sqrt{2}} \sin\dfrac{\pi t}{2T} & 0 \leq t \leq 2T \\ 0 & \text{otherwise} \end{cases} \tag{6.28}$$

rather than rectangular shaping pulse in OQPSK. Therefore, the signal is expressed as

$$s_n(t) = \frac{A}{\sqrt{2}} a_n^I \cos\frac{\pi t}{2T} \cos\left(2\pi f_c t + \frac{\pi}{4}\right)$$

$$+ \frac{A}{\sqrt{2}} a_n^Q \sin\frac{\pi t}{2T} \sin\left(2\pi f_c t + \frac{\pi}{4}\right) \tag{6.29}$$

where $s_n(t)$ is $s(t)$ at time interval $(2n + 1)T \leq t < (2n + 2)T$. This shows that an MSK signal has weighting functions of $\cos(\pi t/2T)$ and $\sin(\pi t/2T)$ for I and Q channels, respectively [11]. This can be rewritten as

$$s_n(t) = \frac{A}{\sqrt{2}} \cos\left(2\pi f_c t + \frac{\pi b_n}{2T} t + \phi_n + \frac{\pi}{4}\right) \tag{6.30}$$

where $b_n = -a_n^I a_n^Q$, $\phi_n(t) = 0$ for $a_n^I = 1$, and $\phi_n(t) = \pi$ for $a_n^I = -1$. This shows that the MSK signal has two frequencies, $f_c + 1/4T$ and $f_c - 1/4T$. Therefore, MSK is also a special form of continuous-phase frequency shift keying (CPFSK) with frequency separation of $1/2\,T$ and modulation index of 0.5. Figure 6.7 shows waveforms modulated by MSK. This has the characteristics of a constant-envelope frequency-modulated signal. These characteristics will be presented again in Section 6.1.6 on continuous phase modulation.

As seen above, the quadrature modulator and demodulator shown in Figure 6.5 can be used for BPSK, QPSK, OQPSK, and MSK. For BPSK, only the upper half is used. The difference in this modulator for these modulation schemes is the pulse shapes and the relative timing between them in the *I* and *Q* channels. In the quadrature demodulator, the carrier recovery circuit regenerates a carrier spectral line component from the received modulated signal by a nonlinear operation such as the fourth power of the QPSK signal. A phase-locked loop (PLL) locks the fourth power harmonic of the received carrier. The received signal is multiplied by these quadrature carriers and processed through the matched filter, which is an integrated-and-dump filter, for example. Timing local clocks are produced from the received signal by the clock recovery circuit.

6.1.4.6 Power Spectral Density

In this section, we consider the spectral characteristics of these modulation schemes in order to compare their bandwidth efficiencies. The spectral characteristics of a stochastic signal can be obtained by calculating the Fourier transform of its autocorrelation functions. The autocorrelation function $R_s(\tau)$ of $s(t)$ is defined as

$$R_s(\tau) = E[s(t)s(t + \tau)]$$
$$= Re[R_u(\tau)e^{j2\pi f_c t}] \tag{6.31}$$

where $E[\ ]$ denotes the statistical average and $R_u(\tau)$ is the autocorrelation function of the equivalent low-pass signal $u(t)$, as defined by

$$R_u(\tau) = \frac{1}{2}E[u^*(t)u(t + \tau)] \tag{6.32}$$

The Fourier transform of $R_s(\tau)$ gives the power spectral density $S_s(f)$ of $s(t)$.

$$S_s(f) = \frac{1}{2}[S_u(f - f_c) + S_u(-f - f_c)] \tag{6.33}$$

where $S_u(f)$ is the power spectral density of $u(t)$. Since $S_s(f)$ can be represented by $S_u(f)$, we will consider only that of the equivalent low-pass signal. Since $u(t)$ is expressed as

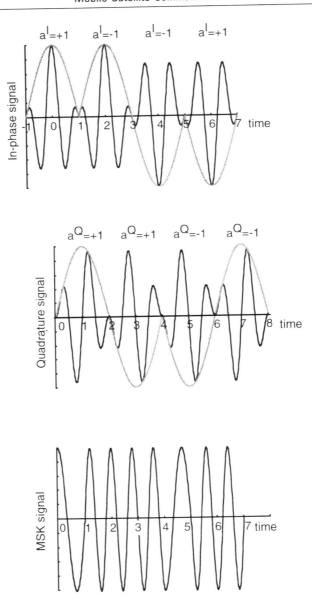

Figure 6.7 MSK modulated signals.

$$u(t) = \sum_{n=-\infty}^{\infty} a_n g(t - nT) \tag{6.34}$$

the autocorrelation function $R_u(t)$ of $u(t)$ can be given by

$$R_u(\tau) = \frac{1}{2} \sum_{n=-\infty}^{\infty} \sum_{m=-\infty}^{\infty} E[a_n^* a_m] g^*(t - nT) g(t + \tau - mT) \tag{6.35}$$

The Fourier transform yields the power spectral density of $u(t)$

$$S_u(f) = \frac{1}{T} |G(f)|^2 S_a(f) \tag{6.36}$$

where $G(f)$ is the Fourier transform of $g(t)$, and $S_a(f)$ is the power spectral density function of the information sequence $\{a_n\}$, which is given by

$$S_a(f) = \sum_{m=-\infty}^{\infty} R_a(m) e^{-j2\pi fmT} \tag{6.37}$$

where $R_a(m)$ is the autocorrelation function defined by

$$R_a(m) = \frac{1}{2} E[a_n^* a_{n+m}] \tag{6.38}$$

When we assume that $R_a(0) = \sigma^2 + \mu^2$ and $R_a(m) = \mu^2$ for $m \neq 0$, we can rewrite $S_a(f)$ as

$$S_a(f) = \sigma^2 + \frac{\mu^2}{T} \sum_{m=-\infty}^{\infty} \delta\left(f - \frac{m}{T}\right) \tag{6.39}$$

Then, $S_u(f)$ is given by [7]

$$S_u(f) = \frac{\sigma^2}{T} |G(f)|^2 + \frac{\mu^2}{T^2} \sum_{n=-\infty}^{\infty} \left|G\left(\frac{n}{T}\right)\right|^2 \delta\left(f - \frac{n}{T}\right) \tag{6.40}$$

For the ideal rectangular pulse $g(t) = A$ for $0 \leq t \leq T$, the power spectral density $S_u(f)$ of $u(t)$ is given by

$$S_u(f) = A^2 T\sigma^2 \left(\frac{\sin \pi fT}{\pi fT}\right)^2 + A^2\mu^2\delta(f) \qquad (6.41)$$

When the information symbols have zero mean (i.e., $\mu = 0$), the second term vanishes. The BPSK and QPSK signals with a rectangular pulse of duration T have the same power spectral densities. However, when T_b is the information bit interval, $T = T_b$ for BPSK and $T = 2T_b$ for QPSK.

The equivalent low-pass signal of OQPSK and MSK is expressed by

$$u(t) = \sum_{n=-\infty}^{\infty} [a_n^I g(t - 2nT) - ja_n^Q g(t - 2nT - T)]e^{j\pi/4} \qquad (6.42)$$

The power spectral density of OQPSK with a rectangular pulse $g(t)$ of duration $2T$ ($T = T_b$) is given by the same power spectral function as that for QPSK, as expressed by

$$S_u(f) = 2T_b\left(\frac{\sin 2\pi fT_b}{2\pi fT_b}\right)^2 \qquad (6.43)$$

On the other hand, the power spectral density of MSK with a sinusoidal pulse is expressed by

$$S_u(f) = \frac{16T_b}{\pi^2}\left(\frac{\cos 2\pi fT_b}{1 - 16f^2 T_b^2}\right)^2 \qquad (6.44)$$

Figure 6.8 compares the spectra of these PSK modulation schemes. The main lobe of MSK is wider than those of QPSK and OQPSK, but the side lobes of MSK decay much faster. As mentioned in Section 6.1.2, this results in the 99% and 35-dB bandwidths of MSK being considerably smaller than those of QPSK, although the null-to-null and half-power bandwidths of MSK are larger than those of QPSK. The former bandwidth efficiency accounts for the popularity of MSK in many digital communication systems. Table 6.3 shows the power spectral density functions and the bit error probability functions for these modulation schemes.

6.1.4.7 Bit Error Performance in Additive Gaussian Noise Channels

In this section, we consider the bit error performance of the PSK family in an additive white Gaussian noise channel with no bandwidth limitation. The received equivalent low-pass signal $v(t)$ can be expressed by

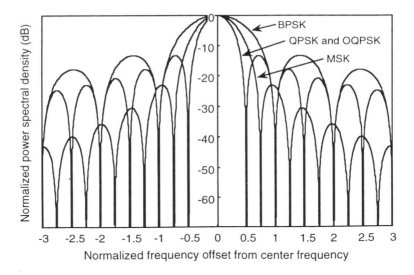

Figure 6.8 Power spectral densities of BPSK, QPSK, OQPSK, and MSK.

$$v(t) = \alpha e^{-j\phi} u_n(t) + n(t) \tag{6.45}$$

where α is a loss factor, ϕ is the phase shift due to time delay, $u_n(t)$ is the nth signal, and $n(t)$ is Gaussian noise. According to the maximum a posteriori (MAP) probability criterion, the demodulator chooses a signal that has the largest posterior probability. This criterion results in finding the largest one of the decision variables I_m, which are defined by [7]

$$I_m = Re\left[e^{j\phi} \int_0^T v(t) u_m^*(t)\, dt \right] \quad m = 1, 2, \ldots, M \text{ for } M\text{-ary PSK} \tag{6.46}$$

Here, we consider coherent demodulation, where we can know the phase ϕ perfectly. If the signal u_1 is assumed to be transmitted for a binary case of $M = 2$, the decision variables can be written as

$$I_1 = 2\alpha E_s + N_1 \tag{6.47}$$
$$I_2 = 2\alpha E_s \rho_{12} + N_2$$

where E_s is the signal energy, ρ_{12} is the real part of the correlation function between u_1 and u_2, and N_m is Gaussian noise, defined by

$$E_s = \frac{1}{2} \int_0^T |u_1(t)|^2\, dt = \frac{1}{2} \int_0^T |u_2(t)|^2\, dt \tag{6.48}$$

Table 6.3
Power Spectral Density Functions and Bit Error Probability Functions for Coherent Detection and Coherent Detection of Differentially Encoded Modulation

Modulation Schemes	Power Spectrum Density	Error Probability for Coherent Detection	Error Probability for Diferentially Encoded
BPSK	$\dfrac{A^2}{R}\left(\dfrac{\sin \pi f/R}{\pi f/R}\right)^2$	$\dfrac{1}{2}\text{erfc}(\sqrt{E_b/N_0})$	$\text{erfc}\sqrt{E_b/N_0}[1 - \frac{1}{2}\text{erfc}\sqrt{E_b/N_0}]$
QPSK, OQPSK, π/4-QPSK	$\dfrac{2A^2}{R}\left(\dfrac{\sin 2\pi f/R}{2\pi f/R}\right)^2$	$\dfrac{1}{2}\text{erfc}(\sqrt{E_b/N_0})$	$2\text{erfc}\sqrt{E_b/N_0} - 2\text{erfc}^2\sqrt{E_b/N_0}$ $+ \text{erfc}^3\sqrt{E_b/N_0} - \frac{1}{4}\text{erfc}^4\sqrt{E_b/N_0}$
MSK	$\dfrac{16A^2}{\pi^2 R}\left(\dfrac{\cos 2\pi f/R}{1 - 16f^2/R^2}\right)^2$	$\dfrac{1}{2}\text{erfc}(\sqrt{E_b/N_0})$	$2\text{erfc}\sqrt{E_b/N_0} - 2\text{erfc}^2\sqrt{E_b/N_0}$ $+ \text{erfc}^3\sqrt{E_b/N_0} - \frac{1}{4}\text{erfc}^4\sqrt{E_b/N_0}$
CPFSK	$\dfrac{4h^2 T}{\pi^2[4(fT)^2 - h^2]} \times$ $\dfrac{(\cos \pi h - \cos 2\pi ft)^2}{1 + \cos^2 \pi h - 2\cos \pi h \cos 2\pi ft}$	$\approx C\exp\left(-d_{min}^2 \dfrac{E_b}{N_0}\right)$	

$$\rho_{12} = Re\left[\frac{1}{2E_s}\int_0^T u_1(t)u_2^*(t)\ dt\right] \tag{6.49}$$

$$N_m = Re[e^{j\phi}\int_0^T n(t)u_m^*(t)\ dt]\quad m = 1,\ 2 \tag{6.50}$$

The probability of decision error is given by the probability that I_1 is less than I_2. Since N_1 and N_2 are still Gaussian noise, a variable I defined as $I = I_1 - I_2$ is also Gaussian with mean μ of $2\alpha E_s(1 - \rho_{12})$ and variance σ^2 of $4E_s N_0(1 - \rho_{12})$, where N_0 is the power of $n(t)$. Therefore, the probability that $I < 0$ is expressed by

$$\begin{aligned} P(I < 0) &= \frac{1}{\sqrt{2\pi\sigma^2}}\int_{-\infty}^0 \exp\left[-\frac{(x - \mu)^2}{2\sigma^2}\right]\ dx \\ &= \frac{1}{2}\text{erfc}\left(\frac{\mu}{\sqrt{2}\sigma}\right) \\ &= \frac{1}{2}\text{erfc}\left(\sqrt{\frac{\alpha^2 E_s}{2N_0}(1 - \rho_{12})}\right) \end{aligned} \tag{6.51}$$

where erfc( ) is the complementary error function given by

$$\text{erfc}(x) = \frac{2}{\sqrt{\pi}}\int_x^\infty \exp(-t^2)\ dt$$

For example, for a BPSK signal, which has $\rho_{12} = -1$ (called antipodal), the bit error probability is given by

$$\begin{aligned} P &= \frac{1}{2}\text{erfc}\left(\sqrt{\frac{\alpha^2 E_s}{N_0}}\right) \\ &= \frac{1}{2}\text{erfc}(\sqrt{\gamma_b}) \end{aligned} \tag{6.52}$$

where $\gamma_b = \alpha^2 E_s/N_0$ is the ratio of energy per bit to noise, or E_b/N_0. Since QPSK is represented by two independent in-phase and quadrature BPSK signals, and Gaussian noises N_1 and N_2 are also independent, the bit error probability is identical to that of BPSK. Moreover, the coherent detection of OQPSK and MSK give the same bit error performance as BPSK and QPSK. On the other hand, the symbol error probability of QPSK is given by $\text{erfc}(\sqrt{\gamma_b}) - (1/4)\text{erfc}^2(\sqrt{\gamma_b})$.

6.1.4.8 Differentially Encoded PSK

For coherent demodulation of PSK, the carrier phase has to be correctly estimated from the received signal and has to be perfectly synchronized in frequency. One method for carrier recovery is to produce the Mth power of the M-ary signal in order to remove its modulated component from the received signal, but such an operation causes phase ambiguity. For example, the BPSK signal, which has two phases, 0 and π, is squared to remove its modulation and becomes an unmodulated carrier, which has two equivalent phases 0 and 2π. However, we have ambiguity of 0 and π in the absolute phase of the received BPSK signal.

One solution is to employ differential encoding, which encodes the signal based on the bit difference between successive bits instead of absolute phase encoding. For example, if the information sequence of $\{\alpha_1 = 0,\ \alpha_2 = \pi\}$ is α_1, α_2, α_1, α_1, $\ldots$ and the initial phase α_1 is arbitrarily chosen as a reference phase, then the differentially encoded sequence $\{A_i\}$ is given by $A_1 = \alpha_1$, $A_2 = A_1 + \alpha_1 = \alpha_1$, $A_3 = A_2 + \alpha_2 = \alpha_2$, $A_4 = A_3 + \alpha_1 = \alpha_2$, $A_5 = A_4 + \alpha_1 = \alpha_2$. At the receiver, if there is no phase shift between successive bits, the bit is estimated to be α_1, and if a phase shift of π is detected, the bit is estimated to be α_2. Then, the decoded sequence can be obtained by $A_{(1)} = \alpha_1$, $A_{(2)} = \alpha_2$, $A_{(3)} = \alpha_1$, $A_{(4)} = \alpha_1$, $\ldots$ at the receiver without knowledge of the absolute signal phase.

There are two types of modulation techniques employing differential encoding. One is a differentially encoded coherent PSK that employs coherent detection of differentially encoded PSK signals. The other is a differentially coherent PSK (DPSK) that employs differential encoding of PSK and makes its decision based on phase differences. Since one error in the decision results in two decoding errors, and two successive errors result in correct bit decision, there will be some degradation in bit error performance relative to the nondifferentially encoded case. When E_b/N_0 is high, the bit error probability of differentially encoded coherent PSK is approximately twice the bit error probability of nondifferentially encoded PSK with absolute phase. The bit error probabilities for coherent detection of differentially encoded BPSK and QPSK are respectively given by [12]

$$P_{\text{BPSK}} = \text{erfc}\sqrt{E_b/N_0}\left[1 - \frac{1}{2}\text{erfc}\sqrt{E_b/N_0}\right]$$

$$P_{\text{QPSK}} = 2\text{erfc}\sqrt{E_b/N_0} - 2\text{erfc}^2\sqrt{E_b/N_0}$$

$$+ \text{erfc}^3\sqrt{E_b/N_0} - \frac{1}{4}\text{erfc}^4\sqrt{E_b/N_0} \tag{6.53}$$

6.1.4.9 Differential PSK (DPSK)

For mobile satellite communications, channel conditions are more severe than additive white Gaussian noise channel due to multipath fading, shadowing, and Doppler effects. One solution to overcome these disturbances is to use differentially coherent detection of differentially encoded signals. When the phase condition is constant for $2T$ seconds, the DPSK demodulator can obtain the optimum a posteriori probability. Moreover, when the phase condition stays constant for nT seconds, where $n > 2$, an alternate differentially coherent detection scheme, which uses not only the following two phases but also the previous $(n - 1)$ phases, is applicable.

Here, we consider the bit error probability of differentially coherent detection of differentially encoded BPSK. The equivalent low-pass signal in the mth interval is given by

$$v_m(t) = \alpha e^{j(\theta_m - \phi)}u_m(t) + n_m(t) \tag{6.54}$$

In the previous interval, the signal is given by

$$v_{m-1}(t) = \alpha e^{j(\theta_{m-1} - \phi)}u_{m-1}(t) + n_{m-1}(t) \tag{6.55}$$

When the signal is passed through a matched filter and sampled, the decision variable for binary DPSK is expressed as [7]

$$I_m = Re[v_m v_{m-1}^*] \tag{6.56}$$

Without loss of generality, we assume the phase difference $\theta_m - \theta_{m-1} = 0$ to calculate the error probability. Therefore, the bit error probability is the probability that $I_m < 0$, which is given by

$$P_{\text{BDPSK}} = \frac{1}{2}\exp[-\gamma_b] \tag{6.57}$$

where γ_b is the signal-to-noise density ratio per bit, E_b/N_0.

The error probability of M-ary DPSK needs complicated calculations, but for large E_b/N_0, several approximations have been proposed. For example, the symbol error probability for M-ary DPSK is approximately given by [12]

$$P_{\text{MDPSK}} \cong \text{erfc}(\xi) + \frac{\xi \exp(-\xi^2)}{4\sqrt{\pi}(1/8 + R_d)} \qquad (6.58)$$

where

$$\xi = \sqrt{2R_d} \sin\left(\frac{\pi}{2M}\right) \qquad (6.59)$$

and $R_d = (\log_2 M)(E_b/N_0)$. Figure 6.9 shows symbol error probability performance for coherent and differentially coherent detection of differentially

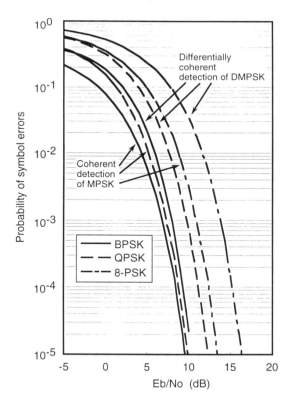

Figure 6.9 Symbol error probability performance for coherent and differentially coherent detection of differentially encoded MPSK.

encoded MPSK. At symbol error probabilities less than 10^{-3}, BDPSK is inferior to coherently detected PSK by almost 3 dB, but at 10^{-5}, the difference is less than 1 dB. Four-phase DPSK is 2.3 dB worse than QPSK at large E_b/N_0.

6.1.4.10 Noncoherent Detection

In this section, we consider the noncoherent detection technique, which is useful when the carrier phase is difficult to estimate at the receiver. In mobile satellite communication channels, for example, there are the following situations. First, the transmitting oscillator that generates the carrier cannot be completely stabilized. Second, a mobile satellite link has a low signal-to-noise ratio. For coherent detection, the bandwidth of the carrier-tracking loop at the receiver must be decreased to increase the *S/N* and obtain a good phase reference. Since mobile satellite propagation characteristics such as multipath fading and the Doppler effect cause rapid phase variation, the carrier phase of the received signal does not remain fixed long enough to be estimated. One solution is the method referred to as noncoherent detection, which does not use knowledge of the carrier phase.

The equivalent low-pass signal is given by

$$v(t) = \alpha e^{-j\phi_n} u_n(t) + n(t) \tag{6.60}$$

Since we do not have any information about the phase ϕ_n, we assume it has random values, which are mutually independent and uniformly distributed.

The decision variables of noncoherent detection for the optimum demodulator are expressed by [7]

$$I_m = \left| \int_0^T v(t) u_m^*(t) \, dt \right| \quad m = 1,2 \tag{6.61}$$

Assuming that the signal of $m = 1$ is transmitted, we can calculate I_m as

$$I_1 = |2\alpha E_s e^{-j\phi_1} + N_1|$$
$$I_2 = |2\alpha E_s \rho_{12} e^{-j\phi_1} + N_2| \tag{6.62}$$

where ρ_{12} is the correlation function between u_1 and u_2, and N_m is defined by

$$N_m = \int_0^T n(t) u_m^*(t) \, dt \quad m = 1, 2 \tag{6.63}$$

Then, N_m are complex-valued Gaussian random variables with zero mean and variance $\sigma^2 = 2E_sN_0$. Therefore, I_1 and I_2 are random variables that are represented by a Rician probability distribution. The bit error probability is the probability that $I_2 > I_1$, which is represented by

$$P = Q(A, B) - \frac{1}{2}\exp\left[- \frac{(A^2 + B^2)}{2} \right] I_0(AB) \qquad (6.64)$$

where

$$A = \sqrt{\gamma/2\left(1 - \sqrt{1 - |\rho_{12}|^2} \right)}$$

$$B = \sqrt{\gamma/2\left(1 + \sqrt{1 - |\rho_{12}|^2} \right)}$$

$$Q(A, B) = e^{-(A^2 + B^2)/2} \sum_{k=0}^{\infty} \left(\frac{A}{B}\right)^k I_k(AB) \qquad (6.65)$$

Here, $Q(A, B)$ is the Q function and $I_0(x)$ is the modified Bessel function of order zero.

For a binary orthogonal signal $\rho_{12} = 0$, such as orthogonal FSK, the bit error probability, which has the minimum value at $P_{12} = 0$, is given by

$$P = \frac{1}{2}\exp\left(- \frac{\gamma_b}{2} \right) \qquad (6.66)$$

On the other hand, for BPSK of $|\rho_{12}| = 1$, the error probability becomes $P = \frac{1}{2}$.

6.1.5 Modulation Techniques—π/4-QPSK and GMSK

6.1.5.1 π/4-QPSK

Recently, π/4-QPSK or π/4 shift QPSK has become very popular for mobile satellite communications as well as for terrestrial mobile communications because it has a compact spectrum with small spectrum restoration due to nonlinear amplification and can perform differential detection. Figures 6.10 and 6.11 show the signal constellation and the block diagram of a conceptual modulator and a differential detector for π/4-QPSK, respectively.

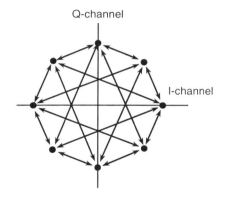

Figure 6.10 Signal constellation for unfiltered $\pi/4$-QPSK.

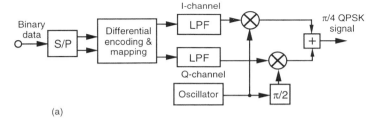

(a)

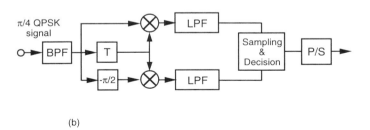

(b)

Figure 6.11 Block diagram of (a) a modulator and (b) demodulator IF-band differential detector employing delay line and mixers for $\pi/4$-QPSK.

Let $u(t)$ and $v(t)$ denote the unfiltered nonreturn to zero (NRZ) pulses in the I and Q channels, respectively, and let u_k and v_k, which are the pulse amplitudes for $kT \le t < (k + 1)T$, be determined by the previous pulses and the current information symbol [13]:

$$\begin{pmatrix} u_k \\ v_k \end{pmatrix} = \begin{pmatrix} \cos\ \theta_k & -\sin\ \theta_k \\ \sin\ \theta_k & \cos\ \theta_k \end{pmatrix} \begin{pmatrix} u_{k-1} \\ v_{k-1} \end{pmatrix}$$

where $\theta_k = \pi/4, 3\pi/4, -3\pi/4, -\pi/4$ correspond to, for example, the information 11, 01, 00, 10, respectively. This equation represents the rotation of the vector (u_{k-1}, v_{k-1}) by an angle θ_k. If the initial phase point in the u-v plane is assumed to be 0 at $t = 0$ (i.e., (1, 0)), then the phase point (u_k, v_k) at $t = kT$ for even numbers of k is one of $\{(1, 0), (0, 1),(-1, 0), (0, -1)\}$, and (u_k, v_k) for odd k is one of $\{(1/\sqrt{2}, 1/\sqrt{2}), (-1/\sqrt{2}, 1/\sqrt{2}), (1/\sqrt{2}, -1/\sqrt{2}), (-1/\sqrt{2}, -1/\sqrt{2})\}$. Therefore, the phase point always shifts its phase over successive time intervals by $\pm\pi/4$ or $\pm 3\pi/4$. The spectrum of a $\pi/4$-QPSK signal is the same as that of a QPSK one that undergoes an instantaneous $\pm\pi/2$ or $\pm\pi$ phase transition, and the same as that of a OQPSK one that undergoes $\pm\pi/2$ phase transitions. However, the $\pi/4$-QPSK signal can reduce the envelope fluctuation due to band-limited filtering or nonlinear amplification more than QPSK can. This is because differential detection can be used since $\pi/4$-QPSK is not an offset scheme.

For mobile satellite communication channels, a strong line-of-sight signal can be expected even in the Rician fading channel, so coherent demodulation is desirable for improved power efficiency. The bit error performance of coherently detected $\pi/4$-QPSK is the same as that of QPSK.

On the other hand, differential detection is also desirable for simple hardware implementation. The bit error probability for differentially coherent detected $\pi/4$-QPSK is given by

$$P_{\pi/4-\text{QPSK}} = e^{-2\gamma_b} \sum_{k=0}^{\infty} (\sqrt{2} - 1)^k I_k(\sqrt{2}\,\gamma_b) - \frac{1}{2} I_0(\sqrt{2}\,\gamma_b) e^{-2\gamma_b} \quad (6.67)$$

where γ_b is E_b/N_0 and I_k is the kth-order modified Bessel function of the first kind. The differential detection is about 2 to 3 dB inferior to the coherent detection in AWGN and fading channels. Noncoherent detection is also applicable. Because of these advantages, $\pi/4$-QPSK has been chosen as the standard modulation technique for several systems of terrestrial and satellite-based mobile communications. Some experimental results show that in fully saturated amplifier systems, $\pi/4$-QPSK still has significant spectral restoration. To reduce this restoration, $\pi/4$-controlled transition QPSK (CTPSK) uses both sinusoidal shaping pulses and timing offsets of the phase transition between I and Q channels [10].

6.1.5.2 Gaussian-Filtered MSK (GMSK)

Instead of using the NRZ pulse used in the conventional MSK, Gaussian-filtered minimum shift keying (GMSK) uses premodulation baseband filtering with a Gaussian-shaped frequency response [14]. Compared to MSK, which has continuous but sharp phase transitions, GMSK has continuous and smooth phase changes, which can reduce spectral side lobes. The pulse shape, which is passed through the Gaussian premodulation filter, can be expressed by

$$g(t) = \frac{1}{2T}\left[Q\left(2\pi B\frac{t - T/2}{\sqrt{\ln 2}}\right) - Q\left(2\pi B\frac{t + T/2}{\sqrt{\ln 2}}\right)\right] \qquad (6.68)$$

where B is the 3-dB bandwidth of the premodulation low-pass filter, which is a variable parameter to be selected for the design of the power spectral efficiency, and

$$Q(t) = \frac{1}{\sqrt{2\pi}}\int_t^\infty e^{-\tau^2/2}d\tau \qquad (6.69)$$

The 99% bandwidths of the GMSK signal for BT = 0.2, 0.25, and 0.5 are 0.79, 0.86, and 1.04, respectively. For example, the case of BT = 0.25 corresponds to $1/T$ = 8 kHz and B = 2 kHz. On the other hand, the 99% bandwidths of MSK and Tamed FM are 1.2 and 0.79, respectively.

Although the bit error rate (BER) performance of GMSK with $BT = \infty$ is the same as that of MSK and QPSK, the smaller BT is, the greater the degradation of the BER performance is. However, experimental results [14] showed that the degradation in the BER performance of GMSK with BT = 0.25 relative to the conventional MSK is only approximately 1.0 dB at BER = 10^{-3} to 10^{-5}. The bit error probability of GMSK in a static, nonfading condition can be approximated as

$$P_{\text{GMSK}} \cong \frac{1}{2}\text{erfc}\left(\sqrt{0.68\left(\frac{E_b}{N_0}\right)}\right) \qquad (6.70)$$

6.1.6 Modulation Technique—Continuous Phase Modulation (CPM)

In this section, we consider several examples of a large class of continuous phase modulation (CPM), which is attractive because it has both good power and bandwidth efficiencies [15–18]. MSK is a special case of CPM, especially CPFSK, and is also referred to as fast frequency shift keying (FFSK). Several methods of improving MSK by achieving a narrower spectrum, smaller side lobes, and better bit error performance have been studied. In this section, we consider a CPFSK scheme.

6.1.6.1   Continuous-Phase Frequency Shift Keying (CPFSK)

For M-ary PSK, the modulator maps a sequence of information-bearing binary bits into a set of discrete phases $\{2\pi(m-1)/M, m = 1, 2, \ldots, M\}$ of a carrier. When the modulator maps the sequence into a set of discrete frequencies $\{f_m, m = 1, 2, \ldots, M\}$ by switching different oscillators tuned to the frequencies $\{f_m\}$, this modulation method is called M-ary FSK. Such an FSK signal may have large spectral side lobes due to a discontinuous phase jump caused by switching independent oscillators. An FSK signal whose phase is controlled to change continuously at symbol boundaries can significantly reduce its side lobes. This is called continuous-phase FSK (CPFSK).

The equivalent low-pass signal $u(t)$ of CPFSK is expressed by [7]

$$
\begin{aligned}
u(t) &= A \, \exp\left(j2\pi h \int_{-\infty}^{t} d(\tau)\,d\tau \right) \\
&= A \, \exp\left[j2\pi h \sum_{n=0}^{\infty} a_n q(t - nT) \right]
\end{aligned}
\tag{6.71}
$$

where

$$
\begin{aligned}
d(t) &= \sum_{n=0}^{\infty} a_n g(t - nT) \\
q(t) &= \int_{-\infty}^{t} g(\tau)\,d\tau
\end{aligned}
\tag{6.72}
$$

where we assume $q(+\infty) = 1/2$. The 2^k-ary symbols $\{a_i\}$ bear information by mapping k-bit blocks, and $g(t)$ is a pulse over time interval $0 \le t \le LT$. The parameter h is called the modulation index. The rate of the phase change is represented by the value of h. When the value of h varies from one symbol to another, the CPM signal is called multi-h modulation [19,20]. If $g(t) = 0$ for $T \le t$ (i.e., $L \le 1$), we have a full-response CPM; if $g(t) \ne 0$ for $T < t$ (i.e., $L > 1$), we have a partial-response CPM. By choosing different pulse shapes for $g(t)$, different h, and different symbol sizes, we get a variety of CPM signals [15]. MSK is CPM with $h = 1/2$.

6.1.6.2   Power Spectral Density of CPFSK

Since it is not easy to calculate the power spectrum of CPM, we refer to the closed-form expression from [7] as follows. We assume that the pulse $g(t)$ is a full-response rectangular pulse. Then, the resulting power spectrum for binary CPFSK is given by

$$S_u(f) = \frac{4h^2 T}{\pi^2 [4(fT)^2 - h^2]^2} \frac{(\cos \pi h - \cos 2\pi fT)^2}{1 + \cos^2 \pi h - 2 \cos \pi h \cos 2\pi fT} \tag{6.73}$$

Figure 6.12 shows the power spectral density of CPFSK for modulation index $h = 0.5$, 0.6, 0.7, and 0.8. For $h = 0.5$, the power spectrum of CPFSK has a single peak in the main lobe at the center of the normalized frequency and smooth tails. Since the power spectrum becomes much broader for $h > 1$, it is considered that CPM of $h < 1$ is suitable for the modulation scheme of mobile satellite communications.

As mentioned in Section 6.1.4, MSK is a special case of binary CPFSK with $h = 0.5$. From (6.73), the power spectral density of MSK is given by

$$S_u(f) = \frac{16T}{\pi^2} \left(\frac{\cos 2\pi fT}{1 - 16(fT)^2} \right)^2 \tag{6.74}$$

This equation is the same as (6.44).

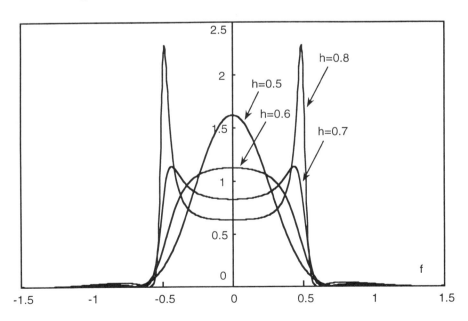

Figure 6.12 Power spectral density of CPFSK for various values of *h*.

6.1.6.3 Bit Error Performance

The phase of the CPM signal $u(t)$ is given by

$$\phi_n(t) = 2\pi h \int_{-\infty}^{t} \sum_{k=0}^{\infty} a_k g(\tau - kT) d\tau = 2\pi h \sum_{k=n-L+1}^{n} a_k q(t - kT) + \pi h \sum_{k=0}^{n-L} a_k$$

$$nT \leq t \leq (n+1)T \quad (6.75)$$

This shows that the phase at $t = nT$ is determined by the data symbol a_n and by many of the previous symbols. For a rational h ($= m/p$), where m and p have no common factor, the number of phase states is finite. For example, for $L = 1$, when m is even, the number of phase states is p, since the second term of (6.75) has p different values in modulo 2π. When m is odd, there are $2p$ phase states. The finite state transitions of the CPM signal for an arbitrary data sequence can be represented by a continuous-phase trajectory or phase trellis. Therefore, maximum likelihood sequence detection (the Viterbi algorithm) can be applied by the trellis description, which will be described in Section 6.2. The bit error performance with maximum likelihood detection is determined by the minimum Euclidean distance $d_{\min}$ between all possible signal sequences. For ideal coherent detection in an AWGN channel at high E_b/N_0, the bit error probability is approximately given by [15]

$$P_{CPM} \approx K \exp\left(-d_{\min}^2 \frac{E_b}{N_0}\right) \quad (6.76)$$

where K is a constant,

$$d_{\min}^2 = \min\left\{\frac{\log_2 M}{T} \int_0^{NT} [1 - \cos\{\phi(t, a) - \phi(t, a')\}] dt\right\}$$

and NT is the receiver observation interval length. Figure 6.13 shows the trade-off between power and bandwidth efficiencies for several CPM schemes. The vertical axis indicates the minimum Euclidean distance $d_{\min}^2$ relative to that of MSK ($d_{\min}^2 = 2$), and the horizontal axis indicates the 99% bandwidth $2BT_b$, where $2B$ is the double-sided bandwidth of CPM signals and $T_b = T/\log_2 M$ for M-ary CPM. The CPM schemes that are shown in the upper left with respect to MSK are superior to MSK in both bandwidth and power efficiency. More efficient CPM schemes than MSK need larger L for

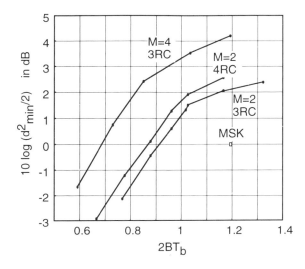

Figure 6.13 Power bandwidth trade-off for CPM schemes using rectangular pulses. (*After:* [15]).

partial response and larger M, which can be achieved by introducing additional memory and increasing the alphabet size as well. Many studies on improving the performance and reducing system complexity, for example, by combining convolutional encoding with CPM schemes, are still in progress.

6.1.7 Modulation Technique—OFDM

Orthogonal frequency division multiplexing (OFDM) has recently received considerable attention for mobile communications over Rayleigh fading channels [21,22] and for digital audio terrestrial/satellite broadcasting [23], as well as for terrestrial digital television and HDTV broadcasting. For possible future applications for mobile and personal satellite communications, this section presents the OFDM scheme [24].

The OFDM scheme is based on the idea of transmitting the information bits in parallel streams, each at a low transmission rate, so that the fading effect is spread out over many bits by using FDM with overlapping subchannels to combat impulsive noise and multipath fading. OFDM is a family of discrete multitone DMT, multichannel modulation, and multicarrier modulation, and it uses carriers that are mutually orthogonal in the FDM scheme. Several methods for separating subbands have been proposed: filtering, staggered QAM, discrete Fourier transform (DFT), and so forth.

In this section, we consider OFDM using DFT, which is achieved at the baseband of parallel data. The DFT spectra of each parallel data stream is the sinc function (i.e., $\sin x/x$), and is allocated to separate frequencies. At the receiver, orthogonal signals can be separated by correlation techniques.

6.1.7.1 Signal Representation of OFDM

For the information-bearing block of complex symbols $\{a_1, a_2, \ldots, a_N\}$, DFT is achieved, and we have N complex variables:

$$D_m = \sum_{n=1}^{N} a_n \exp(-j2\pi nm/N) = \sum_{n=1}^{N} a_n \exp(-j2\pi f_n t_m) \quad \text{for } m = 1, 2, \ldots, N$$

$$(6.77)$$

where $f_n = n/(NT) = n/T_B$, T is the symbol duration, $t_m = mT$, and $T_B = NT$ is the block duration of N symbols. Suppose that these sampled components are applied to a low-pass filter at time interval T. Then the signal is approximately expressed as

$$u(t) = \sum_{n=1}^{N} a_n \exp(-j2\pi f_n t) g(t)$$

$$(6.78)$$

where $g(t)$ is a rectangular pulse: $g(t) = 1$ for $0 \leq t \leq TB$ and $g(t) = 0$ otherwise. This signal can be considered to be the frequency division multiplexed signals, whose frequencies are assigned to be $f_n(= n/T_B)$ for $n = 1, 2, \ldots, N$. The Fourier transform of $u(t)$ is given by

$$V(f) = \sum_{n=1}^{N} a_n G(f - f_n)$$

$$(6.79)$$

where the Fourier transform $G(f)$ of the pulse $g(t)$ is given by $T_B(\sin \pi f T_B / \pi f T_B) \exp(-j\pi f T_B)$, which has nulls at $f = 1/T_B$, $2/T_B$, $3/T_B$, $\ldots$. The frequency spacings of $G(f - f_n)$ for $n = 1, 2, \ldots, N$ are also the same as the spacing of the nulls of $G(f)$. This frequency separation makes the OFDM subchannels orthogonal. Since OFDM converts serial data into parallel data for each subchannel and then into a sampled OFDM signal, the symbol rate in each subchannel can be greatly reduced compared to the conventional serial modulation schemes.

Figure 6.14(a) shows the spectrum of an OFDM subchannel using a rectangular pulse and Figure 6.14(b) shows the total OFDM spectrum. Due to the carrier spacing of $1/T_B$, $2/T_B$, $\ldots$, the OFDM subchannels are mutually orthogonal and the total spectrum is approximately flat.

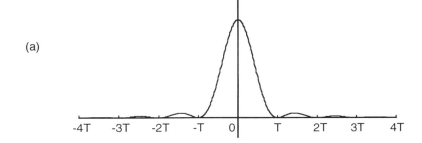

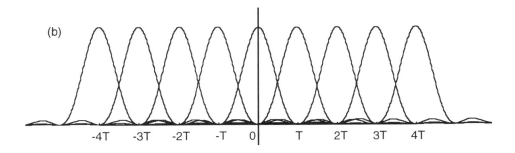

Figure 6.14 (a) Power spectrum of an OFDM subchannel and (b) power spectrum of an OFDM signal, which is the sum of power spectra of subchannels and is approximately flat.

One of the advantages of OFDM is that it can convert a wideband frequency-selective fading channel into a series of narrowband and frequency-nonselective fading subchannels. The number of subcarriers is determined by the channel bandwidth, data throughput, and useful symbol duration. The modulation scheme of each OFDM subchannel can be selected arbitrarily based on the required bandwidth and power efficiencies. Usually, 16-QAM or QPSK modulation schemes are selected for mobile applications. Trellis-coded modulation (TCM) with frequency and time interleaving is applicable, and it is the most effective way to combat frequency-selective fading [24].

OFDM is not a constant-envelope modulation, and the peak power is much larger than the average power. The ratio of peak power to average power depends on the signal constellation and the rolloff factor β of the pulse-shaping filter. For example, for $\beta = 11.5\%$, the peak-to-average power ratio is about 7 dB for 99.99% of the time.

Since OFDM uses many carriers, nonlinear amplification can impair the BER performance. Couasnon and others [25] reported that a 9-dB output backoff of a nonlinear amplifier resulted in negligible BER degradation and adjacent channel interference. OFDM is now under research and development for several applications, such as terrestrial TV broadcasting. For future new applications in mobile and personal satellite communications, such as multimedia communications with audio/video transmission, OFDM will be an attractive candidate for the modulation scheme.

6.1.8 Modulation Techniques—Others

There is a family of nonconstant-envelope modulation schemes [10] that provides small spectral side lobes even when amplification is nonlinear. Examples include quadrature-overlapped raised cosine (QORC) [26], quadrature-overlapped squared raised cosine (QOSRC) [27], quasi-band-limited (QBL) double-interval pulse [6], and quadrature-quadrature PSK (Q^2PSK) [28,29] modulations. The modulation schemes are basically a type of QPSK with various types of pulse shapes to reduce the sidelobe tail efficiently, although the bit error performance has some degradation compared to the unfiltered QPSK. However, simulation results for a nonlinear channel show that QORC has comparable bit error performance to QPSK. QOSRC, which consists of overlapping squared raised cosine pulses with a time offset between quadrature channels, has less envelope fluctuation than QORC, so it has better bit error performance than QORC on a hard-limited channel in the presence of AWGN and intersymbol interference. Q^2PSK with sinusoidal and cosinusoidal shapes in both shaping filters and signal components in I and Q channels has twice the bandwidth efficiency of MSK and has degradation of 1.6 dB in E_b/N_0 at BER = 10^{-5}. However, a simple block coding can reduce the degradation to 0.6 dB. Progress in digital signal processing and device technologies will make these modulation schemes very attractive for future mobile satellite communications.

6.1.9 Carrier Recovery

For carrier recovery or carrier synchronization, there are two main methods: One is pilot-tone or symbol-aided carrier recovery, and the other is carrier recovery directly from the modulated signal without a pilot signal.

In the former method, an unmodulated carrier pilot tone or symbol is transmitted with the information-bearing signal without mutual interference, and the carrier is recovered at the receiver without phase ambiguity. In the latter method, the receiver derives the carrier phase directly from the modulated signal without any redundant carrier tone; this method has the distinct advantage of power efficiency, since it can concentrate the transmitting power to only the signal. In this section, we consider several carrier recovery methods.

In the mobile satellite communications channels, multipath fading and Doppler frequency shift cause irreducible error floors in the bit error performance. Acceptable system performance can be obtained by using one of the following antifading techniques, which use 1) a directional antenna to reduce multipath components, 2) Doppler compensation modems, 3) a tone- or symbol-aided modulation scheme, or 4) differential detection. In tone-aided carrier recovery, since the pilot tone suffers from the same fading as the signal, the receiver can remove the phase fluctuation by tracking the pilot with a PLL. On the other hand, the pilot tone requires additional power and bandwidth, so it decreases nonlinear immunity due to an increase in envelope fluctuations.

Several attempts at differential detection in differentially encoded modulation schemes have been carried out to reduce the inherent degradation in the bit error performance, including the multiple-symbol differential detection (MSDD) method [30–34] and the improved decision feedback scheme with an infinite impulse response (IIR) filter [35]. For mobile satellite communications channels, which have more severe propagation disturbances (such as multipath fading and shadowing) than the AWGN channel, differential detection is very attractive because there is no need for carrier recovery or fast synchronization after shadowing.

6.1.9.1 Pilot-Aided Carrier Recovery

The pilot-aided modulation technique of continuously calibrating the fading channel with a pilot signal [36–41] is known by various names such as feed-forward signal regeneration (FFSR) [42], transparent tone in band (TTIB) [43], and tone-calibration technique (TCT) [44]. The pilot signal provides explicit amplitude and phase information for the receiver to perform coherent demodulation and suppress the error floor in the bit error performance. There are two types of pilot-aided technique, depending on the allocation of the pilot signal. One allocates a pilot tone in the transmitting signal spectrum and the other allocates the pilot symbols, which are periodically inserted in the signal frame, in the time domain. First, we consider the former tone-calibration technique.

6.1.9.2 Tone-Calibration Technique

The allocation of the pilot tone in the transmitting signal spectrum is arbitrary, but the most suitable location is a spectral null of the signal to prevent signal interference. A number of methods to create the desired spectral null have been proposed. Figure 6.15(a) shows an example of TCT using the Manchester encoding technique [44] in BPSK, which creates a null at the center of the

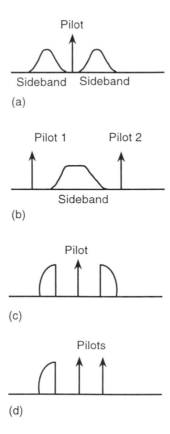

Figure 6.15 Spectrum of pilot-aided modulations: (a) tone-calibrated technique (TCT), (b) dual-tone-calibrated technique (DTCT), (c) transparent tone in band (TTIB), and (d) dual-tone single sideband (DTSSB).

signal spectrum. The system configuration is relatively simple, and TCT can improve the error floor in the BER performance over the Rician fading channel, although it requires twice the bandwidth due to the Manchester encoding and the Doppler effect may impair the performance because of the narrow spectral null. Figure 6.15(b) shows the dual-tone-calibration technique (DTCT) [45], which uses two pilot tones, one at each edge of the signal spectrum, to improve the bandwidth efficiency and the signal-to-noise ratio. However, the processing noise increases and the two pilot tones are easily distorted by filters due to the locations of two tones. Figure 6.15(c) shows the TTIB technique [42]. Since the pilot tone is located at the center of the signal spectrum, TTIB has better bandwidth efficiency and less filter distortion than DTCT, but it needs a complicated receiver configuration, and noise prevents the PLL from having ideal performance. Figure 6.15(d) shows the dual-tone single sideband

(DTSSB), which uses two pilot tones at one side of the higher sideband signal. DTSSB has the advantage of good bandwidth efficiency and simple implementation, but it has an unbalanced spectrum and twice as much noise because it uses two pilot tones.

In these tone-aided modulation schemes, an important role of the pilot tones for MSS modems is to monitor propagation channel conditions and compensate for multipath fading. For coherent demodulation, the pilot tones should match the frequency and phase with the carrier. The insertion of pilot tones increases the total signal power and causes amplitude variation in constant-envelope modulation signals. Moreover, pilot tones at the edge of the spectrum may cause interference in adjacent channels, while those at the spectrum center require symbol reshaping, which generally causes longer spectral tails of the signal. However, due to their simple implementation, robustness against fast fading, and good power and bandwidth efficiencies, tone-aided modulation techniques are useful for coherent demodulation in channels with severe fading.

6.1.9.3 Pilot Symbol-Assisted Technique

Given pilot symbols are inserted periodically into the information symbol sequence at the transmitter [46,47]. An example of the frame format in pilot symbol-assisted modulation (PSAM) is shown in Figure 6.16. Both the pilot symbol and the information symbol will suffer the same multipath fading. The receiver splits the signal into two streams: one to estimate the fading and the other to compensate for it by using the fading time correlation in order to detect the data information. Like TCT, PSAM can suppress the error floors in the BER performance over multipath fading channels, and PSAM does not need to change the transmitted signal spectrum, which can improve the power efficiency and the signal envelope fluctuations compared to TCT. However, PSAM has two disadvantages: the delay caused by the fading estimation and the reduction in information symbol rate due to the insertion of pilot symbols.

In Figure 6.16, each frame has one pilot symbol and n data symbols. To compensate for data symbols in the second frame, for example, pilot symbols

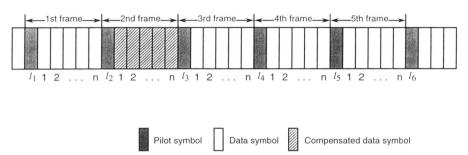

Figure 6.16 Frame format in pilot symbol-assisted modulation.

of l_1, l_2, l_3, and l_4 in neighboring frames are used to accomplish the fading estimation. Usually, data symbols are estimated from pilot symbols using the linear mean-square estimation methods. The second disadvantage of PSAM (i.e., the reduction of information symbol rate due to the insertion of pilot symbols) can be mitigated by selecting a larger frame length and smaller ratio of pilot-to-data symbols. However, to estimate the fading, the pilot symbols should be located within correlated time range of the fading. The other disadvantage of PSAM is the time delay for conducting the fading compensation process. For example, for a Doppler shift of 80 Hz and symbol rate of 2,400 bps, the processing delay is only 6.4 msec for the above scheme [47]. To evaluate the fading correlation in real channels, a training symbol sequence will be employed.

6.1.9.4   Carrier Recovery Without Pilot Tone

In this section, we briefly survey the carrier recovery methods without pilot tone, which are used in many digital communications systems.

6.1.9.5   *M*th Power Loop or Squaring Loop

We consider the carrier recovery technique for the *M*-ary PSK, which is expressed by

$$s(t) = A \cos\left(2\pi f_c t + \frac{2\pi}{M}(m - 1)\right) \qquad \text{for } m = 1, 2, \ldots, M \quad (6.80)$$

The output of the *M*th power law device of this signal has the signal component of frequency Mf_c, in which the information is removed, because $2\pi(m - 1) = 0 \bmod 2\pi$. When this signal is divided by M in frequency, the result is the unmodulated carrier component $\cos(2\pi f_c t + \theta)$ with the phase ambiguity of $2\pi/M$. This ambiguity can be removed by differential encoding.

6.1.9.6   Costas Loop

The phase of the received signal is generated by multiplying the output of the voltage controlled oscillator (VCO) and a $\pi/2$ phase shift of that voltage, as illustrated by the block diagram, which is called the Costas loop, shown in Figure 6.17. Assuming that the signal and the additive narrowband noise are given by

$$s(t) = A \cos(2\pi f_c t + \theta(t))$$
$$n(t) = n_I(t)\cos(2\pi f_c t + \theta(t)) - n_Q(t)\sin(2\pi f_c t + \theta(t)) \qquad (6.81)$$

the received signal $s(t) + n(t)$ is multiplied by $\cos(2\pi f_c t + \hat{\theta})$ and $\sin(2\pi f_c t + \hat{\theta})$, which are outputs from VCO. Then, these two products are given by

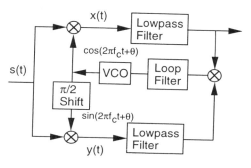

Figure 6.17 Costas loop.

$$x(t) = \frac{1}{2}[A + n_I(t)]\cos(\hat{\theta}(t) - \theta(t)) + \frac{1}{2}n_Q(t)\sin(\hat{\theta}(t) - \theta(t))$$

$$y(t) = \frac{1}{2}[A + n_I(t)]\sin(\hat{\theta}(t) - \theta(t)) - \frac{1}{2}n_Q(t)\cos(\hat{\theta}(t) - \theta(t)) \quad (6.82)$$

where $\hat{\theta}$ is the estimated phase of θ. All double-frequency terms are eliminated by the lowpass filters. The product of $x(t)$ and $y(t)$ generates an error signal, which has the term of $A^2 \sin 2(\hat{\theta}(t) - \theta(t))$. For small phase error, this term is approximated by $2A^2(\hat{\theta}(t) - \theta(t))$, which makes the output frequency and phase of the VCO coincide with the frequency and phase of the received signal. This operation is the same as that of the PLL. Since the output of the VCO has an ambiguity of π, differential encoding and decoding are also required.

6.1.9.7 Decision Feedback Loop

The decision feedback loop removes the modulation and extracts the carrier phase by using the demodulated signal, as shown in Figure 6.18. The lower loop is, for example, an integrate-and-dump circuit followed by a hard limiter,

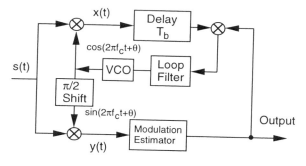

Figure 6.18 Decision feedback loop.

where the reference signal for demodulation is obtained from the decision feedback loop itself. If there is no decision error, then the error signal, which is the input of the loop filter, is given by

$$\epsilon(t) = \frac{1}{2}A^2 \sin(\hat{\theta}(t) - \theta(t)) + \frac{1}{2}An_I(t) \sin(\hat{\theta}(t) - \theta(t)) \qquad (6.83)$$

$$- \frac{1}{2}An_Q(t)\cos(\hat{\theta}(t) - \theta(t))$$

where all the double-frequency terms are neglected. Like the squaring and Costas loops, the decision feedback loop also exhibits a phase ambiguity of π. For BER $< 10^{-2}$, the decision feedback PLL is superior in performance to both the squaring and Costas loops [7].

6.2 Error-Control Techniques

In this section, we introduce error-control techniques for mobile satellite communication channels. Error correction can be achieved by automatic repeat request (ARQ), forward error correction (FEC), or a combination of ARQ and FEC. In the ARQ method, the transmitting terminal separates the information bits into blocks of a finite length, which are encoded for error-control purposes and transmitted. If the receiving terminal does not detect the presence of errors, it sends an ACK (acknowledgment) signal to the transmitting terminal to notify it that the data block was correctly received. If the receiving terminal does detect errors, it sends a NACK (negative acknowledgment) signal to notify it that the block was not correctly received and to request retransmission of the same block. The following three main types of ARQ are shown in Figure 6.19: (a) stop-and-wait ARQ, (b) go-back-N ARQ, and (c) selective-repeat ARQ [9]. In the stop-and-wait ARQ, the transmitting terminal sends one block and delays the transmission of the next block until the ACK or NACK signals are received from the receiving terminal. Since the double-hop delay of about 0.5 sec decreases rapidly the throughput as the transmission rate increases, the stop-and-wait ARQ system is not suitable to satellite communication links, although this ARQ system has advantages of the simplicity of its operation and the small buffer sizes. In the go-back-N ARQ, the blocks are continuously transmitted without waiting for ACK or NACK signals. When the NACK signal is received , the transmitting terminal transmits the erroneous block and all the following blocks, some of which may have already been sent. N refers to the number of blocks repeated. Figure 6.19(b) shows a go-back-3 ARQ scheme. In this system, buffering is required at the transmitting terminal,

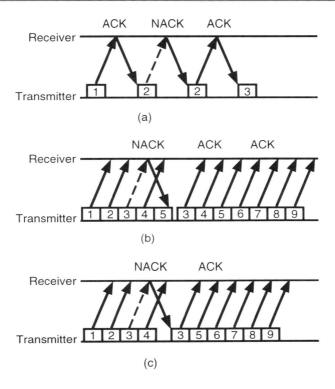

Figure 6.19 Basic ARQ methods: (a) stop-and-wait ARQ, (b) go-back-N ARQ, and (c) selective-repeat ARQ.

and the buffer size depends on the transmission rate and round-trip delay. In the selective-repeat ARQ, the transmitting terminal sends the blocks continuously, and when the NACK signal for the nth block is received, the terminal send only the erroneous nth packet. This method is most superior over the stop-and-wait and go-back-N ARQ methods in terms of throughput for satellite communications. Several variations of these ARQ schemes have been proposed to satisfy specific system requirements. For example, in adaptive ARQ, the transmission rates, the block length, or the coding rates are adaptively changed to improve the throughput in response to channel conditions. Moreover, hybrid ARQ, which is a combination of ARQ and FEC, is used to reduce the BER and the number of retranslated blocks.

To detect the presence of received errors in ARQ systems, encoding for error detection is needed at the transmitting terminal. In this section, we present the basic concept of error-correcting code.

For error-correcting codes, the minimum distance between two arbitrary code words plays an important role in determining the error-correction capability, and several defined distances, depending on the channel characteristics, are used. When the number of alphabets for representing codes is q, the code is called as q-ary code. In this section, we mainly treat binary code (i.e., $q = 2$). There are two different types of codes. One is a block code, whose encoder breaks the continuous sequence of information bits into blocks, and k information bits are encoded in a block of n symbols, $n > k$. The other is a convolutional code, in which the information bits are encoded continuously without their sequence being broken.

6.2.1 Hamming Distance

The Hamming distance between two code words of length n, $\mathbf{a} = (a_1, a_2, \ldots, a_n)$ and $\mathbf{b} = (b_1, b_2, \ldots, b_n)$ is defined as

$$d_H(\mathbf{a}, \mathbf{b}) = \sum_{i=1}^{n} d_H(a_i, b_i) \qquad (6.84)$$

where

$$d_H(a, b) = \begin{cases} 0, & a = b \\ 1, & a \neq b \end{cases}$$

The minimum value of the Hamming distances between arbitrary different code words is called the minimum Hamming distance, which is denoted $d_H^{\min}$. If a single error occurs on a communication channel, the Hamming distance between the transmitted code word and the received code word must be one. Therefore, if the minimum Hamming distance is $d_H^{\min} = 2t + 1$, a t-tuple error can be corrected and a $2t$-tuple error can be detected. Suppose that such code words are composed of k information bits and $(n - k)$ redundant bits for error correction; then, the code is called $(n, k, d_H^{\min})$ code. The ratio of n and k is called code rate $r = k/n$.

For linear codes, a code word $\mathbf{a} = (a_1, a_2, \ldots, a_n)$ satisfies the following linear equation:

$$\sum_{j=1}^{n} a_j h_{ij} = 0 \quad \text{for } i = 1, 2, \ldots, n - k$$

This equation can be rewritten using a matrix $\mathbf{H}$ that has h_{ij} in the ith row and jth column:

$$\mathbf{aH}^T = 0 \tag{6.85}$$

Equation (6.85) is called a parity check equation and matrix $\mathbf{H}$ is called a parity check matrix, which has $(n - k)$ rows and n columns. For any received vector $\mathbf{a}'$, the $(n - k)$-component vector $\mathbf{s} = \mathbf{a}'\mathbf{H}^T$ is called the syndrome. If and only if its syndrome is zero, vector $\mathbf{a}'$ is a code word.

Linear codes with the parity check matrix of $n = 2^m - 1$ columns and $k = n - m$ are called Hamming codes. For example, the (7, 4) code, whose parity check matrix is given by (6.86), is a Hamming code:

$$H = \begin{bmatrix} 1 & 1 & 1 & 0 & 1 & 0 & 0 \\ 0 & 1 & 1 & 1 & 0 & 1 & 0 \\ 1 & 1 & 0 & 1 & 0 & 0 & 1 \end{bmatrix} \tag{6.86}$$

This matrix consists of $2^m - 1$ ($= 7$) different column vectors except the zero vector, and is written by

$$H = [P : I]$$

where P is a $m \times (2^m - 1 - m)$ matrix, and I is a $m \times m$ unit matrix. Then, the generator matrix is defined as

$$G = [I : P^T]$$

$$= \begin{bmatrix} 1 & 0 & 0 & 0 & 1 & 0 & 1 \\ 0 & 1 & 0 & 0 & 1 & 1 & 1 \\ 0 & 0 & 1 & 0 & 1 & 1 & 0 \\ 0 & 0 & 0 & 1 & 0 & 1 & 1 \end{bmatrix} \tag{6.87}$$

and code words $\mathbf{a}$ are expressed as

$$\mathbf{a} = \mathbf{u}\, G \tag{6.88}$$

where $\mathbf{u}$ is a data vector, which consists of $(2^m - 1 - m)$ information bits. The first $k(= 2^m - 1 - m)$ bits of code words are identical to those of the information bits and the successive $(n - k)$ bits are parity check bits. Such a code is called a systematic code. The parity check bits are given by

$$c_1 = a_1 \oplus a_2 \oplus a_3$$
$$c_2 = a_2 \oplus a_3 \oplus a_4$$
$$c_3 = a_1 \oplus a_2 \oplus a_4$$

where $\oplus$ denotes an exclusive OR operation. Since $d_H^{\min} = 3$ for (n, k) Hamming codes, a single error can be corrected. When the received code is expressed as

$$\mathbf{a}' = \mathbf{a} + \mathbf{e}$$

where $\mathbf{e}$ is an error vector and at most one element is not zero. The column of $\mathbf{H}$, which has the same elements as the syndrome $\mathbf{a}'\mathbf{H}^T$, indicates the position of an error.

The Hamming codes are simple configuration, but application areas are limited. Next, in order to deal with more complicated block codes such as BCH code and Reed-Solomon code, we briefly describe the algebraic concept for error-correction coding.

6.2.2 Galois Fields $GF(q^m)$

The coding theory is constructed by using finite field arithmetic. A finite field of q elements, which is referred to as a Galois field after Evariste Galois, is denoted as $GF(q)$. A Galois field is a finite set of elements where addition and multiplication are defined with commutative, associative, and distributive properties as well as including zero element and one element. When q is a prime integer, the integer set $\{0, 1, 2, \ldots, q-1\}$ is a Galois field with modulo-q operation. The set of polynomials of degree $(m - 1)$ or less over $GF(q)$ is shown to be a Galois field $GF(q^m)$ by performing polynomial addition and multiplication with polynomial modulo $p(x)$, which is an irreducible polynomial of degree m over $GF(q)$.

The smallest nonzero integer e such that $\alpha^e = 1$ for an element α of $GF(q)$ is called the order of α. When the order of α is $q - 1$, α is called a primitive element. Then, $(q - 1)$ consecutive powers of α, $(1, \alpha, \alpha^2, \alpha^3, \ldots, \alpha^{q-2})$, are different nonzero elements in $GF(q)$. If $\alpha^i = \alpha^j$ for $0 \leq i < j < q - 1$, we have $\alpha^{(j-i)} = 1$ for $0 < (j - i) < q - 1$. Since this is inconsistent with the assumption that the order of α is $q - 1$, the $(q - 1)$ consecutive powers of α are different from each other. Therefore, an arbitrary nonzero element in $GF(q)$ can be represented by α^i for an integer i ($< q - 1$).

If $p(x)$ is irreducible on $GF(q)$ and $p(\alpha) = 0$ for a primitive element α of $GF(q^m)$, then $p(x)$ is called a primitive polynomial. The polynomial and vector representations of elements in $GF(q^m)$ can be derived from the above exponential representation by the modulo operation of the primitive polynomial.

For example, consider $GF(2^4)$ for a primitive polynomial $p(x) = x^4 + x + 1$. Any elements of $GF(2^4)$ can be represented by powers of a primitive element α, which is also the root of $p(x)$. Using

$p(\alpha) = \alpha^4 + \alpha + 1 = 0$, α^i can be given by linear combination of α^3, α^2, α, and 1, as shown in Table 6.4. For example, $\alpha^5 = \alpha^1 \alpha^4 = \alpha(\alpha + 1) = \alpha^2 + \alpha$. Moreover, any elements in $GF(2^4)$ can be also expressed in vector representation by coefficients of the polynomials. Table 6.4 shows exponential, polynomial, and vector representations for $GF(2^4)$ with a primitive polynomial $p(x) = x^4 + x + 1$.

Using this table, we can calculate addition and multiplication of two elements in $GF(2^4)$. For example, the addition of (0011) and (1010), which is defined as exclusive-OR addition of each bits, results in (1001). The multiplication of (0011) and (1010), which is the multiplication of α^4 and α^9, results in $\alpha^4 \alpha^9 = \alpha^{13} = (1101)$.

6.2.3 Cyclic Codes

If $(c_0, c_1, \ldots, c_{n-1})$ is a code word, and its cyclically shifted word $(c_1, c_2, \ldots, c_{n-1}, c_0)$ is also a code word, then the code is called a cyclic code. Cyclic code is represented by the following code polynomial over $GF(q)$:

$$c(x) = c_0 + c_1 x + c_2 x^2 + \ldots + c_{n-1} x^{n-1} \tag{6.89}$$

Shifting this code word cyclically i times is represented by $x^i c(x)$ modulo $(x^n + 1)$. For example, one cyclic shift of $c(x)$ is expressed by

Table 6.4
Exponential, Polynomial, and Vector Representations for $GF(2^4)$ With a Primitive Polynomial $p(x) = x^4 + x + 1$

0	0	0000
1	1	0001
α	α	0010
α^2	α^2	0100
α^3	α^3	1000
α^4	$\alpha + 1$	0011
α^5	$\alpha^2 + \alpha$	0110
α^6	$\alpha^3 + \alpha^2$	1100
α^7	$\alpha^3 + \alpha + 1$	1011
α^8	$\alpha^2 + 1$	0101
α^9	$\alpha^3 + \alpha$	1010
α^{10}	$\alpha^2 + \alpha + 1$	0111
α^{11}	$\alpha^3 + \alpha^2 + \alpha$	1110
α^{12}	$\alpha^3 + \alpha^2 + \alpha + 1$	1111
α^{13}	$\alpha^3 + \alpha^2 + 1$	1101
α^{14}	$\alpha^3 + 1$	1001

$$xc(x) = c_0x + c_1x^2 + c_2x^3 + \ldots + c_{n-1}x^n$$
$$= c_0x + c_1x^2 + c_2x^3 + \ldots + c_{n-1}(x^n - 1) + c_{n-1}$$
$$= c_0x + c_1x^2 + c_2x^3 + \ldots + c_{n-1} \quad \text{modulo } (x^n + 1)$$

Therefore, if $c(x)$ is a code polynomial, $x^ic(x)$ is also a code polynomial, and its linear combination $\Sigma a_i x^i c(x)$ modulo $(x^n + 1)$ is also a code polynomial.

Let $m = (m_0, m_1, \ldots, m_{k-1})$ be k information symbols; then, a code polynomial is always given by a generator polynomial of degree $n - k$, $g(x) = 1 + g_1x + \ldots g_{n-k-1} x^{n-k-1} + x^{n-k}$.

$$c(x) = m(x)g(x) \qquad \text{modulo } (x^n + 1) \qquad (6.90)$$

where $m(x) = m_0 + m_1x + \ldots + m_{k-1}x^{k-1}$. Such a generator polynomial provides (n, k) cyclic codes. The cyclic code given by (6.90) is not a systematic code. To construct systematic cyclic codes, the following modification is needed:

$$c(x) = x^{n-k}m(x) - r(x) = m_{k-1}x^{n-1} + \ldots \qquad (6.91)$$
$$+ m_0x^{n-k} + r_{n-k-1}x^{n-k-1} + \ldots + r_0$$

where $r(x)$ is the remainder when dividing $x^{n-k}m(x)$ by the generator polynomial $g(x)$. Then, $r(x)$ is the parity check for the cyclic code. The encoding can be achieved by the $(n - k)$-stage shift-register circuit shown in Figure 6.20 [7]. When k information bits have completed to enter the shift register, the $n - k$ digits stored in the register are the parity bits. The code word consists of the k information bits and $n - k$ parity bits.

6.2.4 BCH Codes

The Bose-Chaudhuri-Hocquenghem (BCH) codes are the best constructive codes for burst error correction. They are cyclic codes constructed by a generator

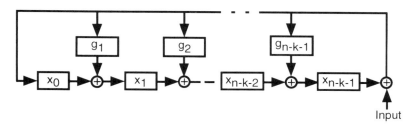

Figure 6.20 $(n - k)$-stage shift-register encoding circuit for the cyclic code generated by $g(x) = 1 + g_1x + \ldots g_{n-k-1} x^{n-k-1} + x^{n-k}$.

polynomial, after independent studies by Bose and Ray-Chaudhuri [48,49], and Hocquenghem [50].

Let β be an element of $GF(q^m)$ and a root of a polynomial $g(x)$ of degree m with coefficients in $GF(q)$; then, β^q is also a root of $g(x)$, since $[g(x)]^q = g(x^q)$. Therefore, $\beta^{q^2}, \beta^{q^3}, \ldots, \beta^{q^{m-1}}$ are also roots of $g(x)$. These roots are called the conjugate roots of β [51]. The generator polynomial for BCH codes is derived from this property of conjugate roots.

For $d-1$ consecutive roots of $\beta, \beta^2, \ldots, \beta^{d-1}$, the generator polynomial $g(x)$ is given by

$$g(x) = \text{LCM}[m_\beta(x), m_{\beta^2}(x), \ldots, m_{\beta^{d-1}}(x)] \qquad (6.92)$$

where LCM is lowest common multiple and $m_\beta(x)$ is the polynomial of least degree that has a root of β. If β is primitive, then $d_{\min}^H = d$; otherwise, $d_{\min}^H \geq d$.

6.2.4.1 Example Binary BCH Code

For $d = 5$, the (15, 7) binary BCH codes can be derived as follows. Let α be a primitive root in $GF(2^4)$. Then, four consecutive roots are $\alpha, \alpha^2, \alpha^3$, and α^4.

The conjugate roots of α are $\alpha, \alpha^2, \alpha^4, \alpha^8, \alpha^{16}(=\alpha)$. The conjugate roots of α^3 are $\alpha^3, \alpha^6, \alpha^{12}, \alpha^{24}(=\alpha^9), \alpha^{18}(=\alpha^3)$. Therefore, $g(x)$ is given by $\text{LCM}[m_\alpha(x)\, m_{\alpha^3}(x)]$ and, using Table 6.4,

$$m_\alpha(x) = (x - \alpha)(x - \alpha^2)(x - \alpha^4)(x - \alpha^8)$$
$$= x^4 + x + 1$$

$$m_{\alpha^3}(x) = (x - \alpha^3)(x - \alpha^6)(x - \alpha^{12})(x - \alpha^9)$$
$$= x^4 + x^3 + x^2 + x + 1$$

Then, $g(x)$ is given by

$$g(x) = x^8 + x^7 + x^6 + x^4 + 1 \qquad (6.93)$$

which is the generator polynomial of a (15, 7) double error-correcting BCH code.

Generator polynomials for BCH codes with different n, k, and d are shown in a text written by Peterson and Weldon [51].

6.2.5 Reed-Solomon Codes

Reed-Solomon code [52] is a nonbinary BCH code with code words from $GF(q)$. For example, $GF(256)$ is used in many applications because the 256 elements in $GF(256)$ can be represented by an 8-bit sequence or 1 byte. The generator polynomial for Reed-Solomon codes is given by

$$g(x) = \prod_{j=1}^{d-1} (x - \alpha^j) \qquad (6.94)$$

where $\alpha, \alpha^2, \ldots, \alpha^{d-1}$ are the consecutive roots and all minimum polynomials have degree of 1. For the (n, k, d_{min}^H) Reed-Solomon codes, we have $n = q - 1$, $k = n - (d - 1)$, and $d_{min}^H = d$.

For example, we will consider the $(7, 5, 3)$ Reed-Solomon code in $GF(8)$. The generator polynomial is expressed as

$$g(x) = (x - \alpha)(x - \alpha^2)$$

As shown in Table 6.4, exponential and vector representations for $GF(2^3)$ with a primitive polynomial $p(x) = x^3 + x + 1$ can be also given by

$$
\begin{array}{ll}
\alpha^1 & 010 \\
\alpha^2 & 100 \\
\alpha^3 & 011 \\
\alpha^4 & 110 \\
\alpha^5 & 111 \\
\alpha^6 & 101 \\
\alpha^7 & 001
\end{array}
$$

Then, the generator polynomial can be rewritten by

$$g(x) = (001)x^2 + (110)x + (011) = x^2 + 6x + 3$$

When the five information symbols are (00073), the information polynomial is expressed by $m(x) = 7x + 3 = (111)x + (011) = \alpha^5 x + \alpha^3$. For systematic codes, the parity check bits are calculated by $x^2 m(x)$ modulo $g(x)$, and it results in $\alpha^4 x + \alpha = (110)x + (010) = 6x + 2$. Therefore, the output code is given by (0007362).

Since valid code polynomials must be a multiple of the generator polynomial and must have the same roots of $(d - 1)$ consecutive roots as those of $g(x)$, we can know whether a received word is a code word.

6.2.6 Convolutional Codes

The convolutional codes are generated through linear operations of a finite-state shift register [7,53]. Figure 6.21 shows an example of a convolutional encoder where three output bits are generated from two input bits (i.e., $n = 3$ and $k = 2$). This code has the rate $r = k/n = 2/3$. The relationship between the input and the output is given by

$$y_i^{(1)} = x_i^{(1)} \oplus x_i^{(2)} \oplus x_{i-1}^{(1)} \oplus x_{i-1}^{(2)}$$
$$y_i^{(2)} = x_i^{(2)} \oplus x_{i-1}^{(1)}$$
$$y_i^{(3)} = x_i^{(1)} \oplus x_i^{(2)} \oplus x_{i-1}^{(2)} \tag{6.95}$$

There are three methods of representing the convolutional code: the tree diagram, the finite-state diagram, and the trellis diagram. Figure 6.22 shows the tree diagram of the above example. Suppose that the shift register is initially filled with an all-zero sequence and the input bits are denoted by $a = 00$, $b = 10$, $c = 01$, $d = 11$, and the corresponding output bits are 000, 101, 111, 010. The four possible states of the shift register are determined by a, b, c, and d, which are the input bits at the previous time. Figures 6.23 and 6.24 show the finite-state diagram and the trellis diagram for the same convolutional codes, respectively. In the finite-state diagram, the branch variables show three output bits. In the trellis diagram, there are four transitions from each state, which correspond to four possible inputs $x^{(1)}$ and $x^{(2)}$, and three output bits in these four transitions are shown in trellis branches. In this convolutional code for rate 2/3, the redundancy of one symbol increases the minimum Hamming distance; in this case, $d_{min}^{H} = 3$.

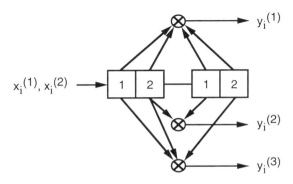

Figure 6.21 Convolutional encoder for rate 2/3, $n = 3$, and $k = 2$.

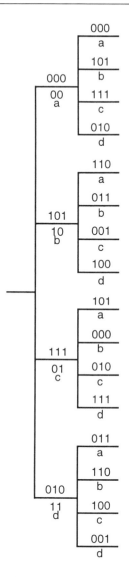

Figure 6.22 Tree diagram for rate 2/3, $n = 3$, and $k = 2$.

6.2.7 Decoding of Convolutional Codes—Viterbi Algorithm

For decoding convolutional codes, the most popular decoding algorithm is the maximum-likelihood decoding developed by Viterbi [54,55]. This algorithm has been used for MSS modems. The decoding process is to determine the transmitted path through the trellis by comparing the Hamming distances

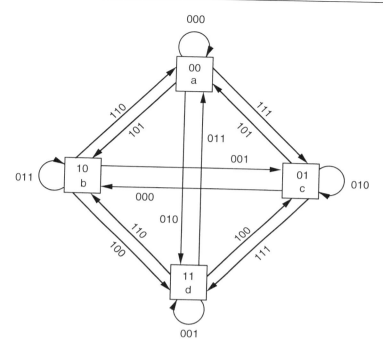

Figure 6.23 Finite-state diagram for rate 2/3, *n* = 3, and *k* = 2.

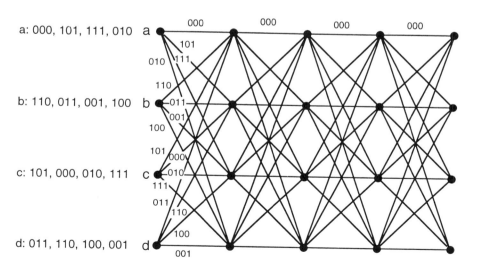

a: 000, 101, 111, 010

b: 110, 011, 001, 100

c: 101, 000, 010, 111

d: 011, 110, 100, 001

Figure 6.24 Trellis diagram for rate 2/3, *n* = 3, and *k* = 2.

between the received bit sequence and all the bit sequences assigned to the trellis branch. For example, assuming that the correct path is 000-101-110, which corresponds to the state transitions *a-b-a*, and the received sequence is 010-101-110 due to an error in the second bit, then we calculate the Hamming distance, referred to as the path metric, between the received sequence and any trellis path in the initial state *a*. Figure 6.25 shows the path metric of each branch for the above convolutional code. There are four paths that reach state *a* at the second branch: *a-a*, *b-a*, *c-a*, and *d-a*, which have the metric of 3, 5, 2, and 2, respectively. Therefore, two paths, *c-a* and *d-a*, survive due to the minimum path metric. At state *b* of the second branch, *a-b*, *b-b*, *c-b*, and *d-b* have the metric of 1, 5, 4, and 2, respectively. Therefore, path *a-b* survives.

This process is written as follows [53]. We have the received sequence (r_i^1, r_i^2, r_i^3) for $i = 1, 2, \ldots$, and calculate the following $M(S_i)$, $S_i = a, b, c, d,$ which is the accumulated metric for state S_i at interval i:

$$M(S_i) = \min_{(y^1 y^2 y^3)} \sum_{k=1}^{i} \sum_{j=1}^{3} y_k^j \oplus r_k^j \tag{6.96}$$

If we know $M(S_i)$, we can estimate $M(S_i + 1)$ at the next interval $i + 1$, as follows:

$$M(S_i + 1) = \min_{(y_i^1 y_i^2 y_i^3)} \left[M(S_i) + \sum_{j=1}^{3} r_{i+1}^j \oplus y_{i+1}^j \right] \tag{6.97}$$

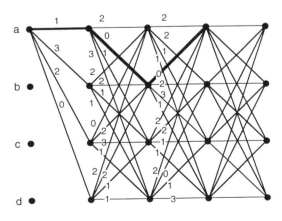

Figure 6.25 Path metric.

When the surviving paths merge, we compare their metrics and select paths having the smallest metric at this stage. Finally, we find the best surviving path, which has the smallest metric among all the survivors.

For constraint length K, which is the number of the shift register stages, we have 2^{K-1} states and hence 2^{K-1} surviving paths at each state and 2^{K-1} metrics. Therefore, the number of computations at each stage increases exponentially with K. This causes a long decoding delay, and the memory required to store the surviving paths and sequences becomes large and expensive. In practice, the path metrics over the trellis interval Ds, which is called the decision depth of the decoder, are stored in the memory of the decoder. Ds is usually selected to be three times the number of states.

6.3 Coded Modulation

In this section, we introduce coded modulation, which is combined with modulation and error-correcting codes without degrading the power or bandwidth efficiencies. Coded modulation was first described in two significant studies: TCM by Ungerboeck [56] and BCM by Imai and Hirakawa [57].

Using forward error correction, such as block codes and convolutional codes, the bit error performance is improved by expanding the required bandwidth. For example, the coding gain for convolutional coding and soft-decision Viterbi decoding with rate 1/2 and a constrain length 7 is about 5.1 dB at a bit error rate of 10^{-5}. Obtaining the power efficiency requires twice the bandwidth of the original (uncoded) signal because of the increase in the symbol rate of the modulation, and a complex implementation.

This can also be done by increasing the number of phases in PSK modulation without expanding its signal bandwidth. When we use the rate 2/3 convolutional codes and 8-PSK, the 8-PSK signal has the same bandwidth as uncoded 4-PSK. However, the bit error performance degrades by about 4 dB due to an increase in phase. If the coding gain becomes more than 4 dB, this idea will be acceptable.

6.3.1 Trellis-Coded Modulation (TCM)

The term trellis in trellis-coded modulation [56,58,59] arises from the use of a state transition diagram that is similar to the trellis diagram of the convolutional codes, as shown in Figure 6.26. However, in TCM, modulation signals are assigned to each trellis branch, although binary code symbols are assigned in the convolutional codes. The Hamming distance is the most important measure for good code design of the convolutional codes, but in TCM the

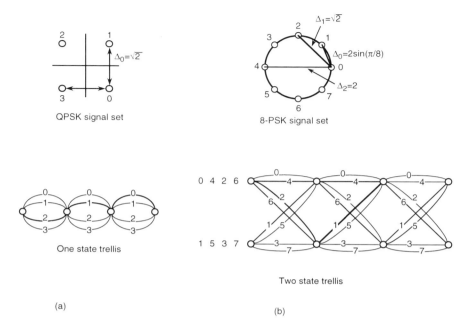

Figure 6.26 (a) Uncoded 4-PSK and (b) two-state trellis-coded 8-PSK.

definition of measuring the distance between modulation signals assigned to each trellis branch is more important. For an AWGN channel, the Euclidean distance (ED) is usually used. The mapping of code symbols optimized for Hamming distance into modulation signals does not guarantee that a good Euclidean distance structure will be obtained. The modulation signal assignment in TCM should be designed based on the Euclidean distance rather than on the Hamming distance.

As an example of TCM schemes, we will consider the transmission of two information bits per symbol with 8-PSK. Figure 6.26(b) shows the signal constellation of 8-PSK modulation with a two-state trellis, and Figure 6.26(a) shows uncoded 4-PSK modulation for reference. Starting from any state, four transitions can occur to send two information bits per symbol in both cases. The uncoded 4-PSK has four parallel transitions, and the minimum Euclidean distance between 4-PSK signals is $\sqrt{2}$, denoted by Δ_0 in Figure 6.26(a).

In Figure 6.26(b), signals {0, 2, 4, 6} are used on the upper state in the trellis diagram, and {1, 3, 5, 7} is used on the lower state. For example, the Euclidean distance between two paths 0-0 and 2-1 is given by

$$d = \sqrt{d^2(0, 2) + d^2(0, 1)} = \sqrt{\Delta_1^2 + \Delta_0^2} = 1.608 \qquad (6.98)$$

where $d(a, b)$ is the Euclidean distance between signal points a and b. In this case, this distance d is the minimum ED. This minimum ED of 8-PSK is referred to as the free ED d_{free}, and the minimum ED of uncoded 4-PSK denotes d_{min}. Therefore, the coding gain, which is defined as $d_{\text{free}}^2/d_{\text{min}}^2$ in decibels, is 1.1 dB.

Figure 6.27 shows four-state and eight-state trellis-coded 8-PSK schemes. The four-state trellis has pairs of two parallel transitions, and two signals whose distance has the largest ED of Δ_2 are assigned. In four transitions starting from one state or merging into one state, the signals with the distance of Δ_1 ($= \sqrt{2}$) are assigned. Since the distance between any paths is larger than the ED for the parallel transitions, the free ED of this four-state trellis is Δ_2 and the coding gain over uncoded 4-PSK is 3.0 dB.

The eight-state trellis has no parallel transitions. Either subset {0, 2, 4, 6} or {1, 3, 5, 7} is assigned to the four transitions from or into one state. Therefore, the paths originating from one state and merging into the same state after two transitions are separated by at least $\sqrt{2}\Delta_1$ in the ED. Considering paths that give the minimum distance, we have $\sqrt{\Delta_1^2 + \Delta_0^2 + \Delta_1^2} = 2.141$ as the free ED and the coding gain over 4-PSK is 3.6 dB. For good performance of the coded modulation, the design of signal assignment to the trellis is very important. Next, we consider set partitioning proposed by Ungerboeck [56].

First, we consider two paths that split from the same node and merge initially at the same node after L intervals. The path of $L = 1$ without splitting is a parallel path. Our problem is how to assign eight signals on branches in order to maximize the free distance. We can imagine that a good assignment is for branches that originate from one node or branches that terminate at the same node to have the largest possible distance.

Let the signals of 8-PSK be expressed by $S = (y_2, y_1, y_0)$. Figure 6.28 shows the set partitioning of an 8-PSK signal constellation into subsets with an increase in the minimum subset distances. The first partitioning into subsets Y_0 and Y_1 is achieved by the value of $y_0 = 0$ and 1, respectively. The minimum ED between signals within Y_0 and Y_1 is Δ_1. At the second partitioning, which is achieved by the value of y_1, Y_0 becomes Y_{00} and Y_{10} for $y_1 = 0$ and 1, respectively, and Y_1 becomes Y_{01} and Y_{11} for $y_1 = 0$ and 1, respectively. Then, the minimum ED within Y_{00}, Y_{10}, Y_{01} and Y_{11} is the same as Δ_2. Finally, for 0 or 1 of y_2, one signal is determined within Y_{00}, Y_{10}, Y_{01} and Y_{11}. This mapping is achieved by successively partitioning a signal set into subsets while increasing the minimum ED $\Delta_0 < \Delta_1 < \Delta_2$ between the signals within the same subset.

Next, we consider a family of TCM codes known as Ungerboeck codes. Figure 6.29 shows a block diagram of an Ungerboeck encoder. At time n, we

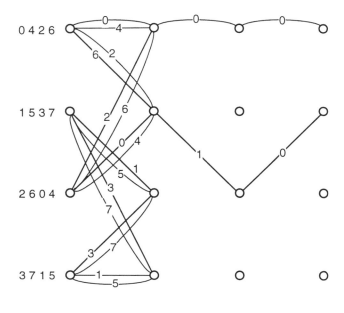

(a)

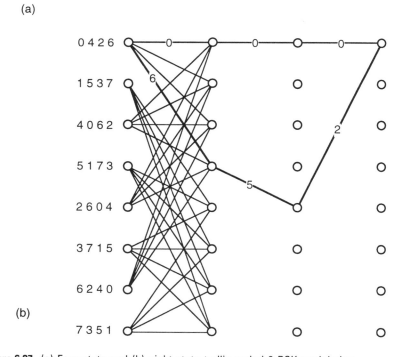

(b)

Figure 6.27 (a) Four-state and (b) eight-state trellis-coded 8-PSK modulation.

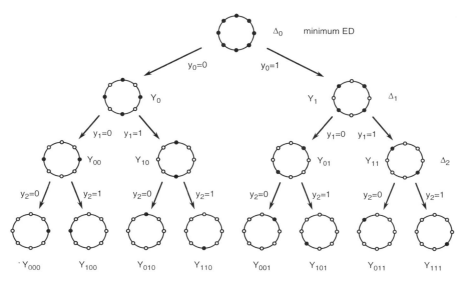

Figure 6.28 Set partitioning of 8-PSK signals into subsets.

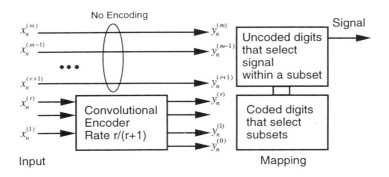

Figure 6.29 Block diagram of an Ungerboeck encoder.

have a sequence of m binary digits $x_n^{(1)}, x_n^{(2)}, \ldots, x_n^{(m)}$ as input to the encoder. The convolutional encoder receives r binary digits of $x_n^{(1)}, x_n^{(2)}, \ldots, x_n^{(r)}$ and generates $r + 1$ binary digits of $y_n^{(0)}, y_n^{(1)}, \ldots, y_n^{(r)}$, which are used to select a subset of the set partitioning. The remaining $(m - r)$ digits of $x_n^{(r+1)}, \ldots, x_n^{(m)}$ are not encoded, and these digits correspond to $(m - r)$ parallel paths in a trellis diagram.

As an example of this encoder, Figure 6.30 shows an 8-PSK TCM encoder with $m = 2$ binary input, rate-1/2 convolutional encoder, and constraint length

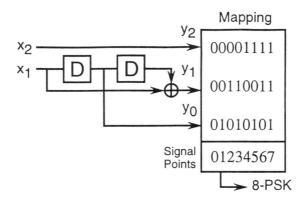

Figure 6.30 8-PSK TCM encoder.

of 2, corresponding to four states. Using the values of y_0 and y_1, the subsets of Y_{y1y0} are selected, and using the value of y_2, a signal Y_{y2y1y0} is selected from Y_{y1y0}.

Figure 6.31 shows the trellis diagram of this TCM encoder. We consider the ED between two paths that start from state 00 at time 0 and reach state 01 at time 3. The signal constellation, shown in Figure 6.28, indicates Δ_1

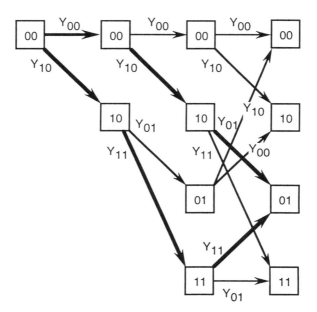

Figure 6.31 Trellis diagram for 8-PSK TCM.

between Y_{00} and Y_{10}, Δ_0 between Y_{10} and Y_{11}, and Δ_1 between Y_{01} and Y_{11}. Therefore, the ED between these two paths is given by

$$d = \sqrt{\Delta_1^2 + \Delta_0^2 + \Delta_1^2} \qquad (6.99)$$

Then, the free ED d_{free} is defined as

$$d_{\text{free}} = \min_{a \ne a'} \left[\sum_n d^2(a_n, a_n') \right]^{1/2} \qquad (6.100)$$

between all pairs of signal sequences **a** and **a'**. In this trellis, the distance between parallel paths is the minimum ED, and $d_{\text{free}} = 2$. Therefore, coding gain of 3 dB is achieved.

If maximum-likelihood soft decoding is applied over an AWGN channel at a high signal-to-noise ratio, the error-event probability will asymptotically approach the lower bound

$$P(\epsilon) \ge N(d_{\text{free}}) Q\left(\frac{d_{\text{free}}}{2\sigma} \right) \qquad (6.101)$$

where $N(d_{\text{free}})$ is the multiplicity of error events with distance d_{free}, $Q(x)$ is the following Gaussian error probability function, and σ^2 is the Gaussian noise power:

$$Q(x) = \frac{1}{\sqrt{2\pi}} \int_x^\infty e^{-y^2/2} dy$$

6.3.1.1 A Modem for Land Mobile Satellite Communications

A trellis-coded 8-DPSK modem with a rate 2/3 and 16 states for 4,800 bits/sec has been implemented in NASA's MSAT-X experimental program for developing a land mobile satellite communications service [60]. For the design of this modem, channel separation of 5 kHz was selected in FDMA, and the required bit error rate is 10^{-3} at E_b/N_0 of 11 dB. The assumed channel model is Rician fading with a Rician factor of 10 dB. Vegetative shadowing is modeled by log-normal distribution statistics.

This TCM encoder is followed by a 128-symbol block interleaver for burst error protection, and the interleaver output is differentially encoded. The pulse-shaping filter is a 100% rolloff root raised-cosine filter in order to produce two intersymbol interference-free points per symbol for Doppler shift estimation [61] and for matched filtering.

Simulation results show that the performance coding gain, which is determined as the reduction in required E_b/N_0 relative to uncoded QPSK at BER of 10^{-3} over the Rician channel with $K = 10$ dB, is 1.6 dB without interleaving. With 128-symbol interleaving, the performance coding gain increases to 3.1 dB.

6.3.2 Block-Coded Modulation (BCM)

Trellis-coded modulation uses the combination of convolutional coding and expanded signal sets of 8-PSK to transmit two information bits per symbol. Instead of convolutional encoding, one could use short binary block codes, which could be simpler and faster to decode. This coded modulation is referred to as block-coded modulation (BCM). Here, we present an example of an 8-PSK BCM scheme with length n and rate r.

First, we assign an array of n rows and three columns of binary digits, as shown below:

$$
\begin{array}{ccc}
a_{11} & a_{12} & a_{13} \\
a_{21} & a_{22} & a_{23} \\
a_{31} & a_{32} & a_{33} \\
\cdots & \cdots & \cdots \\
a_{n1} & a_{n2} & a_{n3}
\end{array}
\qquad (6.102)
$$

Each row of the array corresponds to one signal point $Y_{a_i a_j a_k}$ of 8-PSK. Let the number of information bits in the ith column be k_i. We have

$$k_1 + k_2 + k_3 = 3rn \qquad (6.103)$$

For the ith column, we perform binary block coding [57,62,63] of (n, k_i, d_i), where d_i is the minimum Hamming distance. Assuming two different code words of only the first column, the squared Euclidean distance d^2 between the corresponding signal points is expressed by $d^2 \geq \Delta_2^2 d_1$. Similarly, for two different code words of only the second column, we have $d^2 \geq \Delta_1^2 d_2$. Likewise, for the third column, we have $d^2 \geq \Delta_0^2 d_3$. Therefore, the minimum Euclidean distance between the code words of 8-PSK signal points is expressed by

$$d_{min}^2 = \min(\Delta_2^2 d_1, \Delta_1^2 d_2, \Delta_0^2 d_3) = \min(4d_1, 2d_2, 0.586d_3) \qquad (6.104)$$

The coding gain in the above case is $10 \log(d^2_{\min}/2)$ in dB. Sayegh [62] evaluated several cases of the above block-coded modulation scheme. For instance, 8-PSK with $r = 2/3$ and $n = 7$ has a coding gain of 3 dB over QPSK having the same bandwidth and data rate, and with BCH codes of length 32, it achieves 6 dB gain over QPSK. On the other hand, the convolutional codes require 2^{10} states to achieve coding gain of about 6 dB.

For example, the BCM for information bits a_1, a_2, a_3, . . ., a_{14} with rate 2/3 is considered. For economical assignment, the contributions from each column of the signal array should be balanced at each argument of min( ) in (6.104). We will design a BCM with $d_1 = 1$, $d_2 = 2$, and $d_3 = 7$ by using the following signal assignment:

$$
\begin{array}{ccc}
a_8 & a_2 & a_1 \\
a_9 & a_3 & c_1 \\
a_{10} & a_4 & c_2 \\
\cdots & \cdots & \cdots \\
a_{14} & c_7 & c_6
\end{array}
\qquad (6.105)
$$

The first column is a (7, 7, 1) code, the second column is (7, 6, 2) with a parity check bit of c_7, and the third column is (7, 1, 7), where $c_1 = c_2 = c_3 = \ldots = c_6 = a_1$ for $d_3 = 7$. Therefore, from (6.104), the minimum squared Euclidean distance is

$$
d^2_{\min} = 4 \qquad (6.106)
$$

Thus, this BCM achieves coding gain of 3 dB.

As BCM decoding algorithms, Euclidean decoding and Berlekamp-Massey decoding are well known. Since for error-correction capability t, the complexity of implementation has the order of $t(\log t)^2$, the implementation is not so hard even when t is about 100. On the other hand, the maximum-likelihood decoding algorithm is not well known. However, the BCM of (6.105) can be represented as a trellis based on the binary lattice, as shown in Figure 6.32. Each branch in the trellis corresponds to the signal subsets Y_{00}, Y_{10}, Y_{01}, and Y_{11}. Therefore, each branch has a parallel path. When $a_1 = 0$, seven signals of (6.105), which correspond to seven rows, are all selected from the subsets Y_{00} or Y_{10}. When $a_1 = 1$, the seven signals are selected from Y_{01}, and Y_{11}. This results in two parallel two-state trellises, shown in Figure 6.32 [64].

Using this trellis diagram, we can perform the Viterbi decoding by the conventional metric calculation. The performance coding gain, which is deter-

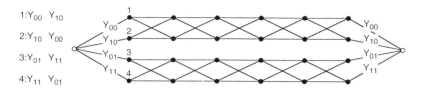

Figure 6.32 Trellis diagram for the BCM scheme.

mined as the reduction in required E_b/N_0 relative to uncoded 4-PSK at a BER of 10^{-4}, over the AWGN channel is 1.4 dB. Performance coding gains of 2.3 dB at a BER of 10^{-3} and 3.7 dB at a BER of 10^{-4} over the Rician fading channel with $K = 10$ dB can be achieved by BCM and a fading compensation method using inner correlation of fading [64].

6.3.2.1 Multiple TCM

To improve the performance of TCM, we can increase the number of states or modify the signal constellation. For example, in Ungerboeck's scheme, a four-state encoder achieves coding gain of 3 dB, and 8-, 16-, 32-, 64-, and 128-state TCMs can achieve coding gains of 4.0, 4.8, 4.8, 5.4, and 6.0 dB, respectively. However, when the number of states exceeds a certain value, the coding gain increases more slowly and implementation complexity increases. On the other hand, several efforts to modify the signal constellation have been carried out. One solution is provided by multidimensional signals. Another is provided by a technique referred to as multiple coded modulation, which gives us multiple TCM (MTCM) and multiple BCM (MBCM).

In the MTCM technique, which was originally proposed by Divsalar and Simon [65], mk binary input bits and $(m + 1)k$ binary encoder output symbols are mapped into k M-ary symbols for $M(= 2^{m+1})$ PSK modulation. This parameter k is referred to as the multiplicity of the code. In the trellis diagram, k symbols are allocated to each branch.

For example, Figure 6.33(a) shows the two-state trellis diagram for conventional rate 1/2 trellis-coded QPSK, and Figure 6.33(b) shows the multiple trellis diagram for the same modulation with $m = 1$ and $k = 2$. Thus, two QPSK symbols are assigned to each branch. For the parallel path between successive zero states, the squared Euclidean distance is given by

$$d^2 = d^2(0, 2) + d^2(0, 2) = 8 \tag{6.107}$$

For the error event path shown by the bold lines in Figure 6.33(b), the squared Euclidean distance is given by

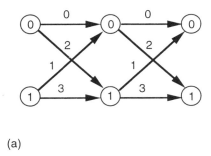

(a)

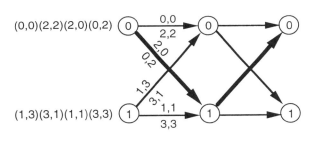

(b)

Figure 6.33 Trellis diagram for MTCM for $k = 2$: (a) trellis diagram for conventional rate 1/2 QPSK TCM and (b) trellis diagram for rate 1/2 QPSK MTCM.

$$d^2 = d^2(0, 0) + d^2(0, 2) + d^2(0, 1) + d^2(0, 3) = 8 \qquad (6.108)$$

Since the squared free distance for the conventional TCM shown in Figure 6.33(a) is 6, the coding gain of this two-state MTCM is $10 \log(8/6) = 1.2$ dB. Performance gains relative to conventional TCM ($k = 1$) and to uncoded BPSK are 1.3 and 3.0 dB, respectively.

6.3.2.2 Multiple BCM

In this section, we extend the discussion of MTCM to MBCM [66,67]. Figure 6.34 shows the signal array of 4 rows and 7 columns for 19 information bits, a_1, a_2, a_3, . . . , a_{19} (for convenience, the rows and columns in the array of the above BCM have been exchanged). Four rows are labeled by four coding levels, l_1, l_2, l_3, and l_4. The l_1 level is encoded by the ($n = 7$, $k = 1$, $d = 7$) codes, where n is the length of the codes, k is the number of information bits,

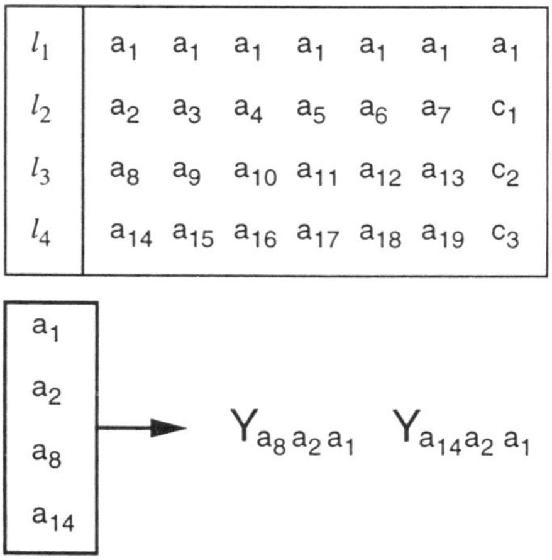

l_1	a_1	a_1	a_1	a_1	a_1	a_1	a_1
l_2	a_2	a_3	a_4	a_5	a_6	a_7	c_1
l_3	a_8	a_9	a_{10}	a_{11}	a_{12}	a_{13}	c_2
l_4	a_{14}	a_{15}	a_{16}	a_{17}	a_{18}	a_{19}	c_3

$$\boxed{\begin{array}{l} a_1 \\ a_2 \\ a_8 \\ a_{14} \end{array}} \longrightarrow Y_{a_8 a_2 a_1} \quad Y_{a_{14} a_2 a_1}$$

Figure 6.34 Signal structure for the proposed MBCM.

and d is the minimum Hamming distance. The l_2, l_3, and l_4 levels are all encoded by (7, 6, 2) using the parity check bits c_1, c_2, and c_3, which are given by

$$c_1 = a_2 \oplus a_3 \oplus a_4 \oplus a_5 \oplus a_6 \oplus a_7$$
$$c_2 = a_8 \oplus a_9 \oplus a_{10} \oplus a_{11} \oplus a_{12} \oplus a_{13}$$
$$c_3 = a_{14} \oplus a_{15} \oplus a_{16} \oplus a_{17} \oplus a_{18} \oplus a_{19} \tag{6.109}$$

The minimum Hamming distances of the four coding levels are given by

$$d_H^{\min}(\ell_1) = 7$$

$$d_H^{\min}(\ell_2) = d_H^{\min}(\ell_3) = d_H^{\min}(\ell_4) = 2 \tag{6.110}$$

Four bits of each column are mapped into two symbols in the signal constellation of 8-PSK, which is shown in Figure 6.28. For example, for the first column, (a_1, a_2, a_8, a_{14}) is mapped to the two symbols $Ya_8 a_2 a_1$ and $Ya_{14} a_2 a_1$. Therefore, the four bits of each column in the signal array are mapped to the same subset, $Ya_2 a_1$.

Figure 6.35 shows the corresponding two-parallel four-state trellis for the MBCM proposed in Figure 6.34. In each branch, two symbols of 8-PSK are assigned. Between each pair of states, there are two parallel branches. For

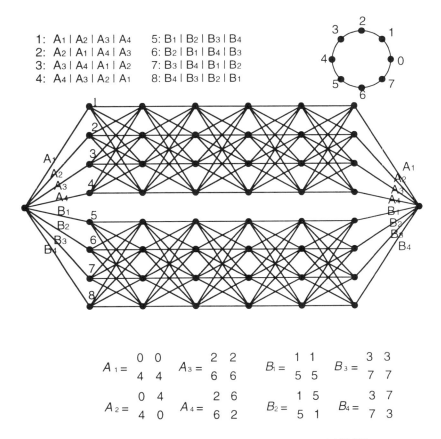

1: A_1 | A_2 | A_3 | A_4 5: B_1 | B_2 | B_3 | B_4
2: A_2 | A_1 | A_4 | A_3 6: B_2 | B_1 | B_4 | B_3
3: A_3 | A_4 | A_1 | A_2 7: B_3 | B_4 | B_1 | B_2
4: A_4 | A_3 | A_2 | A_1 8: B_4 | B_3 | B_2 | B_1

$$A_1 = \begin{matrix} 0 & 0 \\ 4 & 4 \end{matrix} \quad A_3 = \begin{matrix} 2 & 2 \\ 6 & 6 \end{matrix} \quad B_1 = \begin{matrix} 1 & 1 \\ 5 & 5 \end{matrix} \quad B_3 = \begin{matrix} 3 & 3 \\ 7 & 7 \end{matrix}$$

$$A_2 = \begin{matrix} 0 & 4 \\ 4 & 0 \end{matrix} \quad A_4 = \begin{matrix} 2 & 6 \\ 6 & 2 \end{matrix} \quad B_2 = \begin{matrix} 1 & 5 \\ 5 & 1 \end{matrix} \quad B_4 = \begin{matrix} 3 & 7 \\ 7 & 3 \end{matrix}$$

Figure 6.35 Two-parallel four-state trellis diagram for the proposed MBCM.

example, a branch variable A_1 corresponds to a pair of two symbols (0, 0) and (4, 4), and A_2 corresponds to (0, 4) and (4, 0). Both branch variables A_1 and A_2 belong to the same subset of Y_{00} because $a_1 = 0$ and $a_2 = 0$ in the case of the first column. According to the values of a_8 and a_{14}, a pair of symbols is selected from (0, 0), (4, 4), (0, 4) and (4, 0).

The signals assigned to each branch are indicated in Figure 6.35. For example, the four branches starting from node 1 have A_1, A_2, A_3, and A_4, respectively, and at the four branches from node 2, A_2, A_1, A_4, and A_3 are assigned, respectively.

The mapping of each symbol to a signal point in 8-PSK BCM is used by Ungerboeck's set partition in Figure 6.28. The minimum squared Euclidean distance (MSED), $d_E^2(\ )$, of different signal subsets in each level can be given by

$$d_E^2(\ell_1) = \Delta_0^2 = 4\sin^2(\pi/8) = 0.586$$

$$d_E^2(\ell_2) = \Delta_1^2 = 2$$

$$d_E^2(\ell_3) = \Delta_2^2 = 4$$

$$d_E^2(\ell_4) = \Delta_2^2 = 4 \tag{6.111}$$

Considering two different code words in only the first row, we can calculate the MSED between neighboring paths in its trellis, which can be given by $D_E^2 = d_E^2 * d_H$. Moreover, considering two differential code words in only the second, third, and fourth rows, we can calculate the MSED of each case by the following equations:

$$D_E^2(\ell_1) \cong 2 \times 7 \times 0.586 \cong 8.2$$

$$D_E^2(\ell_2) = 2 \times 2 \times 2 = 8$$

$$D_E^2(\ell_3) = 4 \times 2 = 8$$

$$D_E^2(\ell_4) = 4 \times 2 = 8 \tag{6.112}$$

Thus we have the MBCM with $d_{\text{free}}^2 = 8.0$. Since the above-mentioned BCM with one symbol per branch has $d_{\text{free}}^2 = 4.0$, the MBCM with two symbols per branch has coding gain of 3 dB relative to the conventional BCM. The performance coding gain in BER can be obtained by computer simulation, as shown in Figure 6.36. Using the trellis, the Viterbi decoding can be used as well. This two symbols per branch MBCM has performance coding gains of 1.1 dB and 2.2 dB at BER = 10^{-3} relative to the BCM 8-PSK and the uncoded QPSK, respectively.

6.3.3 Summary

In this section, we presented several requirements for MSS modulation schemes such as good power and bandwidth efficiencies, immunity to nonlinearity, robust synchronization under multipath fading and shadowing, and simple implementation. Several modulation techniques such as BPSK, QPSK, OQPSK, MSK, $\pi/4$-QPSK, GMSK, CPM, and OFDM were reviewed together with power spectral characteristics, bit error performance, several detection methods, and carrier recovery methods for designing mobile satellite communications systems.

In the early stages, satellite communication systems were limited by power rather than by bandwidth, and the BPSK, QPSK, and OQPSK schemes and

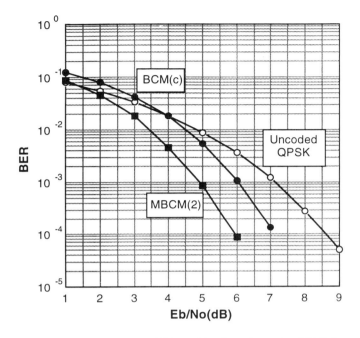

Figure 6.36 BER performance under additive white Gaussian noise BCM(c): conventional BCM, BCM(2): two symbols per branch MBCM.

forward error-correcting techniques such as convolutional encoding and Viterbi decoding were used due to their simple hardware implementation. However, with the recent increase in traffic demand for satellite communication services, mobile satellite communication systems have come to require modulation schemes that are efficient in both power and bandwidth. MSK has been used for aeronautical and maritime communication experiments using Japan's Engineering Test Satellite Five (ETS-V)[68,69]. The current mobile satellite communication systems and ones in the planning stage use the new $\pi/4$-QPSK, MSK, and so forth. Moreover, recent progress in digital signal-processing technology, including both hardware and software, is making possible complicated modulation schemes such as the CPM family, the nonconstant-envelope modulation schemes, and the trellis- or block-coded modulation schemes. These advanced modulation schemes will be very attractive for future mobile and personal satellite communication systems.

6.4 Digital Speech Coding

This section briefly describes typical examples of digital speech-coding techniques for mobile satellite communications, which require high-quality speech coding at bit rates as low as possible and that they be robust under sever propagation conditions such as multipath fading and shadowing. Speech-coding techniques can be classified into two broad groups: waveform coding and parametric coding. Waveform coding is achieved by matching as closely as possible the waveforms of the original and the reconstructed signals. Parametric coding represents the speech signal using a model for speech production, for example, time-varying linear predictive filters, and transmits only parameters of the filters. At the decoder, the speech signal is reconstructed by the inverse prediction filters using received parameters. Several hybrid coding methods also have been proposed as a combination of waveform coding and parametric coding. In the hybrid coding, both the filter parameters and the quantized residual sample, which is the signal that remains after filtering the speech signal, are transmitted.

Figure 6.37 shows bit rates versus speech quality for several speech-coding schemes [70]. A typical example of waveform coding is ADPCM, which produces high-quality speech at bit rates between 32 Kbps and 64 Kbps. At lower bit rates between 4.8 Kbps and 16 Kbps, several linear predictive coding (LPC) methods have been studied as the parametric or hybrid coding, such as multipulse-excited LPC (MPC or MPE-LPC), regular pulse-excited LPC (RPE-LPC), and codebook-excited linear prediction (CELP). Due to high quality at low bit rates, the CELP family, such as low-delay CELP (LD-CELP) [71], conjugate structure algebraic CELP (CS-ACELP), VSELP, multimode-learned CELP (M-LCELP), and pitch-synchronous innovation CELP (PSI-CELP), is a possible candidate for mobile satellite communication systems. Most of these LPC codecs are the hybrid coding methods since they transmit both the model parameters and residual errors.

Here, as typical examples of these speech-coding methods, we describe ADPCM as waveform coding, and two LPC methods of MPC and CELP as parametric coding.

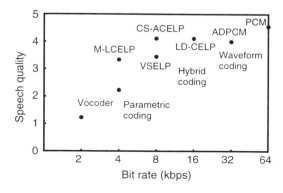

Figure 6.37 Bit rates vs. speech quality for several speech-coding schemes. (*After:* [70]).

6.4.1 ADPCM

Pulse code modulation (PCM) is the simplest method of waveform coding, and the speech signal $s(t)$ is quantized to one of 2^k amplitude levels, where k is the number of binary digits used to represent each sample. When the sampling rate is chosen to be several times the Nyquist rate, adjacent samples become highly correlated, and the signal does not change rapidly from sample to sample. Using high correlation between adjacent speech samples, differential PCM (DPCM) quantizes the differences between the input sample and the predicted value, which is estimated by a linear predictor. Since such differences are smaller than the sampled amplitudes themselves, fewer bits are required to represent the speech signals. DPCM with fixed predictors can provide from 4- to 11-dB improvement over direct quantized PCM. Adaptive differential PCM (ADPCM) uses both adaptive quantization and adaptive prediction to reduce coding errors. Figure 6.38 shows the configuration of an ADPCM encoder with feed-forward adaptive quantization where the quantizer step size is proportional to the variance of the input of the quantizer. ADPCM with feed-forward adaptive prediction provides about 10- to 11-dB improvement in signal to noise ratio (SNR) over PCM. Several adaptation algorithms have been proposed [72]. However, since ADPCM uses a scalar quantization, it is difficult to reduce the bit rate to less than 8 Kbps.

6.4.2 Linear Predictive Coding (LPC)

The linear predictive coding method synthesizes the speech signal using a linear predictive filter, which is excited by appropriate signals such as an impulse sequence for voiced speech or a random noise for unvoiced speech. The LPC codec transmits only the parameters of the linear predictive filter and the index of its selected excitation signals. The speech signal is assumed to be produced by an all-pole filter whose transfer function $H(z)$ is expressed as

$$H(z) = \frac{G}{1 - \sum_{k=1}^{p} a_k z^{-k}} \tag{6.113}$$

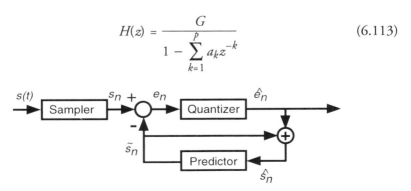

Figure 6.38 Configuration of an ADPCM encoder.

When the excitation sequence is denoted as $\{u_n\}$, the output sequence $\{s_n\}$ of the filter is described by

$$s_n = \sum_{k=1}^{p} a_k s_{n-k} + G u_n \qquad (6.114)$$

The basic idea behind the LPC approach is that a speech signal can be approximated by a linear combination of the previous speech samples. A linear predictor with prediction coefficients a_k estimates the nth sample s_n using p previous speech samples, s_{n-k} by

$$\tilde{s}_n = \sum_{k=1}^{p} \alpha_k s_{n-k} \qquad (6.115)$$

The prediction error e_n is defined as

$$e_n = s_n - \tilde{s}_n = s_n - \sum_{k=1}^{p} \alpha_k s_{n-k} \qquad (6.116)$$

Therefore, the prediction error sequence is the output of a system whose transfer function is expressed as

$$Y(z) = 1 - \sum_{k=1}^{p} \alpha_k z^{-k} \qquad (6.117)$$

If $\alpha_k = a_k$, then $e_n = G u_n$. Then, $H(z) = G/Y(z)$, and $Y(z)$ is an inverse filter of $H(z)$. Since the speech signal is a time-varying statistical process, the prediction coefficients should be estimated from short segments of speech signal.

The basic configuration [73] of an analysis-by-synthesis predictive coder is shown in Figure 6.39. The filter $1/A(z)$ models the spectral envelope of the speech segment being analyzed, and $A(z)$ is called the short-term predictor:

$$\frac{1}{A(z)} = \frac{1}{1 - \sum_{k=1}^{p} \alpha_k z^{-k}} \qquad (6.118)$$

where $\{\alpha_k\}$ are the predictor coefficients to be transmitted, and p is the order of the predictor.

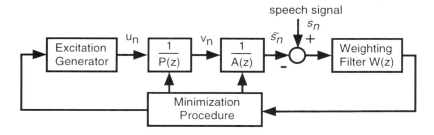

Figure 6.39 Configuration of a linear predictive codec.

The filter $1/P(z)$ models the spectral fine structure of the speech signal, and $P(z)$ is called the long-term predictor:

$$\frac{1}{P(z)} = \frac{1}{1 - \sum\limits_{k=-q}^{r} \beta_k z^{-(D+k)}} \tag{6.119}$$

where $\{\beta_k\}$ are the predictor coefficients, and the delay D shows the pitch period in samples. The number of coefficients, for example, are from 1 ($q = r = 0$) for a one-tap predictor to 3 ($q = r = 1$) for a three-tap predictor.

The short-term predictor $A(z)$ removes the redundancy in the speech signal by the predicted value using the past p samples. However, some periodicity, which is related to the pitch period of the original signal in 50- to 400-Hz pitch frequency, still remains. This residual signal is removed by the long-term predictor $P(z)$ (also called the pitch predictor), and turns into a noise-like signal. The long-term predictor is not essential for medium bit rate LPC codecs, although it can improve their performance. However, the long-term predictor is very essential for low bit rate LPC codecs such as CELP, which uses the excitation signal modeled by a random Gaussian noise process [74].

6.4.3 Multipulse-Excited Linear Prediction Coding (MPC)

The configurations of both MPC and CELP encoders can be shown in Figure 6.39, but the difference between MPC and CELP is their excitation generators. In MPC [75], the excitation generator produces a sequence of pulses located at nonuniformly spaced intervals with different amplitudes. Both amplitudes and positions of these pulses are determined using a closed-loop analysis-by-synthesis method. As shown in Figure 6.39, the synthesized signal is reconstructed using a sequence of pulses produced by the excitation generator and the long-term and short-term predictors. Using a suitable error criterion, the error between the original and the synthesized signals is minimized. To reduce the quantization noise by a masking effect, the following weighting filter $W(z)$ is used:

$$W(z) = \frac{A(z)}{A(z/\gamma)} \tag{6.120}$$

where $A(z)$ is the short-term predictor as defined in (6.118), and the parameter γ is ranging from 0 to 1, typically $\gamma = 0.9$, to control the error power in a given frequency region. Several analysis-by-synthesis methods have been proposed to make the searching procedures faster.

While MPC assumes that both amplitudes and positions of excitation pulses are initially unknown, the regular pulse-excited LPC [76] assumes that the pulses are regularly spaced but the amplitudes are unknown. Two types of codecs need similar bit rates for the same speech quality because MPC needs less number of excitation pulses due to the optimization of pulse positions, but it needs the transmission of pulse positions as well as their amplitudes. In the GSM system, which is the digital cellular system in Europe, the regular pulse-excited LPC codec with long-term predictor (RPE-LTP) at 13 Kbps is used.

6.4.4 Code-Excited Linear Prediction (CELP)

The multipulse and regular pulse-excited LPC can produce good-quality speech at bit rates as low as 9.6 Kbps, but they cannot maintain their quality below 9.6 Kbps because they have to spend the large number of bits for encoding the excitation pulses. On the other hand, as excitation signals, the CELP coder uses the collection of code vectors, which are previously produced using vector quantization techniques based on Gaussian process and are stored as a large codebook. Using each code vector and the predictors, synthesized speech is produced and then the most suitable code vector that produces the lowest error between the original and the reconstructed signal is selected. The index that is assigned to the code vector, the voice gain G, and the values of the parameters for the short-term and long-term predictors are transmitted.

Recently, many improvements have been added to the conventional CELP to reduce the complexity of excitation codebooks and implementation. For example, VSELP [77] uses the excitation vector that is produced by a linear combination of a number of basis vectors. The sparse excitation codebook, where most of the excitation pulses are set to zero, was proposed by Atal [78]. Algebraic codebook, which is produced by error-correcting codes, was proposed by Adoul [79].

6.4.5 Summary

In this section, we described basic digital speech-coding methods, but detailed information will be obtained from the original papers and many textbooks on digital speech processing. Here, we will survey digital speech codecs for terrestrial mobile communication systems. For the digital cellular system, the Pan-Europe GSM system uses 13-Kbps RPE-LTP, the IS-54 system for North America uses VSELP at 7.95 Kbps, and the PDC system for Japan uses VSELP at 6.7 Kbps and PSI-CELP (half-rate) at 3.45 Kbps. For the personal communication systems (PCS) and future public land mobile telecommunication systems (FPLMTS), several coding methods have been proposed and estimated. For example, multipulse-based CELP (MP-CELP), conjugate structure algebraic CELP (CS-ACELP), and multimode learned CELP (M-LCELP) have been proposed. The planned or currently operating mobile satellite communications systems use compatible speech coding as the terrestrial systems, and the future satellite systems will also select the compatible algorithm.

6.5 Multiple Access Techniques

Multiple access (MA) is a technique to use a satellite communication channel efficiently by sharing satellite resources such as frequency bandwidth, power, time, and space by a large number of mobile user terminals. Three main multiple access techniques have been used: FDMA, TDMA, and CDMA. FDMA and TDMA share the frequency bandwidth and the time of satellite transponders, respectively, as shown in Figure 6.40. In CDMA, mobile users share the resources of both frequency and time using a set of mutually orthogonal codes such as pseudorandom noise (PN) sequence.

Multiple access assignment strategy can be classified into three methods: preassignment (or fixed assignment), demand assignment (DA), and random access. In preassignment, channel plans for sharing the system resources are determined previously, regardless of traffic fluctuations. This scheme is suitable to communication links with a large amount of steady traffic. However, since most mobile users in mobile satellite communications do not communicate continuously, the preassignment method is wasteful for the satellite resources. In demand assignment multiple access (DAMA), satellite channels are dynamically assigned to users according to the traffic requirement. Due to high efficiency and system flexibility, DAMA schemes are suited to mobile satellite communication systems. In random access, a large number of users use the satellite resources in bursts, with long inactive intervals. To increase the system throughput, several methods have been proposed [80]: pure ALOHA, slotted ALOHA, reservation ALOHA, and spread ALOHA.

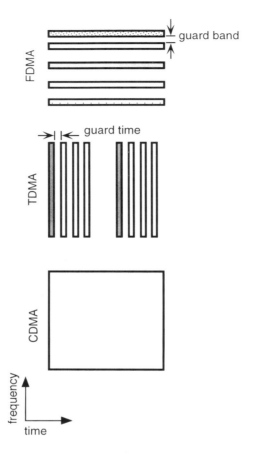

Figure 6.40 Multiple access methods.

6.5.1 Frequency Division Multiple Access (FDMA)

FDMA is the most common multiple access technique for satellite communication systems. Especially, a single channel per carrier (SCPC), in which each carrier bears one voice or data channel, has been used in most operating mobile satellite systems that provide voice services, since this technique allows frequency reallocations according to increases of traffic and future developments in modulation schemes.

Figure 6.41 shows the concept of FDMA, where transmitting signals occupy nonoverlapping frequency bands with guard bands between signals to avoid interchannel interference. When the satellite transponder is operated close to its saturation, nonlinear amplification produces intermodulation (*IM*) products, which may cause interference in the signals of other users. In order

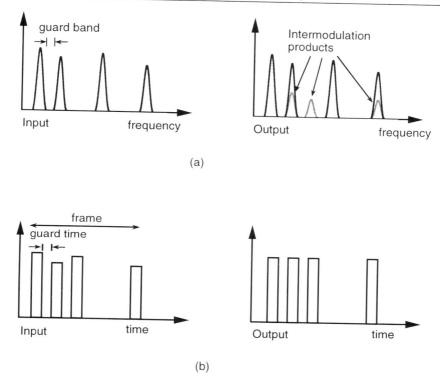

Figure 6.41 Concepts of (a) FDMA and (b) TDMA.

to reduce *IM*, it is necessary to operate the transponder by reducing the total input power, referred to as input backoff. This causes inefficient use of the available transponder power. Several nonlinear models for satellite traveling wave tube (TWT) amplifiers, which have both amplitude and phase nonlinearities, have been proposed to calculate the carrier-to-*IM* versus the input backoff [81–85]. The amplitude nonlinearity is referred to as *AM-AM* conversion, and the phase nonlinearity is called *AM-PM* conversion. Both nonlinearities produce *IM* products. The effect of *IM* in a link budget of the total $(C/N_0)_{\text{Total}}$ is expressed as

$$\left(\frac{C}{N_0}\right)^{-1}_{\text{Total}} = \left(\frac{C}{N_0}\right)^{-1}_{\text{up}} + \left(\frac{C}{N_0}\right)^{-1}_{\text{down}} + \left(\frac{C}{I_0}\right)^{-1}_{IM} \qquad (6.121)$$

where I_0 is the equivalent *IM* spectral density; $I_0 = I/B$ for I of *IM* power and B of signal bandwidth.

The frequencies of the third- and fifth-order *IM*, f_{IM} are expressed as

$$f_{IM^3} = f_i - f_j + f_k, \; 2f_i - f_j \qquad\qquad \text{for } i, j, k, l, m = 1, 2, 3, \ldots$$
$$f_{IM^5} = f_i - f_j + f_k - f_l + f_m, \; 3f_i - 2f_j, \ldots$$
$$\text{(6.122)}$$

where f_i, f_j, f_k, f_l, f_m are the frequencies of assigned signals. Therefore, suitable channel allocation can reduce *IM* falling on the wanted signals. Several frequency allocation methods for avoiding *IM* interference have been proposed, but suitable frequency allocation requires wider frequency bandwidth.

As shown in Chapter 7, in the planned LEO systems, multiple access techniques have been evolved from FDMA to TDMA to CDMA.

6.5.2 Time Division Multiple Access (TDMA)

Figure 6.41 shows the concept of TDMA, where each mobile terminal transmits a data burst with a guard time to avoid overlaps between TDMA bursts. Since only one TDMA burst occupies the full bandwidth of the satellite transponder at a time, input backoff, which is needed to reduce *IM* interference in FDMA, is not necessary in TDMA. TDMA permits the satellite amplifier to be operated in full saturation. This results in a significant increase in channel capacity. Another advantage over FDMA is its flexibility [9]. Time-slot assignments are easier to adjust than frequency channel assignments.

Figure 6.42 shows a typical TDMA frame, which consists of a reference burst and data bursts. An accessing signal that occupies an assigned slot in the frame is referred to as a burst. A reference burst is transmitted periodically to

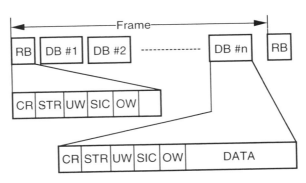

RB: Reference burst
STR: Symbol timing recovery
UW: Unique word
SIC: Station identification code
OW: Order wire
DB: Data burst

Figure 6.42 TDMA frame format.

indicate the start of each frame to control the transmission timing of all data bursts. To improve the imperfect timing of TDMA bursts, several synchronization methods of random access, open-loop, and closed-loop have been proposed [9]. The Earth station that transmits the reference burst is called the reference station.

A preamble is the initial part of a burst and consists of carrier recovery (CR) for coherent demodulation, symbol-timing recovery (STR), a unique word (UW), a station-identification code (SIC), and control symbols. In order to achieve fast carrier recovery and clock recovery from received TDMA bursts, an unmodulated carrier and clock recovery symbols, which are modulated by reference clocks, are included. The data part may contain the time-multiplexed data for all destinations of the burst. At a receiver, these high-speed subbursts are provided to the appropriate elastic buffers, where the data are read at lower speed.

As an example of operational TDMA systems, the INMARSAT uses TDMA for supporting telex from maritime earth stations to the ground station. The transmission rate of the TDMA bursts is 4,800 bps. The frame length is 1.74 sec and the guard time is 40 msec using the open-loop burst synchronization method.

6.5.3 Code Division Multiple Access (CDMA)

In CDMA, the resources of both frequency bandwidth and time are shared by all users using orthogonal codes. Spread spectrum multiple access (SSMA), which is achieved by a PN sequence generated by irreducible polynomials, is the most popular CDMA method. A spread spectrum method, which uses low-rate error-correcting codes including orthogonal codes with Hadamard or waveform transformation, has also been proposed [86].

The spread spectrum techniques can be classified into two categories: direct sequence (DS) method and frequency hopping (FH) method. In DS, the modulated signal is multiplied by PN codes with a chip rate R_c much larger than an information bit rate R_b. The resulting signal has wider frequency bandwidth than the original modulated signal. Figure 6.43 shows a typical DS transmitter and receiver. The transmitting signal $s(t)$ can be expressed as

$$s(t) = Ad(t)c(t)\cos(2\pi f_c t) \tag{6.123}$$

where $d(t)$ is the input data signal and $c(t)$ is a spreading PN sequence. At a receiving terminal, the received signal is despread by using the same PN sequence $c(t)$, as shown by

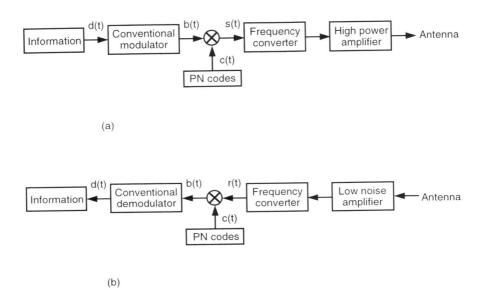

Figure 6.43 Direct sequence spread spectrum (a) transmitter and (b) receiver.

$$r(t) = Ad(t)c(t)\ \cos(2\pi f_c t) \times c(t)$$
$$= Ad(t)\ \cos(2\pi f_c t) \qquad (6.124)$$

where $\{c(t)\}^2 = 1$. Only the same PN code can achieve the despreading of the received signal bandwidth. In this process, the interference or jamming spectrum is spread by the PN codes and other users' signals spread by different PN codes are not despread. Therefore, interference or jamming power density in the bandwidth of the received signal decreases from their original powers. The most widely accepted measure of interference rejection is the processing gain G_p, which is given by the ratio of R_c/R_b, typically $G_p = 20$ to 60 dB [9]. The input and output signal-to-noise ratios are related as

$$\left(\frac{S}{N}\right)_{\text{output}} = G_p \left(\frac{S}{N}\right)_{\text{input}} \qquad (6.125)$$

In the forward link, the hub station transmits the spread spectrum signals, which are spread with synchronized PN sequence to different mobile users. Since orthogonal codes can be used, the mutual interference in the network is negligible and the channel capacity is close to that of TDMA. This is referred to as a synchronous SSMA. Conversely, in the return link the signals transmitted from different mobile users are not synchronized, and they are not orthogonal.

This is referred to as asynchronous SSMA. The nonorthogonality causes the interference due to the transmissions of other mobile users in the network. As the number of simultaneously accessing users increases, the communication quality gradually degrades, in a process called *graceful degradation.*

In an FH spread spectrum system, a bandwidth spreading effect is achieved by pseudorandom frequency hopping. The hopping pattern and hopping rate are determined by the PN code and code rate, respectively. Figure 6.44 shows an FH spread spectrum transmitter and receiver. At the receiver, when the frequency hopping of the frequency synthesizer is synchronized with that in the transmitter, the frequency hopping of the received signal is removed at the demodulator. The processing gain for FH is given by

$$G_p = \frac{W}{\Delta f} \tag{6.126}$$

where W is the frequency bandwidth and Δf is the bandwidth of the original modulated signal. A combined system of DS and FH is called a hybrid system, and the processing gain can be improved without increases of chip rate. The hybrid system has been used in the Joint Tactical Information Distribution

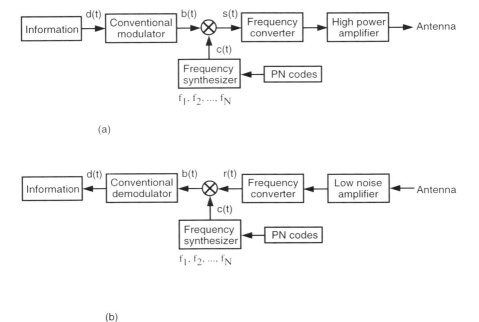

Figure 6.44 Frequency hopping spread spectrum (a) transmitter and (b) receiver.

System (JTIDS) and OmniTracs [87], which is the Ku-band mobile satellite communication system.

6.5.4 Summary

The best choice of multiple access schemes depends on the characteristics of the traffic to be transmitted in the networks, such as regularity and duration of calls, required data rates, as well as propagation characteristics in mobile satellite communications systems.

FDMA systems have relatively simple network control and system configurations, and flexibility for future extension of network configurations and improvements of modulation schemes. The digital satellite network via very small aperture terminal (VSAT) will still use FDMA, although nonlinearity of amplifiers may restrict the efficiencies of bandwidth and power.

TDMA requires system complexity for burst synchronization, and higher EIRP for mobile transmitters, which transmit the TDMA bursts at higher symbol rates than the original information rate. However, due to high efficiency and flexibility, TDMA will be used in personal satellite communications via LEOs and MEOs as well as in trunk channels having large capacities.

CDMA has advantages on anti-interference, sharing with other radio networks, frequency reuse in onboard multibeam satellite systems, and the capability of random access. CDMA will stand out as a strong candidate to choose multiple access schemes in future mobile and personal satellite communication systems.

References

[1]   Ghais, A., G. Berzins, and D. Wright, "INMARSAT and the future of mobile satellite services," *IEEE J. of Selected Areas in Communications*, Vol. SAC-5, No. 4, May 1987, pp. 592–600.

[2]   Taira, S., et al., "Experiments on ACSSB land mobile satellite communications," *Proc. IEEE Vehicular Technology Conf.*, May, 1990, pp. 695–700.

[3]   Jones, L. T., and W. A. Kissick, "ACSB—a minimum performance assessment," *Proc. Mobile Satellite Conf.*, Pasadena, CA, May 1988, pp. 351–358.

[4]   Sydor, J. T., "A study on the co- and adjacent channel protection requirements for mobile satellite ACSSB modulation," *Proc. Mobile Satellite Conf.*, Pasadena, CA, May 1988, pp. 359–364.

[5]   Lodge, J. H., "Mobile satellite communications systems: toward global personal communications," *IEEE Communications Magazine*, Vol. 29, No. 11, Nov. 1991, pp. 24–30.

[6]   Amoroso, F., "The bandwidth of digital data signals," *IEEE Communications Magazine*, Vol. 18, No. 6, Nov. 1980, pp. 13–24.

[7] Proakis, J. G., *Digital Communications*, New York: McGraw-Hill, 1989.

[8] Spilker, J. J., Jr., *Digital Communications by Satellite*, Englewood Cliffs, NJ: Prentice Hall, 1977.

[9] Bhargava, V. K., et al., *Digital Communications by Satellite*, New York, NY: John Wiley & Sons, 1981.

[10] Xiong, F., "Modem techniques in satellite communications," *IEEE Communications Magazine*, Vol. 32, No. 8, Aug. 1994, pp. 84–98.

[11] Ziemer, R. E. and C. R. Ryan, "Minimum-shift keyed modem implementations for high data rates," *IEEE Communications Magazine*, Vol. 21, No. 7, Oct. 1983, pp. 26–37.

[12] Lindsey, W. C., and M. K. Simon, *Telecommunication Systems Engineering*, New York, NY: Dover Press, 1973.

[13] Liu, C.-L., and K. Feher, "$\pi/4$-QPSK modems for satellite sound/data broadcast systems," *IEEE Trans. on Broadcasting*, Vol. 37, No. 1, March 1991, pp. 1–8.

[14] Murota, K., and K. Hirade, "GMSK modulation for digital mobile radio telephony," *IEEE Trans. on Communications*, Vol. COM-29, No. 7, July 1981, pp. 1044–1050.

[15] Sundberg, C.-E., "Continuous phase modulation," *IEEE Communications Magazine*, Vol. 24, No. 4, April 1986, pp. 25–38.

[16] Anderson, J. B., T. Aulin, and C.-E. Sundberg, *Digital Phase Modulation*, New York, NY: Plenum Press, 1986.

[17] Aulin, T., and C.-E. Sundberg, "Continuous phase modulation—Part I: Full response signaling," *IEEE Trans. on Communications*, Vol. COM-29, No. 3, March 1981, pp. 196–206.

[18] Aulin, T., N. Rydbeck, and C.-E. Sundberg, "Continuous phase modulation—Part II: Partial response signaling," *IEEE Trans. on Communications*, Vol. COM-29, No. 3, March 1981, pp. 210–225.

[19] Miyakawa, H., H. Harashima, and Y. Tanaka, "A new digital modulation scheme, multi-code binary CPFSK," *Proc. 3rd Int. Conf. on Digital Satellite Communications*, Nov. 1975, pp. 105–112.

[20] Sasase, I., and S. Mori, "Multi-h phase-coded modulation," *IEEE Communications Magazine*, Vol. 29, No. 12, Dec. 1991, pp. 46–56.

[21] Cimini, L. J., "Analysis and simulation of a digital mobile channel using orthogonal frequency division multiplexing," *IEEE Trans. on Communications*, Vol. COM-33, July 1985, pp. 665–675.

[22] Casas, E. F., and C. Leung, "OFDM for data communication over mobile radio FM channels—Part I: analysis and experimental results," *IEEE Trans. on Communications*, Vol. 39, No. 5, May 1991, pp. 783–793.

[23] Floch, B. L., R. Lassalle, and D. Castelain, "Digital sound broadcasting to mobile receivers," *IEEE Trans. on Consumer Electronics*, Aug. 1989.

[24] Zou, W. Y., and Y. Wu, "COFDM: an overview," *IEEE Trans. on Broadcasting*, Vol. 41, No. 1, March 1995, pp. 1–8.

[25] Couasnon, T. D., et al., "Results of the first terrestrial television broadcasting field tests in Germany," *IEEE Trans. on Consumer Electronics*, Vol. 39, Aug. 1993.

[26] Austin, M. C., and M. U. Chang, "Quadrature overlapped raised-cosine modulation," *IEEE Trans. on Communications*, Vol. COM-29, No. 3, March 1981, pp. 237–249.

[27]   Sasase, I., R. Nagayama, and S. Mori, "Bandwidth efficient quadrature overlapped squared raised-cosine modulation," *IEEE Trans. on Communications*, Vol. COM-33, No. 1, Jan. 1985, pp. 101–103.

[28]   Saha, D., and T. G. Birdsall, "Quadrature-quadrature phase shift keying," *IEEE Trans. on Communications*, Vol. 37, No. 5, May 1989, pp. 437–448.

[29]   El-Ghandour, O., and D. Saha, "Differential detection in quadrature-quadrature phase shift keying (Q2PSK) systems," *IEEE Trans. on Communications*, Vol. COM-39, No. 5. May 1991, pp. 703–712.

[30]   Divsalar, D., and M. K. Simon, "Multiple-symbol differential detection of MPSK," *IEEE Trans. on Communications*, Vol. 38, No. 3, March 1990, pp. 300–308.

[31]   Simon, M. K., and D. Divsalar, "On the implementation and performance of single and double differential detection schemes," *IEEE Trans. on Communications*, Vol. 40, No. 2, Feb. 1992, pp. 278–291.

[32]   Edbauer, F., "Bit error rate of binary and quaternary DPSK signals with multiple differential feedback detection," *IEEE Trans. on Communications*, Vol. 40, No. 3, March 1992, pp. 457–460.

[33]   Qu, S., and S. M. Fleisher, "Double differential MPSK on the fast Rician fading channel," *IEEE Trans. on Communications*, Vol. 41, No. 3, Aug. 1992, pp. 278–295.

[34]   Mackenthun, K. M., Jr., "A fast algorithm for multiple-symbol differential detection of MPSK," *IEEE Trans. on Communications*, Vol. 42, No. 2, Feb. 1994, pp. 1471–1474.

[35]   Hamamoto, N., "Differential detection with IIR filter for improving DPSK detection performance," *IEEE Trans. on Communications*, Vol. 44, No. 8, Aug. 1996, pp. 959–966.

[36]   Davarian, F., "High performance digital communication in mobile channels," *Proc. IEEE 34th Vehicular Technology Conf.*, 1984, pp. 114–118.

[37]   Yokoyama, M., "BPSK system with sounder to combat Rayleigh fading in mobile radio communication," *IEEE Trans. on Vehicular Technology*, Vol. VT-34, No. 1, Feb. 1985, pp. 35–40.

[38]   Davarian, F., "Comments on BPSK system with sounder to combat Rayleigh fading in mobile radio communication," *IEEE Trans. on Vehicular Technology*, Vol. VT-34, Nov. 1985, pp. 154–156.

[39]   Rafferty, W., et al., "Laboratory measurements and a theoretical analysis of the TCT fading channel radio system," *IEEE Trans. on Communications*, Vol. COM-35, No. 2, Feb. 1987, pp. 172–180.

[40]   Korn, I., "Coherent detection of M-ary phase-shift keying in the satellite mobile channel with tone calibration," *IEEE Trans. on Communications*, Vol. COM-37, No. 10, Oct. 1989, pp. 997–1002.

[41]   Leung, P.S.K., and K. Feher, "Pilot tone aided coherent GMSK systems for sound/data mobile broadcasting," *IEEE Trans. on Broadcasting*, Vol. 39, No. 2, June 1993, pp. 295–300.

[42]   McGeehan, J. P., and A. J. Bateman, "Theoretical and experimental investigation of feed-forward signal regeneration as a means of combating multipath propagation effects in pilot-based SSB mobile radio systems," *IEEE Trans. on Vehicular Technology*, Vol. VT-32, Feb. 1983, pp. 106–120.

[43]   Bateman, A. J., and J. P. McGeehan, "Phase-locked transparent tone-in-band(TTIB): a

new spectrum configuration particularly suited to the transmission of data over SSB mobile radio networks," *IEEE Trans. on Communications*, Vol. COM-32, No. 1, Jan. 1984, pp. 81–87.

[44]   Davarian, F., "Mobile digital communications via tone calibration," *IEEE Trans. on Vehicular Technology*, Vol. VT-36, No. 2, May 1987, pp. 55–62.

[45]   Simon, M. K., "Dual-pilot tone calibration technique," *IEEE Trans. on Vehicular Technology*, Vol. VT-35, No. 2, May 1986, pp. 63–70.

[46]   Caves, J. K., "An analysis of pilot symbol assisted modulation for Rayleigh fading channels," *IEEE Trans. on Vehicular Technology*, Vol. VT-40, Nov. 1991, pp. 686–693.

[47]   Li, H.-B., Y. Iwanami, and T. Ikeda, "Improvement on the error performance of M-ary PSK under a Rician fading using time domain TCT," *Trans. IEICE*, Japan, Vol. J75-A, Jan. 1992, pp. 108–117 (in Japanese).

[48]   Bose, R. C., and D. K. Ray-Chaudhuri, "On a class of error-correcting binary group codes," *Information and Control*, Vol. 3, March 1960, pp. 68–79.

[49]   Bose, R. C., and D. K. Ray-Chaudhuri, "Further results on error correcting binary group codes," *Information and Control*, Vol. 3, Sept. 1960, pp. 279–290.

[50]   Hocquenghem, A., "Codes correcteurs d'erreurs," *Chiffres*, Vol. 2, 1959, pp. 147–156.

[51]   Peterson, W. W., and E. J., Jr., Weldon, *Error-Correcting Codes*, 2nd ed., Cambridge, MA: MIT Press, 1991.

[52]   Reed, I. S., and G. Solomon, "Polynomial codes over certain finite field," *SIAM J. of Applied Mathematics*, Vol. 8, 1960, pp. 300–304.

[53]   Biglieri, E., et al., *Introduction to Trellis-coded Modulation With Applications*, New York, NY: Macmillan Publishing Company, 1991.

[54]   Viterbi, A. J., and J. K. Ohmura, *Principles of Digital Communication and Coding*, New York, NY: McGraw-Hill, 1979.

[55]   Viterbi, A. J., "Error bounds for convolutional codes and an asymptotically optimum decoding algorithm," *IEEE Trans. on Inform. Theory*, Vol. IT-13, No. 2, April 1967, pp. 260–269.

[56]   Ungerboeck, G., "Channel coding with multilevel/phase signals," *IEEE Trans. on Inform. Theory*, Vol. IT-28, No. 1, Jan. 1982, pp. 55–67.

[57]   Imai, H., and S. Hirakawa, "A new multilevel coding method using error-correcting codes," *IEEE Trans. on Inform. Theory*, Vol. IT-23, No. 3, May 1977, pp. 371–377.

[58]   Ungerboeck, G., "Trellis-coded modulation with redundant signal sets, Part-I: introduction," *IEEE Communications Magazine*, Vol. 25, No. 2, Feb. 1987, pp. 5–11.

[59]   Ungerboeck, G., "Trellis-coded modulation with redundant signal sets, Part-II: state of the art," *IEEE Communications Magazine*, Vol. 25, No. 2, Feb. 1987, pp. 12–21.

[60]   Divsalar, D., and M. K. Simon, "Trellis coded modulation for 4800-9600 bits/s transmission over a fading mobile satellite channel," *IEEE J. of Selected Areas in Communication*, Vol. SAC-5, No. 2, Feb. 1987, pp. 162–175.

[61]   Simon, M. K., and D. Divsalar, "Doppler-corrected differential detection of MPSK," *IEEE Trans. on Communications*, Vol. 37, No. 2, Feb. 1989, pp. 99–109.

[62]   Sayegh, S. I., "A class of optimum block codes in signal space," *IEEE Trans. on Communications*, Vol. COM-34, No. 10, Oct. 1986, pp. 1043–1045.

[63] Forney, G. D., et al., "Efficient modulation for band-limited channels," *IEEE J. of Selected Areas in Communications*, Vol. SAC-2, No. 5, Sept. 1984, pp. 632–647.

[64] Li, H.-B., Y. Iwanami, and T. Ikeda, "Performance of a multidimensional BCM scheme with fading estimation based on time correlation," *IEEE ICC'93*, Geneva, May 1993, pp. 443–447.

[65] Divsalar, D., and M. K. Simon, "Multiple trellis coded modulation (MTCM)," *IEEE Trans. on Communications*, Vol. 36, No. 4, April 1988, pp. 410–419.

[66] Li, H.-B., H. Wakana, and T. Ikegami, "Performance of a multiple block coded modulation scheme," *Proc. 4th IEEE ICUPC*, Tokyo Japan, Nov. 1995, pp. 27–31.

[67] Li, H.-B., and T. Ikegami, "Two symbols/branch MBCM scheme under fading channels," *Proc. 7th IEEE PIMRC*, Taipei, Taiwan, Oct. 1996, pp. 312–316.

[68] Ohmori, S., et al., "Experiments on aeronautical satellite communication using ETS-V satellite," *IEEE Trans. on Aerospace Electronic Systems*, Vol. 28, No. 3, July 1992, pp. 788–796.

[69] Wakana, H., et al., "Experiments on maritime satellite communications using the ETS-V satellite," *J. Communications Research Laboratory*, Vol. 38, No. 2, July 1991, pp. 223–237.

[70] Ozawa, K., "Speech coding technologies and their practical applications," *IEICE Technical Report*, SP95-110, Feb. 1996, pp. 9–16 (in Japanese).

[71] Chen, J-H., et al. "A low-delay CELP coder for CCITT 16 kb/s speech coding standard," *IEEE J. of Selected Areas in Communications*, 1992, pp. 830–849.

[72] Rabiner, L. R., and R. W. Schafer, *Digital Processing of Speech Signals*, Englewood Cliffs, NJ: Prentice-Hall, 1978.

[73] Kroon, P., and E. F. Deprettere, "A class of analysis-by-synthesis predictive coders for high quality speech coding at rates between 4.8 and 16 kbits/s," *IEEE J. of Selected Areas in Communications*, Vol. 6, No. 2, Feb. 1988, pp. 353–363.

[74] Salami, R. A., et al., *Mobile Radio Communications*, R. Steele (ed.), Chapter 3 Speech Coding, London, U.K.: Pentech Press Limited, 1992.

[75] Atal, B. S., "Predictive coding of speech at low bit rates," *IEEE Trans. on Communications*, Vol. 30, April 1982, pp. 600–614.

[76] Kroon, P., E. F. Deprettere, and R. J. Sluyter, "Regular-pulse excitation—a novel approach to efficient multipulse coding of speech," *IEEE Trans. on Acoustics, Speech, and Signal Processing*, Vol. 34, No. 5, Oct. 1986, pp. 1054–1063.

[77] Gerson, I. A., and M. A. Jasiuk, "Vector sum excitation linear prediction (VSELP) speech coding at 8 kbps," *Proc. ICASSP'90*, Albuquerque, NM, April 3–6, 1990, pp. 461–464.

[78] Atal, B. S., and J. R. Remde, "A new model of LPC excitation for producing natural-sounding speech at low bit rates," *Proc. IEEE Int. Conf. Acoustics, Speech, and Signal Processing*, April 1982, pp. 614–617.

[79] Adoul, J-P., and C. Lamblin, "A comparison of some algebraic structures for CELP coding of speech," *Proc. ICASSP'87*, 1987, pp. 1953–1956.

[80] Abramson, N., (ed.), *Multiple Access Communications*, Piscataway, NJ: IEEE Press, 1992.

[81] Berman, A. L., and C. E. Mahle, "Nonlinear phase shift in traveling-wave tubes as applied to multiple access communications satellites," *IEEE Trans. on Communication Technology*, Vol. COM-18, No. 1, Feb. 1970, pp. 37–48.

[82] Shimbo, O., "Effects of intermodulation, AM-PM conversion, and additive noise in multicarrier TWT systems," *Proc. IEEE*, Vol. 59, No. 2, Feb. 1971, pp. 230–238.

[83] Horstein, M., and D. T. Laflame, "Intermodulation spectra for two SCPC systems," *IEEE Trans. on Communications Technology*, Vol. COM-25, No. 9, Sept. 1977, pp. 990–994.

[84] Saleh, A. A. M., "Frequency-independent and frequency-dependent nonlinear models of TWT amplifiers," *IEEE Trans. on Communications Technology*, Vol. COM-29, No. 11, Nov. 1981, pp. 1715–1720.

[85] Wakana, H., "A new method for computing intermodulation products in SCPC systems," *IEEE Trans. on Communications*, Vol. 43, No. 2/3/4, Feb./March/April 1995, pp. 1067–1074.

[86] Wu, W. W., *Elements of Digital Satellite Communication*, Vol. I, Computer Science Press, 1984.

[87] Salmasi, A., and K. Gilhousen, "On the system design aspects of code division multiple access applied to digital cellular and personal communications networks," *Proc. IEEE Vehicular Technology Conf.*, May 1991, pp. 57–62.

7

Operational and Forthcoming Systems

This chapter presents operational or currently planned mobile satellite communications systems [1–4], including low Earth orbit (LEO) and highly elliptical orbit (HEO) satellite systems as well as geostationary satellite systems. The first-generation mobile satellite service (MSS) systems (for example, as in the INMARSAT systems) are characterized by global beam features of geostationary satellites and relatively large user terminals. The second-generation MSS systems are expected to play an important role wherever the terrestrial-based systems are not competitive due to low traffic density in rural and remote areas and to ease the terrestrial system during times of congestion.

The MSS systems are classified into geostationary Earth orbit (GEO) and nongeostationary orbit (NGSO) satellite systems. The NGSO systems are further subdivided into LEO, medium Earth orbit (MEO) (or intermediate circular orbit, ICO), and HEO systems, depending on their altitudes or the shapes of their orbits. The LEO systems are subdivided into big LEO and little LEO systems; big LEOs generally target real-time near-toll-quality voice as well as data, paging, facsimile, and radiodetermination satellite service (RDSS), while little LEOs, also referred to as nonvoice, nongeostationary (NVNG) systems, target low-data-rate (on the order of kilobits per second) data messaging, and RDSS using frequencies below 1 GHz. Recently, Microsoft and McCaw Cellular Communications have proposed the Teledesic *broadband LEO* system, which will use several hundred LEO satellites to provide videoconferencing, interactive multimedia and real-time digital network connection services in the Ka band. A number of global broadband GEO systems using the Ka-band frequency have also been proposed recently to provide high-speed data transmission services. The operational or planned MSS systems for GEOs, little LEOs, big LEOs, broadband LEOs and GEOs, and HEOs are presented in Sections 7.2, 7.3, 7.4, 7.5, and 7.6, respectively.

7.1 Overview

A brief review of the advantages and disadvantages of GEO, LEO, MEO, and HEO systems is shown in Table 7.1 [2]. GEO systems have relatively simple configurations of both space segments and Earth stations, an extremely wide footprint on the ground, time-invariant elevation angles to the satellites, and fixed propagation delay of about 0.25 seconds for a single hop. However, they have power-limited links, excessive propagation delays for voice and automatic repeat request (ARQ)-based packet data transmission, and extremely low elevation angles in high-latitude countries and polar regions.

Compared to GEOs, LEO systems have low propagation delay and loss, use relatively small, low-power and low-cost handheld user terminals thanks to their low propagation loss, and feature global service capability, including even high-latitude areas; moreover, their satellites are easier to launch. However, they need a large number of satellites for continuous communications and have a shorter satellite visibility period, more frequent handoffs, a much larger Doppler shift, and more complex onboard control systems. The altitude of LEOs is chosen to be in the range from 500 to 1,500 km, which is below the two van Allen radiation belts at 1,500 to 5,000 km and 13,000 to 20,000 km. On the other hand, the MEO or ICO satellites, which are positioned at altitudes of about 10,000 km between the outer and inner van Allen belts, have intermediate characteristics between those of GEO and LEO satellites.

The HEO systems feature comparable propagation delays and loss at the apogee altitude to those of GEOs systems, high elevation angles at high-latitude countries like Europe and Canada, and no eclipsing within service areas. However, they require large, tracking onboard antennas, and have larger Doppler shifts, high fuel consumption for satellite altitude controls, and a shorter satellite lifetime as a result of passing through the van Allen radiation belts, which are a major source of potentially damaging ionizing radiation. Selection of the satellite orbit is a trade-off between many factors, including satellite cost; satellite antenna size; battery power, weight, and lifetime; number of satellites and launch flexibility; minimum elevation angles within service areas; uplink and downlink frequencies; effect of van Allen radiation belts; handset power; propagation delay; system reliability; and required service quality.

Table 7.1
Advantages and Disadvantages of GEO, LEO, MEO, and HEO Systems

	GEO	LEO/MEO	HEO
Advantages	—Configuration simplicity —Extremely wide spotbeam footprint —Relatively time-invariant satellite-ground terminal geometry —Simple space segment control system —Fixed propagation delay	—Much lower propagation delays —Much better link margin —Easier launch —Ability to support handheld terminals	—High elevation angles —Flexible system design
Disadvantages	—Power-limited links —Excessive propagation delay for voice and ARQ-based packet data —Inability to cover polar regions	—A large number of satellites —More complex onboard control subsystems —Less satellite dwell time —More frequent handoffs —Much larger Doppler shift	—Lower link margin than MEOs —Large onboard antennas —Large Doppler shift —Shorter lifetime due to periodically crossing through the van Allen radiation belt

7.2 Satellite Systems Using Geostationary Orbits—GEOs

Characteristics of existing and planned GEO MSS systems are presented in Table 7.2.

7.2.1 INMARSAT [5,6]

The first-generation MSS has been provided by the International Maritime Satellite Organization (INMARSAT), which was established in 1979 as an international organization mandated with providing MSS for maritime users, by leasing three Marisat satellites from COMSAT General, and subsequently, two Marecs spacecraft from the European Space Agency (ESA). The INMARSAT Standard A system (now called INMARSAT-A) provides voice, data, and telex on demand between ships and the international telecommunications network via a Coast Earth Station (CES). Table 7.3 shows spacecraft, launch date, initial weight in geostationary orbit, frequencies, capacity, beam coverage, maximum equivalent isotropically radiated power (EIRP), and the launcher of each satellite. The user frequency between mobile terminals and the satellites is the L band at 1.6 and 1.5 GHz, and the feeder link frequency between the satellites and CESs is the C band at 6 and 4 GHz.

In 1985, the INMARSAT extended its service to include aeronautical services, and further to support land mobile services in 1989. In late 1994, the INMARSAT changed its name to the International Mobile Satellite Organization. By September 1994, INMARSAT terminals numbered 31,628 for maritime, 20,506 for land mobile, and 981 for aeronautical applications. To provide these services, the INMARSAT developed third-generation satellites, the INMARSAT-3 satellites. The first of five INMARSAT-3 satellites was launched in April 1996 by the U.S.-based Atlas-Centaur launcher, with the remainder following at regular intervals during 1996 and 1997. Since the INMARSAT-3 satellites are eight times as powerful as the INMARSAT-2 satellites as a result of using the satellite spotbeam technique, smaller and cheaper mobile terminals can be used. Table 7.4 and Figure 7.1 show the locations of the INMARSAT satellites, as of July 1996. Each INMARSAT-3 satellite has a global beam and five spot beams, whose power and bandwidth can be dynamically reallocated between beams, depending on traffic demand. INMARSAT-3 satellites carry navigation payloads as well as mobile-to-mobile communication capability.

The services shown below are provided using different types of INMARSAT standard terminals for maritime, land, and aeronautical mobile satellite communications.

Table 7.2
Characteristics of Existing and Planned GEO Satellites

Name	Owner	Country	No. of Satellites	Latitude (Degrees)	Weight (Kg)	Battery Power (W)	Lifetime (yr)	Launch	Cost to Build ($US)	User Frequency (MHz)	Feeder Link Frequency (MHz)	Service	Multiple Access	Repeater Type
AMSC	American Mobile Satellite Corp.	USA	1 (3)	101° W (62° W, 139° W)	2,500	3,000	12	1995	$500 million	L band (1,646.5–1,660.5/ 1,545–1,559)	Ku band (13/11, 14/12 GHz)	Voice/ data/fax/ messaging	FDMA	Bent pipe
INMARSAT-3	Inmarsat	UK	4 (4 spares)	54° W, 15.5° W, 64° E, 178° E	1,928	2,312	13	1996	$330 million	L band (1,626.5–1,646.5/ 1,530.0–1,545.0)	C band (6.4/3.6 GHz)	Voice/ data/fax/ messaging	FDMA	Bent pipe
MSAT	TMIC	Canada	2	106.5° W	2,500	3,000	12	1996	$500 million	L band (1,646.5–1,660.5/ 1,545–1,559)	Ku band (13/11 GHz)	Voice/ data/fax/ messaging	FDMA	Bent pipe
N-STAR	NTT	Japan	2	132° E, 136° E	2,000	5,000	10	1996	N/A	S band (2.6/2.5 GHz)	C band (6/4 GHz)	Voice/ data/fax/ messaging	FDMA	Bent pipe
Optus	Optus Communications	Australia	2	160° E, 156° E	1,582	3,500	10–15	1992, 1994	N/A	L band (1,646.5–1,660.5/ 1,545–1,559 MHz)	Ku band (14.0115–14.0255/ 12.2635–12.2775 GHz)	Voice/ data/fax/ messaging	FDMA	Bent pipe
EMS/Italsat I-F2	ESA	Europe	1	10.2° E	N/A	Communication: 400W	8	1996	N/A	L band (1,631.5–1,660.5 MHz, 1,530–1,559 MHz)	Ku band (14.231–14.250/ 12.731–12.750 GHz)	Voice/ data/fax/ messaging	CDMA/ FDMA	Bent pipe
LLM/Artemis	ESA	Europe	1	16.4° E	2,600	2,200	5	2000	$400 million	L band (1,631.5–1,660.5 MHz, 1,530–1,559 MHz)	Ku band (14.231–14.250/ 12.731–12.750 GHz)	Voice/ data/fax/ messaging	CDMA/ FDMA	Bent pipe

Table 7.3
INMARSAT Satellites

	Inmarsat-1			Inmarsat-2	Inmarsat-3
Spacecraft and launch date (UT)	Marisat-F1 '76.02.19 Marisat-F2 '76.06.10 Marisat-F3 '76.10.14	Marecs-A '81.12.20 Marecs-B2 '84.11.10	Intelsat V-MCS A '82.09.28 Intelsat V-MCS B '83.05.19 Intelsat V-MCS D '84.03.05	F1 '90.10.30 F2 '91.03.08 F3 '91.12.16 F4 '92.04.15	F1 '96.04.04 F2 scheduled F3 scheduled F4 scheduled F5 scheduled
Initial weight in orbit	330 kg	560 kg	1,870 kg	800 kg	820 kg
Frequency L band					
Satellite to Earth	1,537.0–1,541.0 MHz	1,537.5–1,542.5 MHz	1,535.0–1,542.5 MHz	1,525.5–1,559.0 MHz	1,530.0–1,548.1 MHz
Earth to satellite	1,683.5–1,642.5 MHz	1,638.5–1,644.0 MHz	1,636.5–1,644.0 MHz	1,626.5–1,660.5 MHz	1,626.5–1,649.6 MHz
C band					
Satellite to Earth	4,195.0–4,199.0 MHz	4,194.5–4,200.0 MHz	4,192.5–4,200.0 MHz	3,599.0–3,629.0 MHz	3,600.0–3,623.1 MHz
Earth to satellite	6,420.0–6,434.0 MHz	6,420.0–6,435.0 MHz	6,417.5–6,425.0 MHz	6,425.0–6,454.0 MHz	6,425.0–6,443.1 MHz
Capacity (INMARSAT-A voice)	8 channels	40 channels	50 channels	250 channels	9 × 250 channels
Beam coverage		Global		Global	Global/5-Spot
Max. EIRP L band				39 dBW	40 dBW/Global 48 dBW/Spot
C band				24 dBW	27 dBW
Launcher	Delta	Ariane	Atlas-Centaur	F1 Delta F2 Delta F3 Ariane 4 F4 Ariane 4	F1 Atlas-Centaur F2 F3 F4

Table 7.4
Locations of INMARSAT Satellites

Ocean Region	Spacecraft	Location	Status
Atlantic-West	INMARSAT-2 F4	54° W	Operational
(AOR-W)	Intelsat V MCS-B	31.5° W	Spare
Atlantic-East	INMARSAT-2 F2	15.5° W	Operational
(AOR-E)	Marecs B2	15° W	Spare
Indian	INMARSAT-3 F1	64° E	Operational
(IOR)	INMARSAT-2 F1	65° E	Spare
Pacific	INMARSAT-2 F3	178° E	Operational
(POR)	Marisat F3	178.2° E	Spare
			as of July 1996

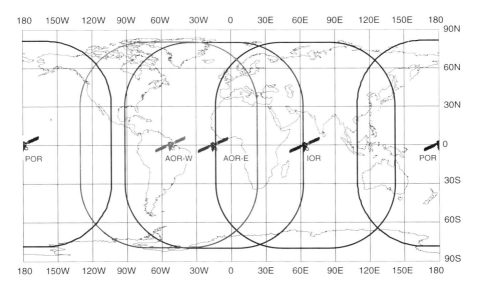

Figure 7.1 Locations of INMARSAT-3.

7.2.1.1 INMARSAT-A

The INMARSAT-A analog system provides direct-dial phone, data, telex, facsimile, and electronic mail. INMARSAT-A parabolic antennas, which are about 1m in diameter, are installed on various types of vessels, including oil tankers, liquid natural gas carriers, seismic survey vessels, fishing boats, and cargo and container vessels. INMARSAT-A can be also used in maritime communications services such as fleet monitoring, videoconferencing with 64-Kbps high-speed data transmission services, electronic data interchange used to submit customs documentation in advance of the ship's arrival in port, and weather forecasting.

For land mobile applications, transportable INMARSAT-A receivers can be packed in one or two suitcase-sized containers with a collapsible antenna. Data services using a voice-band modem can be provided at rates of up to 9.6 Kbps. The high-speed data service is also available in both simplex and duplex modes for applications such as large-volume file exchange, broadcast-quality audio, and store-and-forward video transfer.

7.2.1.2   INMARSAT-B

INMARSAT-B, which is the digital replacement for INMARSAT-A, provides high-quality satellite phone, telex, medium- and high-speed data, and facsimile services with lower communication charges, ranging from $3 to $7 per minute. Requiring less power to operate and being cheaper to use than INMARSAT-A, INMARSAT-B is attractive for all of the applications currently available via INMARSAT-A.

7.2.1.3   INMARSAT-C

INMARSAT-C provides a two-way *store-and-forward* messaging, facsimile, and e-mail service (600 bps) via small terminals weighing only a few kilograms. The smallest available INMARSAT-C terminal weighs around 4 kg, costs $5,000 to $8,000 to buy, and costs about $1 per kilobit to operate. The terminal can be mounted on a truck dashboard or at the helm of a small vessel, or carried around inside a briefcase. INMARSAT-C supports the transmission of data such as location, speed, and heading; fuel stocks; and fuel consumption of vessels. The enhanced group call capability and the international SafetyNETS service allow weather analyses, warnings, and predictions for better safety at sea. INMARSAT-C is also suitable for supervisory control and data acquisition (SCADA). Typical applications include the monitoring of river levels, weather information-gathering, and oil pipeline supervision.

7.2.1.4   INMARSAT-M

Briefcase-type INMARSAT-M communicators provide 6.4-Kbps two-way digital voice and 2.4-Kbps data and facsimile services. The cost of a call is $3 to $6 per minute. In rural or remote areas where the telecommunication infrastructure is poor and there is a lack of satisfactory long-distance telephone links, the combination of INMARSAT-M with the terrestrial wireless system is a very cost-effective way of adding national and international reach to these areas.

7.2.1.5   INMARSAT-Aero [7]

The INMARSAT-Aero system provides interconnection service between jet aircraft and the public switched telephone network (PSTN) with 9.6-Kbps digital voice and data services up to 9.6 Kbps using high-gain (12 dBi) steerable antennas. A system using an omnidirectional antenna provides a low-bit-rate (600 bps) data service for aircraft fleet management. By 1994, Lufthansa, Swissair, Air Canada, British Airways, United Airlines, and KLM had all outfitted jumbo jets with INMARSAT-Aero terminals.

Recently, INMARSAT aeronautical services were classified into four categories: INMARSAT-Aero-C, Aero-H, Aero-I, and Aero-L.

INMARSAT-Aero-C service is a messaging and data reporting service providing aircraft with store-and-forward satellite communications at 600 bps. Potential applications are weather and flight plan updates, flight plan submission, position reporting during flight, and communications for smaller aircraft operating in remote areas. INMARSAT-Aero-C is suitable for smaller aircraft in remote regions that do not need the full telephone and data capability provided by Aero-L and -H.

INMARSAT-Aero-H service provides aircraft with two-way digital voice, G3 facsimile at 4.8 Kbps, and real-time data transmission at up to 10.5 Kbps to meet the requirements of flight crew, cabin staff, and passengers. Potential applications are real-time engine and airframe monitoring, weather and flight plan updates, maintenance and fuel requests, interactive passenger services, and catering information and crew scheduling.

INMARSAT-Aero-I operates via the spot beams of the INMARSAT-3 satellites, which provide telephony, G3 facsimile at 2.4 Kbps, and real-time packet-mode data transmission at 4.8 Kbps. Potential applications are news and weather broadcasts, point-to-multipoint data broadcast, and interactive passenger services.

INMARSAT-Aero-L provides aircraft with a two-way and real-time data communication capability at 600 or 1,200 bps using a low-gain antenna. It has been developed to meet the needs of aircraft operators requiring highly reliable data links for their flight crew, cabin crew, and passengers. This system supports worldwide automatic position reporting and polling for air traffic control (ATC) and operations in management and communication, and supports Acars/Aircom-type messaging in the world. Potential applications are real-time engine and air frame monitoring, weather and flight plan updates, maintenance and fuel requests, and catering information and crew scheduling.

The numbers of INMARSAT standard terminals are 25,611 (17,948 are for maritime application) for INMARSAT-A, 1,947 (902) for INMARSAT-B, 9,003 (1,675) for INMARSAT-M, 23,520 (14,751) for INMARSAT-C, and 797 for INMARSAT-Aero, as of March 1996. The numbers of INMARSAT-B and -M terminals for land mobile application are increasing, while the number of INMARSAT-A terminals is decreasing as users shift over from type A terminals to types B and M.

7.2.2 The United States' AMSC [8]

In 1985, the Federal Communications Commission (FCC) proposed to establish U.S. domestic MSSs and allocate their spectrum. Twelve applications were filed in response to a notice of proposed rulemaking (NPRM) issued by the FCC. In 1986, the FCC allocated a total of 28 MHz in the L band for MSS systems, but in 1987 the Mobile World Administrative Radio Conference (MOB-WARC-87) did not adopt U.S. proposals to make the L-band multiservice generic MSS allocation. The FCC determined that only one licensee would be viable and ordered all qualified applicants to form a single consortium. In 1988, eight of the twelve applicants formed the American Mobile Satellite Consortium (AMSC), which was licensed by the FCC in 1989. AMSC later changed its name to American Mobile Satellite Corporation. AMSC's major shareholders include Hughes Communication, Inc., AT&T Wireless Services, Inc., and Singapore Telecom.

AMSC is authorized to construct, launch, and operate the U.S. domestic mobile satellite system consisting of three satellites using the L band for mobile service links and the Ku band for feeder links. The orbital locations are 101 degrees west for the central satellite, 62 degrees west for the eastern satellite, and 139 degrees west for the western satellite. The second and third satellites will be launched when the extra capacity is required. The FCC allocated 28 MHz of the L-band spectrum (1,545 to 1559 MHz and 1,646.5 to 1,660.5 MHz). For the feeder link, 200 MHz of the Ku band was allocated. The central satellite uses the 13/11 GHz band, and the other two satellites use 14/12 GHz. The spacecraft uses the three-axis HS-601 bus from Hughes Aircraft Corporation and a communication payload from SPAR Aerospace. The service coverage is all 50 states of America (including Hawaii and Alaska), Puerto Rico, Mexico, and up to 300 km offshore using L-band antennas about 5.5m in diameter, producing six spot beams with EIRP of 56 dBW and $G/T = 3$ dB/K.

A joint operating agreement between AMSC and Telecom Mobile Inc. (TMI), which is the Canadian counterpart to AMSC and operates the MSS in Canada, provides restoration and mutual backup for each satellite. Both AMSC and TMI introduced interim two-way data messaging services in the early 1990s, by leasing capacity from INMARSAT and using INMARSAT-C-type terminals. In April 1995, AMSC's MSAT-2 satellite was launched by an Atlas-II A rocket, and MSS services have been successfully operating since December 1995, in spite of trouble in the onboard hybrid matrix.

AMSC is providing SKYCELL (Satellite Communication Services and Mobile Messaging Service), which is a set of seamless mobile communication services for maritime, aeronautical, emergency restoration and rescue, pipeline maintenance, fleet management, and utility and service vehicle dispatch communications. This multimode communication system, which integrates satellite-based and terrestrial mobile communication systems, can provide commercial trucking fleets with two-way data communications and global vehicle location services, no matter where their vehicles are. The terrestrial system ensures communications while vehicles are traveling in urban areas where tall buildings may block the line of sight to the satellite, while the satellite system ensures communications while vehicles are traveling through remote areas that are not covered by terrestrial services.

7.2.3 Canada's MSAT [9,10]

The MSAT^{TM} (Mobile Satellite) system provides voice, circuit-switched data (up to 4,800 bps), G3 facsimile services, and radio broadcast and dispatch services to land, maritime, aeronautical mobile users in North America, Central America, Mexico, the Caribbean, and up to 400 km offshore via a Canadian geostationary MSAT-1 satellite located at 106.5 degrees west. In the late 1960s, the Canadian government commenced the MSAT program at the Communication Research Centre in Ottawa. In 1988, the MSAT project moved from the government to the private sector under TMI. After going bankrupt in 1993, TMI was reconstructed as TMI Communications & Co. (TMIC), with investment by Bell Canada Enterprise. In April 1996, TMIC's MSAT-1 satellite was launched by an Ariane-4 rocket and MSS services started in June 1996. The MSAT system provides approximately 1,800 demand-assigned 6.4-Kbps digital communication channels as well as voice and data switches, which are managed through the public switched telephone network, public packet-switched data networks, and private networks.

The MSAT provides five types of mobile satellite communications services: telephone calls, group voice calls, facsimile, circuit-switched data calls, and packet-switched data calls. One option is dual mode, which handles both satellite and cellular services with the same handset terminal. The following are examples of several applications.

7.2.3.1 Wide Area Fleet Management (WAFM) [9]

MSAT WAFM service provides voice and data communications between vehicles and central dispatch centers, which allow vehicle operators to compile and send messages, status codes, emergency alerts, and global positioning system (GPS) position reports. The government and commercial fleet managers in the land, maritime, and aeronautical mobile communities require wide area, "seamless" fleet management. MSAT WAFM has applications for courier companies, public safety agencies, and remote school bus fleets; tow-boats, fishing vessels, and ferries; and general aviation.

The largest market for MSAT WAFM is trucking. It enables discussion with dispatch about complex maintenance problems, maintenance and tracking of shipments for shippers and consignees, and assignment of vehicles' routes. Potential users in the maritime community includes tugs, passenger ferries, cargo ferries, coastal freighters, fishing companies, and government fleets for policing, fishery patrols, and coast guard activities. Marine search and rescue (SAR) is primarily undertaken by Canadian Coast Guard vessels.

In rural areas of Canada, some school buses operate on long rural routes that are not served by terrestrial radio systems. Since rural school buses are canceled on extremely cold days, an MSAT-equipped school bus would have the capability of emergency communications and could transfer location information.

7.2.4 Japan's N-Star

In Japan, mobile satellite communication experiments were carried out by the Communications Research Laboratory of the Ministry of Posts and Telecommunications and other several institutes in Japan via the Engineering Test Satellite Five (ETS-V), which was launched in August 1987. These experiments included maritime, aeronautical, and land mobile satellite communications with L-band coverage of the Asia and Pacific region, and passed on various useful results for development of the commercial MSS systems. After the unfortunate failure of the ETS-VI satellite, which carried an experimental payload for S-band mobile satellite communications, to reach geostationary orbit in 1994, two N-Star satellites were launched into geostationary orbital slots of 132 degrees E and 136 degrees E by Ariane rockets in Kourou, French Guiana, on August 29, 1995, and February 5, 1996, respectively. Using two S-band transponders in each satellite, commercial mobile satellite communication services to portable terminals and land vehicles in domestic areas, and to vessels up to 370 km offshore around Japan started on March 29, 1996. Separate multibeam (four-beam) antennas for transmitting and receiving in the S band at 2.6 and 2.5 GHz were installed on the satellites to avoid mutual interference. The feeder link frequency is 6 GHz for uplink and 4 GHz for downlink. Figure 7.2 shows the service coverage of the N-Star satellites.

The payloads of the N-Star satellites are not only for MSS but also for fixed satellite communication services in the Ku band (14/12 GHz) and Ka band (30/20 GHz). Mutual connections among the C band, Ku band, and Ka band are available to provide mutual backup.

The N-Star's mobile satellite communication services, which are provided by a cellular service provider, NTT Mobile Communication Network Inc. (NTT DoCoMo), target the extension of cellular services to areas that are not

Service coverage of terrestrial maritime phone system

Service coverage of terrestrial cellular system

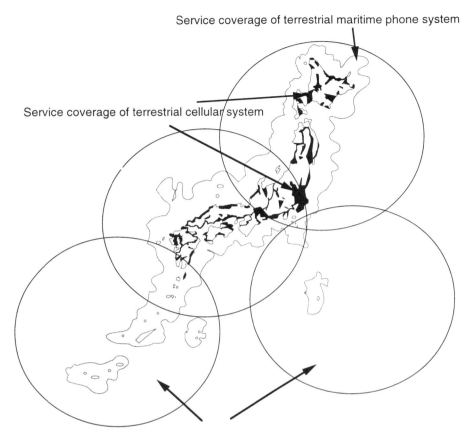

Service coverage of mobile satellite communication services via N-Star

Figure 7.2 Service coverage of the N-Star satellites.

covered by cellular systems and the extension of maritime two-way communications, both onshore and offshore, up to 200 nautical miles (about 370 km) around the coast of Japan. The N-Star's services are twofold, as described below.

Satellite Portable Phone

The satellite portable terminal has two modes of operation: satellite phone and terrestrial cellular digital phone at 800 MHz. When a user is inside a service area of the terrestrial cellular system, this dual-mode phone selects the cellular system, and when the user is traveling through rural or remote areas where cellular services are not available, it selects the N-Star's satellite phone. The portal terminal has a volume of 2,800 cc, weight of 2.7 kg, and transmission power of 2W. A vehicle-mounted terminal consists of a 30-cm-diameter antenna and radio equipment of 2,700 cc in size. The antenna has automatic satellite tracking capability. The transmitting power is also 2W. The service offers digital voice communication at 5.6 Kbps and data communication at 4.8 Kbps.

Potential users are manufacturing companies, mining and exploration companies, construction companies, and reporters operating in mountainous regions and rural and remote areas, where business communications including voice, data, still pictures and video, and dispatch information can be readily transmitted and received using satellite phones. Satellite phones may play an important rule in appropriate life-saving emergency care and data acquisition and transmission in the event of natural disasters.

Maritime Satellite Phone

Maritime phone services are provided only by a single mode of the satellite phone. Two maritime systems are available. Satellite Marine Phone, which is installed on a vessel, consists of a 30-cm-diameter antenna and radio equipment of 2,700 cc in volume. The antenna has automatic satellite tracking capability and the transmitting power is 2W. The other system is the Satellite Portable Marine phone, which is similar in shape to the INMARSAT-M terminal. The volume is 2,800 cc and the weight is 2.7 kg. Potential users include fishing companies, passenger ferries, cargo ferries, coastal freighters, and government fleets for policing and coast guard activities.

7.2.5   Australia's MobileSat [11,12]

Mobile satellite telephone service MobileSat is provided by Optus Communications Pty., Ltd. to Australia and its neighboring waters. In August 1992, the OPTUS-B1 satellite was launched by a Long March rocket in China. After the unsuccessful launch of the B2 in December 1992, the B3 satellite was successfully launched in August 1994. Commercial MobileSat services started in August 1994. Optus's MobileSat service is operated in conjunction with its mobile cellular services, which are a digital Global System for Mobile (GSM) communications system and the resale of the analog American Mobile Phone System (AMPS). While the terrestrial cellular system provides mobile communication services for 85% of the population, it only covers about 5% of the land area. MobileSat, on the other hand, provides 100% coverage of Australia.

The frequencies for the feeder link between the satellites and the gateway station are the Ku band at 14.0115 to 14.0255 GHz for uplink and 12.2635 to 12.2775 GHz for downlink. The mobile users use right-hand circularly polarized L-band signals of 1,646.5 to 1,660.5 MHz for uplink and 1,545.0 to 1,559.0 MHz for downlink. With nominal transmission power of 150W for each of their L-band transmitters, the two Optus B satellites located at longitude 156 degrees east and 160 degrees east cover Australia with EIRP of 46 dBW. Figure 7.3 shows *G/T* of the L-band receiver and EIRP of the L-band transmitter.

The MobileSat ground infrastructure consists of two network management stations (NMSs) in Sydney and Perth. Access to the public switched telephone network is via a gateway station (GW), which is controlled by the

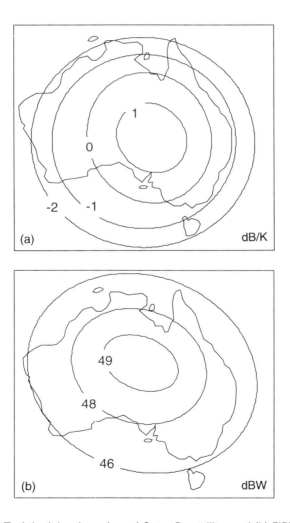

Figure 7.3 (a) G/T of the L-band receiver of Optus-B satellites and (b) EIRP of the L-band transmitter of Optus-B satellites.

NMS. The NMS is also responsible for overall management and coordination of network operations, including the allocation of power and bandwidth during call establishment. Figure 7.4 shows the configuration of the MobileSat network. The gateway station provides echo cancellation and conversion of the circuit-switched signal to 64 Kbps, which is then multiplexed onto an ISDN 30B+D primary rate access. Optus offers its Australian customers a number of rate plans: for example, the voice plan is A$100 (U.S.$75) for a one-time connection fee, A$45 (U.S.$34) per month as a service fee, and A$1.80 (U.S.$1.35) per minute for call charges for all domestic calls within Australia.

The specifications of MobileSat telephones are $\pi/4$ QPSK modulation at 6,600 bps, 4,200 bps improved multiband excited (IMBE) digital voice encoding developed by Digital Voice System Inc., and G3 fax and data transmission at 2,400 bps. As a demand assigned multiple access (DAMA) scheme, the single channel per carrier (SCPC) scheme is used for circuit-switched channels. Outbound signaling from NMS/GW to the mobile terminal uses a $\pi/4$ QPSK time division multiplexing (TDM) signal at 9,600 bps with up to 24 inbound channels operating as slotted ALOHA random access on an aviation BPSK carrier at 2,400 bps. Eight inbound channels are allocated for acknowledging the outbound signaling with a 12-byte signal unit.

The MobileSat system predominantly uses an adjustable mast antenna, which is approximately 1m long and 20 mm in diameter, is azimuthally omnidirectional, has an adjustable elevation pattern with a beamwidth of about 10 degrees, and has a G/T of -18 dB/K. Tables 7.5 and 7.6 show the link budget of the MobileSat systems.

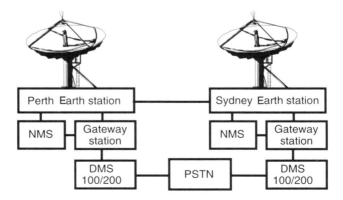

NMS: Network Management Station
PSTN: Public Switched Telephone Network

Figure 7.4 Configuration of the MobileSat network.

Table 7.5
Inbound Link Budget

Mobile terminal EIRP	10	dBW
Propagation loss	188	dB
Satellite G/T	−1	dB/K
Uplink C/N_0	49.6	dBHz
Satellite EIRP	8	dBW
Propagation loss	206	dB
Gateway G/T	38	dB/K
Downlink C/N_0	68.6	dBHz
Total C/I	60	dBHz
Total C/N_0	49.2	dBHz
Req'd C/N_0	43	dBHz
Link margin	6.2	dB

Table 7.6
Outbound Link Budget

Gateway EIRP	42	dBW
Propagation loss	207	dB
Satellite G/T	2	dB/K
Uplink C/N_0	65.6	dBHz
Satellite EIRP	25	dBW
Propagation loss	188	dB
Mobile Terminal G/T	−18	dB/K
Downlink C/N_0	47.6	dBHz
Total C/I	56	dBHz
Total C/N_0	47.0	dBHz
Req'd C/N_0	43	dBHz
Link margin	4.0	dB

The MobileSat nominally aims to support 50,000 users by the year 2000. The early users were mainly mining and exploration, emergency, marine, government, transportation, and support services operating in Australia's rural and remote areas.

7.2.6 OmniTracs and EutelTracs [13]

In the late 1980s, Qualcomm Inc. (joined by Omninet Communication Services) launched Omnitracs in North America to provide a spread spectrum two-way mobile messaging and RDSS via two Ku-band (14/12 GHz) satellites. The target market was the long-distance road haulage industry. Although the mobile service using Ku-band GEO satellites was a commercial pioneer, the high equipment cost and user charge, and the large terminal size, restricted market penetration. In 1991, this service was introduced to the European market as Euteltracs by using Ku-band transponders of the Eutelsat satellites, but it did not experience the success achieved in the United States. At the end of 1993, roughly 45,000 trucks were equipped with Qualcomm terminals, with 90% of sales being in the United States and only 10% in Europe.

The Euteltracs mobile satellite messaging and automatic satellite position reporting (ASPR) service is Europe's first commercially operated mobile satellite service. The organization, established in 1985, promotes satellite networks and services for TV and radio broadcasting, public telephony, videoconferencing, and mobile services.

7.2.7 European Mobile Satellite Services [14,15]

In Europe, ESA promoted land mobile satellite communications technologies via the PRODAT program, which conducted successful field trials with a low-data-rate terminal using the Marecs satellite. In parallel, two L-band mobile satellite communication payloads are being procured by ESA to promote European Mobile Satellite Services (EMSS): the European Mobile System (EMS) payload on the Italsat l-F2 satellite and the L-band Land Mobile (LLM) payload on the Artemis satellite. The EMS allows data and voice communications with a capacity of 300 channels. The available capacity is partially being used to demonstrate and evaluate emerging mobile satellite services. The EMSS operational phase started with the EMS payload in orbit in 1996, and is continuing with the LLM payload, which was planned to be launched in 1997 but will be delayed by a couple of years. The main feature of the LLM is the use of multiple beams, which is an efficient technique for the utilization of spacecraft resources through flexible adaptation to traffic demand.

7.2.7.1 EMS/Italsat l-F2

The EMS payload has an L-band (1.6 and 1.5 GHz) transponder for the user link and a Ku-band (14 and 12 GHz) transponder for the feeder link. The L-band antenna of the Italsat satellite, which is located at 10.2 degrees E in a geostationary orbit, has been designed to optimize the European coverage, and exhibits a minimum edge of coverage gain of 26 dBi. The total L-band EIRP is 42.5 dBW, L-band G/T is −2.0 dB/K, and Ku-band G/T is −1.4 dB/K. Figure 7.5(a) shows regional European beam coverage of EMS.

7.2.7.2 LLM/Artemis

The LLM payload of Artemis, which is a geostationary satellite at 16.4 degrees E, has two transponders for the same frequency bands as those in EMS. The

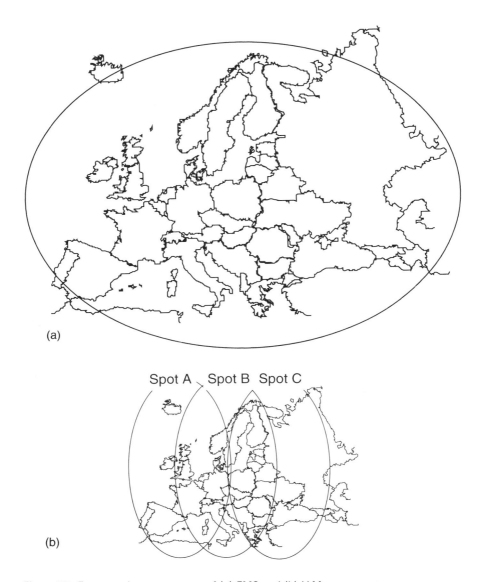

Figure 7.5 European beam coverage of (a) EMS and (b) LLM.

L-band antenna covers four areas: one broad beam covers Europe and North Africa, and three overlapping spot beams with a minimum edge of coverage gain of 29 dBi cover Europe (spot A on northwestern Europe, spot B on central Europe, and spot C on eastern Europe and Asia). Figure 7.5(b) shows the coverage of these three spot beams of the LLM. The available frequency bandwidth is 15 MHz, which is divided into three 4-MHz segments and three 1-MHz segments. The total L-band EIRP is 45.5 dBW in the spot beams and 43.5 dBW on the broadbeam coverage. The L-band G/T is 0 dB/K for the spot beam, -2 dB/K for the broad beam, and the Ku-band G/T is -1.4 dB/K.

The EMSS is partly based on ESA's PRODAT program and mobile satellite business network (MSBN). The features of compact and low-cost mobile terminals for the PRODAT system are as follows:

- TDM/BPSK 1,500 bps at reception;
- CDMA/OQPSK 600 bps at transmission;
- Adaptive ARQ-based block coding;
- A quadrifilar helical antenna 120 mm high or a printed antenna 90 mm in diameter and 20 mm thick;
- RF power of 10W.

The PRODAT was designed in accordance with the X400 ITU-T standard to establish messaging services compatible with terrestrial services. The MSBN was designed to provide voice and data transmission services between L-band mobile terminals and low-cost Ku-band VSAT stations using code division multiple access (CDMA). A quasi-synchronous CDMA (QSCDMA) technique has been developed to increase the maximum number of simultaneously operable channels in a given bandwidth. The main design feature is a decentralized satellite network architecture from fixed VSAT sites.

The EMSS provides an opportunity to demonstrate improved and new mobile satellite communications services. For example, an aeronautical satellite data link system for civil aviation is one candidate for applications of EMSS. A detailed design study is planned to follow, with prototypes being produced to demonstrate in-flight performance. EMSS users and applications include civil protection agencies, disaster response efforts, law enforcement, fire fighting and telemedicine.

7.2.8 Russian Marathon Mobile Satellite Service (MSS) Program [16]

For remote areas with a small population, such as Siberia, Kamchatka, and Tyumen in Russia, the present mobile satellite communication services are still inadequate. The Russian government is developing a fixed satellite communications system called Express, a mobile satellite communications system called Marathon, and a broadcasting satellite system called Gals.

The Marathon MSS system uses five geostationary Arcos satellites and four highly elliptically orbiting Mayak satellites to provide telephone, facsimile, telegraphy, telex, e-mail, and data services. The Arcos and Mayak satellites use the L-band frequencies of 1,631.5 to 1,660.5 MHz for user uplink and 1,530.0 to 1,559.0 MHz for user downlink with steerable multibeam phased array antennas, in which the beam shape can be changed in response to traffic requirements. The frequency of the feeder link is the C band at 6,355.0 to 6,420.0 MHz and 4,030.0 to 4,095.0 MHz. The Arcos satellites will be located at 40.0 degrees east, 90.5 degrees east, 145.5 degrees east, 160.0 degrees west, and 13.5 degrees west. The Mayak satellites have orbit inclination of 64.0 degrees and −84.5 degrees, and a period of 11.96 hours. It was announced that the Marathon system fully meets the requirements for INMARSAT-A, -B, -C, -M, and -Aero standards as well as those for the European land mobile satellite standard. Reliable communications links between mobiles and remote users are expected to make a major contribution to more efficient and safer transportation, continuous environmental monitoring, and the introduction of new advanced economic and production management concepts.

7.2.9 Other Proposed GEO Systems

Recently, several GEO systems, such as Asia Pacific Mobile Telecommunications, Agrani, and Garuda, have been proposed to provide personal satellite communication services using handheld terminals via large satellites. In this section, we briefly introduce these systems.

7.2.9.1 PCSat

Personal Communications Satellite Corp., AMSC's subsidiary corporation, filed an application with the FCC of the United States in April 1994. The PCSat system, which will be the second generation of the AMSC MSAT system, will provide both satellite- and ground-based cellular communication services using the same handsets over North America, operating in cooperation with terrestrial cellular companies by the year 1998. Two PCSat satellites, which will be launched in 1997 to 98 degrees west and 104 degrees west, carry S-band onboard antennas with more than 100 spot beams to provide personal communication services to about 2 million users.

7.2.9.2 CelSat

Celsat Inc. has proposed the CelSat system, which mainly provides North American coverage using three GEO satellites located at 78 degrees west, 89 degrees west, and 109.4 degrees west in cooperation with terrestrial cellular systems, and it filed an application with the FCC in April 1994. The CelSat system uses S-band frequencies of 1,970 to 1,990 MHz and 2,160 to 2,180 MHz with SS/CDMA. The CelSat satellites produce 146 multibeams from 20m diameter antennas and will be launched in 1997 to provide 88,000 voice channels.

7.2.9.3 APMT

Asia Pacific Mobile Telecommunications (APMT) system, which is planned by Singapore Telecom, Singapore Technology Venture Corp., and so forth, will provide personal satellite communication services using handheld terminals as well as land mobile satellite communications services. Two large GEO satellites will be launched in 1998 to provide coverage of East Asia and China using the L-band frequencies. The traffic capacity will be 16,000 simultaneous calls.

7.2.9.4 ASC/Agrani

The ASC system or Agrani system, which has been proposed by the India-based Afro-Asia Satellite Communications Corp, will start operations by the year 1998 using two Hughes HS-601 GEO satellites located at 52 degrees E and 46 degrees E. In 1995, Thailand-based Telecom Asia decided to invest. The system aims to improve the telecommunications infrastructure in India and Africa as well as mobile satellite communication services. The traffic capacity of 40,000 voice calls is provided by using an S-band onboard antenna 12m in diameter.

7.2.9.5 ACeS/Garuda

The Asia Cellular Satellite System (ACeS), which was developed by a consortium including Indonesia's PT Pasifik Satellite Nusantara, Philippine Long Distance Telephone Company, and Thailand's Jasmine International Public Corpora-tion, will provide voice, facsimile, and pager services to handheld and fixed terminals throughout Southeast Asia, India, and China in 1998. ACeS will allow the same telephone handsets to be used for both the ACeS satellite system and GSM-based cellular communications systems over Asia. Lockheed Martin has signed a $650 million contract to build the ACeS system, which includes two AB2100AX satellites. Each satellite will offer at least 11,000 circuits at S-band frequencies. The first ACeS satellite is planned to be launched in 1998. Three national gateways will be constructed in Indonesia, the Philippines, and Thailand to connect the ACeS system with local telephone systems and with network and satellite control centers.

7.3 Little LEO Satellite Systems [3]

The U.S. FCC has licensed three little LEO systems: Orbcomm (owned by Orbital Communications Corporation), Starsys (by GE Americom), and VITA (by a nonprofit group, Volunteers in Technical Assistance, Inc.) in the frequency bands of 137 to 138 MHz and 148 to 149.9 MHz [1]. In November 1994, five additional companies filed applications for licenses for new little LEO systems with the FCC (CTA, LeoOne, Final Analysis, E-Sat, and GE-Americom). These little LEO systems use the very/ultrahigh-frequency VHF/UHF band, because it allows user terminals to be made at lower cost since these frequencies are already used. The drawback is that these frequencies are heavily used worldwide for public and private services. Since there is not enough space to accommodate all the proposed little LEO systems within the limited frequency bands, a rulemaking process will be required to select suitable LEO systems. In the 1995 World Radiocommunication Conference (WRC), proposals to allocate additional frequencies for little LEOs were unsuccessful, except for a small amount of additional spectrum in the 400-MHz band. European systems include ITAMSAT (by Italy), Ariane Radioamateur Satellite Enseignement Espace (ARSENE, by CNES of France), and IRIS-LLMS (by ESA) [1].

Customers for little LEOs are the shipping and container industry for position determination of intermodal containers on trucks, ships, and railway cars, the utilities companies for reading meters and monitoring pipeline flow rates, rental vehicle companies for tracking stolen vehicles and locating misplaced ones, insurance companies for locating stolen vehicles, construction companies, environmental services for monitoring of processing plants, and the government for tracking valuable assets [3]. Table 7.7 shows the characteristics of little LEOs.

7.3.1 Orbcomm [17]

The Orbcomm system has been developed by Orbital Communications Corporation, a subsidiary of Orbital Sciences Corporation, in cooperation with Teleglobe Canada, Inc., to provide low-cost mobile two-way data and message communications. The first two satellites, Flight Model 1 and 2 (FM1, FM2), were successfully launched on April 3, 1995. In the first phase of operation, there is a constellation of 26 satellites, which includes 24 satellites in three orbital planes of eight satellites, each with an inclination of 45 degrees at 775 km altitude. FM1 and FM2 are orbiting in a 739-km circular orbit, inclined at 70 degrees. In the second phase, a fourth plane of satellites at a 45-degree inclination and a second polar orbit with two satellites will be added to increase coverage and capacity. The total number of satellites will eventually be 36. The launch vehicle will be the Pegasus XL.

Table 7.7
Characteristics of Little LEOs

Name	Owner	Country	No. of Satellites	Altitude (km)	Inclination (deg)	No. Orbit Planes	Weight (kg)	Battery Power (W)	Lifetime (y)	Launch	Project Budget (million$)	User Frequency (MHz)	Feeder Link Frequency (MHz)	Service	Multiple Access
Orbcomm	Orbital Comms Corp.	USA	32/4	775/739	45/70	4/2	38.5	160	4 to 6	1995–1997	155	148.905–159.9 (up), 137–138 (down)	149.61/137.56	Data/asset tracking, messaging	FDMA (user) TDMA (feeder link)
Stamet	Starsys Global Positioning, Inc.	USA	24	1,000	53	6	330	N/A	5	1998–2000	197	137.048–137.952, 148–148.905, 150–150.05	400.595–400.645	Messaging, emergency alert services, global RDSS	CDMA
VITASAT	Volunteers in Technical Assistance	USA	2	970	88	1	45	N/A	5	1996–1997	3	148.055–150.395 (up), 137.09–137.9 (down)	400.28–400.91	Disaster, medicine, education	FDMA
E-Sat	E-Sat, Inc.	USA	6	1,262	N/A	1	114	N/A	10	1996–2000	50	148.0–149.9 (up), 137.0–138.0 (down)	399–400.05, 400.15–401	Data	CDMA
Falsat	Final Analysis Communications Services, Inc.	USA	26	1,000	N/A	6	100	N/A	7 to 10	1996–2000	140	137 138, 148–150	400–401	Data acquisition, SCADA, messaging, tracking	FDMA
GE Americom or Eyesat	GE American Communications, Inc.	USA	24	800	98	4	15	N/A	5	1997–2001	190	148.0–149.9	399.9–400.05, 400.15–401	Asset tracking, remote communications	FDMA
GEMNET	CTA Commercial Systems, Inc.	USA	38	1,000	50/90	4/1	45.2	100	5 to 7	1997	183	37–138, 148–150	312–315, 387–390	Asset tracking, meter reading, e-mail	N/A
Leo One USA	LEO One USA Corp.	USA	48	950	50	8	125	N/A	5 to 7	1997–2000	225	148.0–150.05 (up), 137–138 (down)	400.15–401	Smart car, data	FDMA
Eyesat	Eyetel	USA	24	N/A	N/A	N/A	N/A	N/A	N/A	N/A	N/A	148–150.05	399.9–400.05, 400.15–401	Asset tracking, remote communications	N/A
GONETS	Smolsat	Russia	36	N/A	N/A	N/A	250	N/A	N/A	1996–1999	300	312–315, 390	N/A	Fax, e-mail, RDSS	N/A
LEO One Panamericana	LEO One Panamericana	Mexico	12	1,400	N/A	N/A	150	N/A	N/A	1996	N/A	137/148, 387–390, 312–315	N/A	Data, fax, RDSS, e-mail	N/A

Users in the temperate climate zones around the world will have a satellite continuously in view for over 98% of the time, and will have to wait less than two minutes to directly access a satellite the rest of the time. The system is designed to meet the following system performance objectives:

- 98% availability of satellites above 5 degrees in elevation angles;
- 99% of outages < 5 minutes, due to no available satellite being in view;
- 98% of outages < 2 minutes, due to no available satellite being in view.

The position and velocity data transmitted from onboard GPS receivers are used to determine the orbital elements for Orbcomm satellites. The position determination service for user terminals is also available by measuring Doppler frequency shifts of satellite signals. Full-time 24-hour communication services are not available, but the waiting time for access to a satellite is designed to be less than one minute in high-traffic areas. Uplink from the user is a 2,400-bps QPSK signal at 148 MHz, and downlink to the user is a 4,800-bps BPSK signal at 137 MHz. Orbcomm has proposed different types of user terminals, including portable, handheld, and mobile units.

The network control center is situated at the center of a star network of local gateway Earth stations and serves as the interconnection to the public data networks. The main function of the satellite control center is to monitor and control the Orbcomm satellites as well as to archive data, perform systems analysis, and generate reports.

7.3.2 Starsys

The Starsys system was scheduled to be operated from 1996 by Starsys Global Positioning Inc. under the investment from North American Collect Location Satellite Inc., GE Americom, and so forth. It consists of a constellation of 24 little LEO satellites 1,000 km in altitude, with an inclination of 53 degrees. Since both Starsys and Orbcomm use the same frequency bands of 150/140 MHz, the Starsys services will compete with the Orbcomm services, but Starsys is planning to lease their network capacity to network service providers.

7.3.3 VITA

VITA (Volunteers in Technical Assistance, Inc.) is a private, nonprofit organization engaged in international development. Since its establishment in 1959,

VITA has been committed to helping people in developing countries improve the quality of their lives through the collection, distillation, and transfer of information, and is the first organization to pursue the application of advanced microelectronics and space technology to the dissemination of information for development and humanitarian purposes. Information is also disseminated by VITA's weekly *Voice of America* radio program and its electronic newsletter. In order to facilitate the transfer of information, VITA has developed a global communication program, VITACOMM, which employs three low-cost communications technologies: VITASAT for a LEO satellite system, VITAPAC for a terrestrial digital radio system, and VITANET for an electronic message delivery system using existing telephone networks. VITACOMM allows global information resources, including a volunteer roster of over 5,000 experts in a variety of fields, to participate in timely dialogue for development. The projects address a variety of issues such as enterprise development, agriculture, rural rehabilitation, renewable energy, and protection of the environment.

In 1990, the first mission was installed on the UoSat-3 satellite, which was constructed and operated by Surrey Satellite Technology, Ltd., and has a polar, sun-synchronous orbit with perigee of 785 km and apogee of 800 km with a period of 100.8 minutes and an inclination of 98.7 degrees for sending messages of 50 to 250 pages in length. The duration of each pass is no more than 15 minutes at any location. In July 1991, the UoSat-5 (Healthsat-1) satellite was launched by an Ariane rocket. This satellite has a store-and-forward messaging capability with 9.6-Kbps FSK using VHF/UHF bands. In September 1993, the Healthsat-2 satellite, which also uses the VHF/UHF bands, was launched.

Since VITA failed to procure the building cost of about $3 million for the next VITA satellite, it decided to accept CTA's support in constructing the VITA satellite and share the satellite capacity fifty-fifty with CTA. However, due to the unsuccessful launch of the VITASAT-A by the LLV-1 rocket in July 1995, the start of the services will be delayed more than one year.

Table 7.7 lists several little LEO groups, including the five second-round applicants to the FCC license. As an example of a second-round applicant, the GEMnet system developed and operated by CTA Commercial Systems, Inc. is presented in the next section.

7.3.4 GEMnet [18]

The GEMnet (Global Electronics Message network) will provide global digital data communications services using a little LEO system that consists of a constellation of 38 LEO satellites. It is operated by two-way data CTA Commercial Systems, Inc. (a subsidiary of CTA, Inc.), which is one of the second-round applicants that filed after the three organizations, Orbcomm, Starsys, and VITA, which filed in the FCC's first round. It will start services in 1997.

The GEMnet constellation will consist of 38 satellites in low Earth orbits and two on-ground spares. The space segment constellation consists of four planes of eight satellites in 50-degree inclined circular orbits and one plane of six satellites in polar orbits to provide not only optimal coverage in the areas where demand is likely to be greatest, but also global coverage. All spacecraft altitudes are 1,000 km.

The GEMnet ground segment consists of user terminals, gateway stations, a network operations center, and a terrestrial network of dedicated links. Six major services that GEMnet will initially address were identified as 1) asset tracking and monitoring (GEMtrack), 2) utility/meter reading (GEMread), 3) e-mail (GEMmail), 4) global paging (GEMpage), 5) buoy/environmental sensor reading (GEMsense), and 6) direct-to-home communications services.

The required E_b/N_0 of 7 dB for a 10^{-5} bit error rate is met by the GEMnet systems, which use differentially encoded offset-QPSK (OQPSK) modulation with covolutional encoding having a rate of 7/8 and constraint length of 7. The uplink at 148 MHz consists of 11 data channels at 2.4 or 4.8 Kbps, one of which is reserved as the signaling channel with a modified reservation ALOHA protocol.

7.4 Big LEO Satellite Systems

Six main contenders for big LEO systems filed for Federal Communications Commission (FCC) licenses, in 1990 and 1991, to provide global mobile satellite services. These were Iridium by Motorola Satellite Communications, Odyssey by TRW, Globalstar by Loral/Qualcomm Partnership, Ellipso by Mobile Communications Holdings, Aries by Constellation Communications, and AMSC by AMSC Subsidiary Corporation. The big LEO systems provide a wide variety of mobile services, such as voice, data, and facsimile transmission; position location; paging; search and rescue; disaster management; environmental monitoring; cargo tracking; and industrial monitoring. On January 31, 1995, three systems, Iridium, Globalstar, and Odyssey, were awarded licenses by the FCC to operate in the United States. Decisions on the other systems were deferred for one year to allow them to establish financing, which is a critical consideration in the FCC approval process. Moreover, ICO Global Communications, formerly known as INMARSAT-P Affiliate Company, also proposed an MEO (ICO) system. AMSC did not initially file for a big LEO license, opting to defer filing its financial qualifications with the FCC until January 31, 1996.

Industry experts feel that only two or three of these systems can survive past the year 2000, because of the limited amount of financing to support each of these proposed systems, which range in cost from 800 million to $3.37 billion each [3]. The characteristics of big LEOs are shown in Table 7.8.

7.4.1 Iridium [19–22]

The Iridium system, which was proposed, developed, and marketed by Motorola Corporation, is being financed by a private international consortium of telecommunications and industrial companies, and will become operational in 1998. The Iridium network consists of a constellation of 66 satellites, 785 km in altitude, with 6 polar orbital planes inclined 86.4 degrees, each containing 11 satellites. The resulting orbital period is approximately 100 minutes.

To a user terminal stand anywhere on the Earth, a satellite is always visible at least 8 degrees above the horizon. Each satellite uses three L-band antennas to cover the ground with 48 beams, and the diameter of each spot beam is about 600 km. The 66 satellites provide 3,168 cells, of which only 2,150 need to be active to cover the whole surface of the Earth. As the spacecraft moves, the user encounters adjacent beams about once a minute. Adjacent beams use different frequencies, but nonadjacent beams can reuse the frequency already used by another beam. The allocated frequency is divided into 12 subbands, and each subband is reused four times on a single satellite. Since at high northern or southern latitudes some of the outer beams on each satellite are not used, 2,150 spot beams are actually active to cover the globe, so the frequency reuse factor is 2,150/12 = 180. The system is designed for each spot beam to support 80 channels, so the channel capacity worldwide is 2,150 × 80 = 172,000 channels. Uplink and downlink frequencies are identically allocated in the range from 1,610 to 1,626.5 MHz. Using 50-Kbps time division multiple access (TDMA) bursts in uplink and downlink, 4,800-bps voice or 2,400-bps data, full duplex communications services are available [22].

The Iridium system uses only a one-way communication link at a time, which is known as time-duplexing, and the user terminals can rapidly switch modes between receive and transmit. The use of one set of frequencies for uplink and downlink simplifies the user's hardware.

Ka-band intersatellite links with four cross-links on each satellites (front, back, and two in adjacent orbits) provide reliable, high-speed communications between neighboring satellites and connect a subscriber to a gateway station via various possible paths. This flexibility improves call delivery efficiency and system reliability. The network configuration of the Iridium system is shown in Figure 7.6.

Table 7.8
Characteristics of Big LEOs

Name	Owner	Country	No. of Satellites	Altitude (km)	Period (min)	Inclination (deg)	No. Orbit Planes	Weight (kg)	Battery Power (W)	Life Time (yr)	Launch
Constellation	Constellation Communications Inc.	USA	35/11	2,000/2,000	N/A	62/0	7/1	200	N/A	7	1997
Ellipso	Mobile Communications Holdings, Inc.	USA	10/6	7,846*520/ 8,040	N/A	116.5/0	2/1	300	N/A	5	1997
Globalstar	Loral/Qualcomm (LQSS)	USA	48 + 8 spares	1,414	113	52	8	426	1,000	7.5	1997
ICO	ICO Global Communications Ltd.	UK	12	10,355	360	45	2	1,600	5,000	10–12	1998
Iridium	Motorola Inc./ Iridium Inc.	USA	66 + 6 spares	785	100.13	86.4	6	689	1,200	8	1997
Odyssey	TRW Inc.	USA	12	10,354	359.53	55	3	1,334	1,800	15	1997

Table 7.8 (Continued)
Characteristics of Big LEOs

Name	Project Budget (billion$)	User Frequency (MHz)	Feeder Link Frequency (MHz)	Service	Multiple Access	Repeater Type	Inter-satellite Link	Max. Data Rate (Kbps)	Terminal Cost ($)	Rate	No. of Spot Beams
Constellation	1.7	1,610–1,626.5 2,483.5–2,500	5,050–5,225, 6,825–7,025	Voice/data/ fax/paging RDSS	CDMA	Bent pipe	No	9.6	1,500	$30/mon	32
Ellipso	1.1	1,610–1,626.5 2,483.5–2,500	15.4–15.7 GHz, 6.725–7.025 GHz	Voice/data/ RDSS	CDMA	Bent pipe	No	9.6	1,000	$0.5/min, $35/mon.	61
Globalstar	1.9	1,610–1,626.5, 2,483.5–2,500	5,091–5,250, 6,875–7,055	Voice/data/ fax/paging RDSS	CDMA	Bent pipe	No	9.6	750	$0.30– 0.35/min	16
ICO	2.6	1,980–2,010, 2,170–2,200	5,100–5,250, 6,925–7,075	Voice/data	TDMA	Bent pipe	No	2.4	1,500	$2/min	85–150
Iridium	3.37	1,616–1,626.5	29.1–29.3 19.4–19.6 GHz	Voice/data/ fax/paging	TDD/TDMA/ FDMA	Processing	Yes Ka band	2.4	2,000–3,000	$3/min	48
Odyssey	1.35	1,610–1,621.35 2,483.5–2,500	29.1–29.4, 19.3–19.6 GHz	Voice/data/ fax/paging/ messaging	CDMA	Bent pipe	No	9.6	500–600	$1/min	37

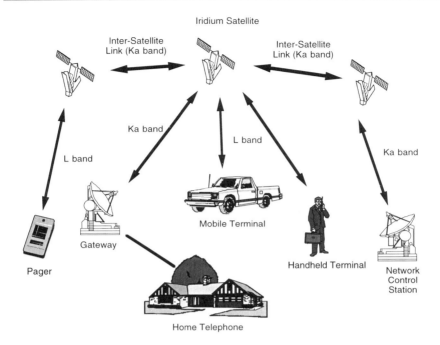

Figure 7.6 Network configuration of the Iridium system.

The constellation of satellites is controlled by the system control segment. Connection to and from the public switched telephone network is achieved through gateway stations, which use high-gain K-band parabolic antennas to track Iridium satellites. An Iridium subscriber unit is a handheld satellite telephone and a message termination device is a pocket-size pager to receive a short message. Each satellite provides 1,100 voice channels with high quality of service. The Iridium call-processing architecture is based on the GSM standard, which is a popular terrestrial digital cellular standard, although the gateway station needs unique management due to the satellite constellations.

The industrial consortium Iridium has signed a $3.4 billion contract to purchase the Iridium space system from Motorola's satellite communications division. Lockheed Corporation designs and constructs the satellite bus and Raytheon corporation designs the phased array antenna for communication between the satellite and the user terminals. The Canadian firm, COM DEV, is responsible for the antennas for intersatellite and gateway links. Telesat, Siemens, Telespazio, and Bechtel are among other key element suppliers of the system. It will cost up to $3 per minute to use the link.

The satellites will be launched by Proton rockets of Khrunichev Enterprise of the Russian Federation, Long March 2C rockets of China Great Wall Industry Corporation, and Delta II rockets of McDonnell Douglas.

The initial Iridium concept was developed in 1987, and its spectrum was allocated in the 1992 World Administration Radio Conference (WARC). By 1994, Iridium, Inc., had all its equity financing in place. On July 19, 1993, Motorola Satellite Communications Division awarded Lockheed Missiles & Space Company (LMSC) a contract valued at more than $700 million to design, develop, and manufacture 125 satellite buses for the Iridium system [21].

7.4.2 Globalstar [23–25]

The Globalstar system uses CDMA with an efficient power control technique, multiple-beam active phased array antennas for frequency reuse, variable rate voice encoding, multisatellite path diversity, and soft handoff beams and satellites. Overall responsibility for the Globalstar system development rests with Loral Qualcomm Partnership L. P.

The Globalstar system has a constellation of 48 satellites in 8 planes with 6 satellites per plane inclined at 52 degrees, which was chosen to provide 100% coverage from 70 degrees south to 70 degrees north. The orbital altitude is 1,414 km and the orbital period is 114 minutes. GPS receivers are installed on the satellites to provide position and attitude control to maintain orbit and to furnish a frequency reference.

The first group of satellites will be launched in mid 1997, interim service will begin in mid 1998, and full services will be started in 1999. The first four satellites will be launched via a Delta rocket from Cape Canaveral, and the subsequent launches will use the Delta, Ukrainian Zenit, and Chinese Long March rockets with up to 12 satellites per launch.

The user link uses the L band of 1,610 to 1,626.5 MHz for uplink and the S band of 2,483.5 to 2,500 MHz for downlink. The feeder link uses the C band of 5 GHz for uplink and 7 GHz for downlink. The L- and S-band satellite antennas are active phased array antennas to divide the user coverage into 16 beams that collectively fill the 5,760-km-diameter circle on the Earth visible to an individual satellite. The channel capacity of one satellite is 2,148 links. The satellite transponders are not regenerative but simple *bent pipe* types without intersatellite capability. The main features are described in more detail below.

7.4.2.1 CDMA With Efficient Power Control

The multiple access scheme for the Globalstar system is modified Qualcomm's CDMA, which is codified in the U.S. EIA/71A IS-95 CDMA standard. In multibeam satellite systems, CDMA can achieve more effective frequency reuse than FDMA or TDMA, since the typical required beam isolation is as low as 2 or 3 dB for CDMA, compared to 18 dB or more for FDMA or TDMA.

As a result of using closed-loop and open-loop dynamic power control techniques for both the gateway stations and the user terminals, all signals arrive at the CDMA receiver at about the same level.

7.4.2.2   Active Phased Array Antennas

Active phased array antennas are installed on the satellite with low-noise amplifiers (LNAs) and monolithic microwave integrated circuit (MMIC) solid-state high-power amplifiers (HPAs) attached at each antenna element. The antennas have 91 radiating elements for transmission in the S band and 61 elements for reception in the L band. The 91 HPAs provide an output power of 4.2W and a gain of about 50 dB, and the 61 LNAs have a noise figure of 1.6 dB and a gain of about 40 dB.

7.4.2.3   Variable Rate Voice Encoding

When speech activity is low, the data transmission rate is reduced, and the transmitting power is also reduced. Four different data rates of voice encoding are available: 1,200, 2,400, 4,800, and 9,600 bps. When no speech activity is detected, the data rate is reduced to 1,200 bps. Preliminary results suggest that the Globalstar system may not require the 9,600-bps rate, but will only need an average rate of 2,400 bps. In addition to voice communications, the Globalstar system will provide position determination to within 300m, paging, and messaging services.

7.4.2.4   Path Diversity to Enhance Communications Reliability

Since the same CDMA signals received at a user terminal from different satellites can be separately demodulated by the user terminal, they can be combined to improve the signal-to-noise ratio. This feature is called path diversity. When buildings, trees, or other obstructions block the direct path between a user and an available satellite, the user's receiver uses the signals from other satellites. This diversity can reduce the possibility of loss of communication paths and improve the communications reliability. A receiver that can track several signals is called a rake receiver.

7.4.2.5   Soft Handoff Between Beams and Satellites

When the user terminal is moving between beams or between satellites, the handoff may take place every two to four minutes or faster. In the Globalstar CDMA system, the gateway measures the return power from a user terminal in an adjacent beam, and then begins to transmit a time-shifted copy of the same signal in a new beam. This enables the rake receiver in the user terminal to perform a soft handoff seamlessly.

7.4.3 Odyssey [26]

TRW filed the MEO system known as Odyssey with the FCC on May 31, 1991, to provide high-quality voice, data, facsimile, paging, and RDSS services with a constellation of 12 satellites orbiting in three 55-degree inclined planes at an altitude of 10,354 km, and established Odyssey Worldwide Service Ltd. with Canada's Teleglobe and so forth. In January 1995, the Odyssey was licensed from the FCC. The Odyssey satellites employ dynamically steerable multibeam antennas and bent-pipe transponders with the spread spectrum CDMA techniques.

The MEO satellite system has several advantages over the LEO system. The in-orbit lifetime of a MEO, 15 years for Odyssey, is longer than that of a LEO, which is 5 to 7 years. Continuous services to several regions can be started with only nine satellites, and the addition of only three satellites ensures that two or three satellites are visible at all times. This will ensure that service can be provided in a short time to market.

User terminals are designed to use dual modes for terrestrial and satellite communications services. If terrestrial cellular service is not available, then the call is routed through the Odyssey system. User circuits established through an Odyssey satellite enter the public switched telephone network at a ground station located within the region served by the satellite. Since the satellite cell coverage, whose diameter is typically 800 km, is relatively large, a user will remain within the same cell area during a call. This precludes the need for handover.

The user frequency is the L band of 1,610 to 1,626.5 (or 1,621.35) MHz for uplink and the S band of 2,483.5 to 2,500 MHz for downlink. The Ka band is used for the feeder links. The gateway station connects between the Odyssey link and the public switched telephone network, and each gateway station is equipped with four 10-ft tracking antennas, which are separated by 30 km. Three antennas simultaneously communicate with as many satellites. The fourth antenna provides diversity during heavy rainfall, since rain cells are typically much smaller than 30 km in diameter.

The capacity of each Odyssey satellite is approximately 2,300 voice circuits. Since the 12 satellites can provide dual satellite coverage of any region, the system capacity for any given region will be 4,600 voice circuits, and 2.3 million subscribers worldwide will be supported.

The cost of the Odyssey service is based on $0.65 per minute, and its main targets are corporate and government users, business travelers, residents of sparsely populated regions, and citizens of nations that lack a communications infrastructure. Launches are scheduled to begin in 1997, and initial operations will start with six satellites in 1998.

7.4.4 Constellation/Aries [3]

Constellation Communications Inc., which was established by investment from CTA/DSI Inc., Pacific Communications Sciences, Inc., and International Microspace, Inc., will develop a constellation known as the "Aries" system. This constellation network will consist of 48 satellites: 12 satellites in each of four polar orbital planes at 1,020 km to provide voice, data, facsimile, and RDSS. Each satellite has seven antenna beams to provide global coverage. User links use L-band and S-band frequencies, and feeder links use the C band of 6 and 5 GHz. The constellation is scheduled to begin full operation in 1997.

7.4.5 Ellipso [27]

Ellipsat Corp. filed the first application with the FCC for a mobile satellite communications system that provides voice services using elliptically orbiting small satellites. Now, the Ellipso mobile satellite system is planned by Mobile Communications Holdings, Inc. to reach its subscribers with terrestrial cellular telephone service at a competitive price (i.e., around U.S. $0.5 per minute and U.S. $35 per month).

Since the land and population of the Earth are asymmetrically distributed between the northern and southern hemispheres, the constellation of the Ellipso satellites is designed to offer efficient coverage. For example, the southern hemisphere contains much less land mass at high latitudes than the northern hemisphere. Europe, the United States, Canada, and Japan lie at 40 degrees north, while southern New Zealand, Tasmania, Argentina, and Chile lie at 40 degrees south, and all the areas further south of 40 degrees south are relatively sparsely populated.

The Ellipso system uses two complementary and coordinated constellations of satellites, Ellipso-Borealis and Ellipso-Concordia.

The Ellipso-Borealis uses 10 satellites in elliptical orbits in two planes inclined at 116.5 degrees to prevent movement of the apogee around the orbital path. The apogee and the perigee are 7,846 km and 520 km, respectively, with a 3-hour orbital period. The Borealis orbits provide northern service for a greater percentage of their orbits. Moreover, the orbits are carefully configured to be sun-synchronous and to optimize the time of day at which the greatest satellite coverage occurs.

The Ellipso-Concordia uses six satellites in an circular equatorial orbit at an altitude of 8,040 km. This constellation provide continuous coverage of all tropical latitudes plus all temperate latitudes to 47 degrees south. Figure 7.7 shows the initial Ellipso constellations: two inclined Borealis orbits and the equatorial orbital plane of the Concordia.

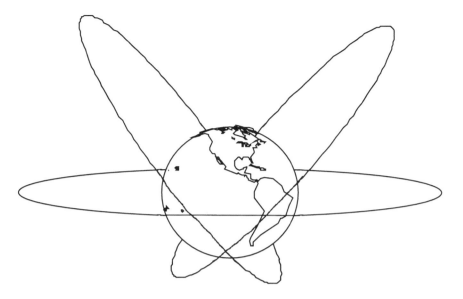

Figure 7.7 Ellipso constellations: Borealis and Concordia orbits.

The Ellipso satellite is three-axis stabilized and its user-link antennas divide the visible Earth into 61 adjoining areas. The 61 beams cover the same 12 MHz within the MSS band. Therefore, users do not need to switch frequencies as beams or satellites serving them change during a call. The user signals in different beams are translated to a different frequency and polarization on the feeder link between the satellites and the ground station.

There will be three types of Ellipso user terminals: mobile or portable, fixed, and handheld. Both the mobile and handheld terminals use antennas having hemispheric coverage, and the fixed terminals use a directional antenna. The Ellipso terminals will be dual-mode terminals for operating selectively in the Ellipso system and the terrestrial cellular system, depending on service availability and the price of calls. The Ellipso system uses digital voice coding provided by code-excited linear prediction (CELP) at 4.15 Kbps and supports data services such as Hayes modem data, facsimile, message forwarding, paging, and position location information at 300 to 9,600 bps. CDMA is used to permit simultaneous access by many Ellipso users. Compared to FDMA and TDMA, the CDMA system can tolerate interference from neighboring beams or satellites, and can use all available time and frequency resources in every beam and in every satellite. The handoffs from beam to beam and from satellite to satellite do not require any frequency or time adjustment at the user terminal, which are known as soft handoffs.

Path diversity is also available. The ground control station will combine all received signals, which are transmitted from a user terminal and passed through different paths, to create a composite signal. If the signal over one path fades due to shadowing or blockage, the ground control station will automatically make adjustments by using the signal propagating along another path. Moreover, the CDMA signals have better resistance to multipath fading, and provide communications security and privacy, because they are difficult to detect without knowledge of the user's spectrum spreading sequence.

7.4.6　ICO [3]

Since 1996, INMARSAT has studied global personal satellite communications system using handheld terminals, and has proposed the INMARSAT-P system as Project 21 using 12 satellites operating in an intermediate circular orbit (ICO) at an altitude of 10,355 km. In January 1995, ICO Global Communications Ltd. was established as a private company to provide personal/mobile satellite communications services, and it will start full commercial services by the year 2000. Its goal is to bring to the global marketplace PCS-class services using pocket-sized handheld terminals.

The main ICO services are high-quality voice with dual modes of cellular and satellite, duplex data at 2.4 Kbps or higher, G3 facsimile, global paging with indoor coverage, and personal navigation services. Since it is very costly to provide large link margins for good-quality voice communications in heavily blocked environments such as inside buildings, the ICO provides a high-margin call alerting function and a very powerful paging service to extend the coverage inside buildings. The handheld terminal has a volume of less than 300 cc and costs less than $1,500 with dual-mode operation. It is designed based on a worst case average transmission power of 0.25W, averaged over six minutes. This power level is lower than or similar to that of cellular and cordless phones. The current plan is to use a transmitting frequency of 1,610 to 1,626.5 MHz and a receiving frequency of 2,483.5 to 2,500 MHz, but this could change if the FPLMTS band becomes available globally before the year 2000. The ICO satellites links with terrestrial networks through ICO's ICONET, which consists of 12 Earth stations or satellite access nodes, which also link with gateways that serve as the primary interface with public switched telephone and mobile networks.

The ICO space segment consists of ten operational satellites and two in-orbit spares at an altitude of 10,354 km. Two orbital planes, inclined 45 degrees to the equator, are used to provide continuous overlapping coverage of the ground. The footprint diameter on the ground with minimum elevation angle of 20 degrees is 10,850 km and the maximum footprint pass time is approximately 97 minutes. Using a digital onboard processor and TDMA, each satellite can handle 4,500 simultaneous telephone calls. The satellites will be launched by Atlas IIA, Delta III, Proton, and Zenit Sea Launch launchers. The first launch is scheduled for 1998.

7.5 Broadband Systems

Optical-fiber networks will provide various advanced multimedia services such as audio, video, and high-resolution graphics. Such advanced networks will soon cover urban areas, but most of the population living in rural and remote areas lacking adequate telecommunications infrastructure will not have the opportunity to participate in such advanced communications because of the high cost of extending the networks. However, using broadband and high-data-rate global satellite systems, which will provide seamless compatibility with terrestrial fiber networks, multimedia services can be extended to remote areas. Tables 7.9 and 7.10 shows the characteristics of several examples of broadband LEO and GEO systems, respectively, which mostly use the Ka band for user service links.

Such new-generation broadband satellites for high-data-rate service in the Ka band of 30/20 GHz have recently attracted great interest in the satellite communication industry, and in the United States, applications for seven global and seven regional broadband satellite systems were submitted to the FCC by September 1995. Applicants included big players such as AT&T for VoiceSpan, Lockheed Martin for Astrolink, Loral for Cyberstar, GE America for GE*Star, Motorola for Millenium, and Hughes for Spaceway. In this section, we present several broadband satellite systems.

7.5.1 Teledesic Satellite System [28]

Teledesic plans to provide fiber-like global broadband services in the Ka band of 20 and 30 GHz using a constellation of 924 LEO satellites orbiting at altitudes between 695 and 705 km. (Recently, the constellation was modified to that of 288 LEO satellites orbiting at altitudes of 1,400 km, but in this text we describe the previous configuration.) There are 21 circular orbital planes at a sun-synchronous inclination of approximately 98.16 degrees with adjacent ascending nodes spaced at 9.5 degrees around the equator. This sun-synchronous orbit allows significant savings in solar power arrays and allows some of the satellite's electronics to be cooled. Each plane contains a minimum of 40 operational satellites plus up to 4 in-orbit spares spaced evenly around the orbit. The satellites in adjacent planes travel in the same direction except at

Table 7.9
Characteristics of Broadband LEO Systems

System Name	Operator	Country	Service	Number of Satellites	Altitude (km)	Inclination (deg)	Number of Orbit Planes	Weight (kg)	Launch	User Frequency (GHz)
Teledesic Network	Teledesic Corporation	USA	Fixed, Voice/data/video	288	1,400	98.142–98.182	21	795	2,000	27.5–30 17.8–18.6 18.8–20.2
M-star System	Motorola Satellite Systems, Inc.	USA	Fixed, Voice/data up to 51.84 Mbps	72	1,350	47	12	2,210 lbs.	2,000	47.2–50.2 (up) 37.5–40.5 (down) 59–64 (intersatellite)
SkyBridge	Alcatel Espace	France	Fixed, Broadband data up to 60 Mbps/video	64	1,457	N/A	N/A	700	2,001–2,002	Ku band

Table 7.10
Characteristics of Broadband GEO Systems—
Ka-band Satellite Applications Accepted for Filing by the FCC of the United States by September 1996

Name	Owner	No. of Satellites	Uplink Frequency	Downlink Frequency	Intersatellite Link Frequency	Coverage	Services
VoiceSpan	AT&T Corporation	12 GEOs (7 locations)	29.0–30.0 GHz (international) 28.35–28.6, 29.25–30.0 GHz (domestic U.S.)	19.2–20.2 GHz	59–64 GHz	Global coverage	Electronic messaging and mailboxes, multimedia bridging, software distribution, and voice
Millenium	Comm, Inc.	4 GEOs	28.35–28.60 GHz for service links, 29.5–30.0 GHz for uplinks	18.55–18.80 GHz for service links, 19.70–20.20 GHz for downlinks	59.5–60.5 GHz, 62.5–63.5 GHz	The Western Hemisphere	Video, image, fax, audio, and computer data
EchoStar	EchoStar Satellite Corporation	2 GEOs	29.5–30.0 GHz	19.7–20.2 GHz	59 GHz and 61 GHz	Full CONUS	High-speed switched data, video, and video telephone
GE*Star	GE American Communications, Inc.	9 GEOs (at 5 locations)	28.35–28.6 GHz, 29.25–30.0 GHz	18.55–18.80 GHz, 19.45–20.2 GHz	N/A	Global coverage	High-speed digital communications, video, audio, videotelephony, videoconferencing
Galaxy	Hughes Communications Galaxy, Inc.	20 GEOs (at 15 locations)	2.5 GHz band in the Ka band	2.5 GHz band in the Ka band	22.55–23.55, 32.0–33.0, 54.25–58.25, 59–64 GHz	Global coverage	Two-way interactive services, direct-to-home and other forms of video distribution
KaStar	KaStar Satellite Communications Corp.	2 GEOs	29.0–29.5 GHz, 29.5–30.0 GHz	19.2–19.7 GHz, 19.7–20.0 GHz	61.9–62.3 GHz	CONUS, Alaska, and Hawaii	High-speed switched data, video, and video telephone
Astrolink	Lockheed Martin Corporation	9 GEOs (5 locations)	29.5–30.0 GHz for service links, 28.35–28.6, 29.25–29.5 GHz for gateways	19.7–20.2 GHz for service links, 18.55–18.8, 19.45–19.7 GHz for gateways	54.25–58.2, 59.0–64.0 GHz	Global coverage	Voice, data, and video transmission

Table 7.10 (Continued)
Characteristics of Broadband GEO Systems—
Ka-band Satellite Applications Accepted for Filing by the FCC of the United States by September 1996

Name	Owner	No. of Satellites	Uplink Frequency	Downlink Frequency	Intersatellite Link Frequency	Coverage	Services
CyberStar	Loral Aerospace Holdings, Inc.	3 GEOs	28.35–28.6 GHz, 29.5–30.0 GHz	18.95–20.20 GHz	60 GHz	North America, Asia, Europe	Video telephony, videoconferencing, medical and technical teleimaging, and high-data-rate computer transmission
Morning Star	Morning Star Satellite Co., L.L.C.	4 GEOs	30 GHz	20 GHz	N/A	The Western U.S., Canada, Hawaii and Alaska, Asia, Europe	N/A
NetSat 28	NetSat 28	1 GEO	28.35–28.6, 29.25–29.5 GHz	18.55–18.8, 19.45–20.2 GHz	N/A	CONUS	N/A
Orion F-6	Orion Asia Pacific Corporation	1 GEO	Ka band	Ka band	N/A	Asia Pacific region	High-speed data transmissions
Orion F-2	Orion Atlantic, L.P.	1 GEO	Ka band	Ka band	N/A	North America, South America, and West Africa	High-speed data transmissions
Orion F-2, F-8, and F-9	Orion Network Systems, Inc.	3 GEOs	28.35–28.60, 29.25–29.5, 29.5–30.0 GHz	18.55–18.80, 19.45–19.7, 19.70–20.20 GHz	N/A	CONUS, Alaska, Hawaii, Puerto Rico, Virgin Islands, Southern Africa, Australia, Southeast Asia, China	Video, audio, voice, and multimedia services
PAS-10 and PAS-11	PanAmSat Corporation	2 GEOs	Ka band	Ka band	N/A	The U.S., Latin America, Europe, and Western Africa	Video programming, data, voice
VisionStar	VisionStar, Inc.	1 GEO	28.35–28.6, 29.25–29.5, 29.5–30.0 GHz	18.55–18.8, 19.45–19.7, 19.7–20.2 GHz	N/A	The United States	Video programming, interactive services, distance learning, video conferencing

the constellation seams, where ascending and descending portions of the orbits overlap. Teledesic Corporation submitted an application to the FCC in March 1994.

The satellite footprint consists of contiguous cells, which is analogous to a terrestrial cellular system. High-gain satellite antennas produce small basic cells of 53.3 km^2. The Earth's surface is covered by approximately 20,000 super cells, each consisting of 9 basic cells. The footprint sweeps over the Earth's surface at approximately 25,000 km/hr. As a satellite passes overhead, it steers its antenna beams to the fixed cell locations within its footprint. This beam steering compensates for the satellite's motion as well as the Earth's rotation.

The Teledesic network uses fast packet-switching technology similar to the asynchronous transfer mode (ATM) technology. All communication links transport data and voice as fixed-length (512 bits) packets. Each packet contains a header for address and sequence information, an error-control section used to verify the integrity of the header, and a payload section that carries digital audio, video, and data. The fast packet-switching network combines the advantage of a circuit-switched network and a packet-switched network. The basic unit of channel capacity is the basic channel, which supports a 16-Kbps payload data rate and an associated 2-Kbps D channel for signaling and control. Basic channels can be aggregated to support higher data rates. For example, eight basic channels can be aggregated to support the equivalent of a 2B+D ISDN link.

The Teledesic network provides a quality of service comparable to terrestrial communication systems, including fiber-like delays, bit error rates less than 10^{-10}, and a link availability of 99.9% over most of the United States. The system provides 24-hour seamless coverage to over 95% of the Earth's surface and almost 100% of the Earth's population. All configurations operate at multiples of the 16-Kbps basic channel rate up to 2.048 Mbps. The terminals use antennas with a diameter from 16 cm (8 cm for mobile) to 1.8m with a transmission power of 0.01 to 4.7W, depending on antenna size, transmission rate, and so forth. Within its service area, each satellite can support 100,000 simultaneous basic channels. The Teledesic system also supports a smaller number of fixed-site GigaLink terminals that operate at the OC-3 rate (155.52 Mbps) and multiples of that rate up to OC-24 (1.24416 Gbps). Each satellite can support up to 16 GigaLink terminals within its service area.

Intersatellite links, which use the 60-GHz band, operate at the OC-3 rate and multiples of this rate up to OC-24, depending on the instantaneous capacity requirement.

The satellite antennas are advanced active-element phased array systems using GaAs MMIC amplifiers and beam-steering circuits, which provide dynamic control of gain, beam shape, and power level. The octagonal baseplate also supports eight pairs of intersatellite link antennas.

The Teledesic network will provide "fiber-like" service quality, including low transmission delay, high data rates and low bit error rates, to fixed and mobile users around the world starting in 2001. Figure 7.8 shows Teledesic's network configuration.

7.5.2 M-Star

Motorola, who developed the Iridium system, has started a new broadband satellite project, called M-Star, to provide voice, data, and video services using 72 LEO satellites. M-Star is expected to take four years from getting FCC approval and $6.1 billion to construct a 72-satellite constellation with data rates of about 1 Gbps for optical intersatellite links and 155 Mbps for satellite-to-ground transmission using millimeter waves of 59 to 64 GHz. The M-Star's service coverage is not worldwide, but is between ±57 degrees of latitude, to extend broadband wireless connectivity to mobile users.

7.5.3 Broadband GEO Systems

Table 7.10 shows characteristics of broadband GEO systems, which were submitted to the FCC by September 1995 [29].

AT&T VoiceSpan Digital Satellite Services use 12 satellites on seven orbital slots: one satellite will be located at 103 degrees west for North America,

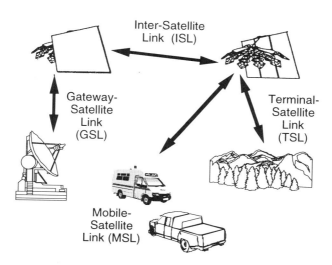

Figure 7.8 Teledesic's LEO network configuration.

two will be located at 93 degrees west for the United States, one will serve the Caribbean and South America at 54 degrees west, two will cover Europe at 42 degrees west, two will cover Africa at 1 degree west, two will cover Asia and India at 92 degrees west, and two will be located at 116 degrees west for Asia and Australia. The first satellite will be launched in 2000, and full deployment will occur in 2003.

Motorola proposed the Millenium system as a GEO system as well as the M-Star system as a LEO system to provide global broadband services. Motorola submitted an application of the Millenium to the FCC, under the name of Comm. Inc., to provide broadband capacity to third-party retailers in the North, Central, and South American markets. GE American Communications Inc. has prepared GE*Star system, which uses nine GEO satellites on five orbital positions of 17 degrees east, 56 degrees east, 114.5 degrees east, 106 degrees west, and 82 degrees west. The first satellite, which will be launched in 2000, will begin to provide the service in the North American market, and the service will be expanded as each satellite is launched.

Astrolink, which was proposed by Lockheed Martin Telecommunication, consists of nine GEO satellites on five orbital positions to provide global broadband services. CyberStar was proposed by Loral Space and Communications Ltd. to provide global services with one satellite over America at 110 degrees west, one over Europe at 28 degrees east, and one over Asia at 105.5 degrees east. The first satellite is planned to be launched in 1999. Spaceway was proposed by Hughes Communications Inc. (HCI), which was the first proposed Ka-band GEO system. HCI has expanded the number of satellites from the original proposal, and now 15 orbital positions are proposed. These positions will be used for both the Spaceway and Galaxy constellation, which will be operated by HCI and PanAmSat, respectively. PanAmSat Corporation was formed in May 1997 by the merger of the previous PanAmSat Corporation and the Galaxy Satellite Services division of HCI.

These Ka-band satellite systems were recently proposed in the United States, but for several countries in Asia and Europe the Ka-band system is not new. In Japan, the first experimental satellite CS (Sakura), which was launched in 1977, installed six Ka-band transponders. The succeeding series CS-2, CS-3 and N-Star also carry the Ka-band transponders to provide satellite communication services to the fixed Earth stations. Another satellite series using Ka-band are the Superbird satellites, which are operated by Space Communications Corporation (SCC). As experimental satellites, ETS-VI (Engineering Test Satellite Six), which was launched in 1994 but could not reach the geostationary orbit due to failure of an apogee motor, and COMETS (Communications Engineering Test Satellite), which will be launched in 1997, have Ka-band transponders as well as millimeter-wave transponders.

7.6 HEO Systems

While geostationary satellites cannot provide mobile satellite communications service with good elevation angles in high-latitude areas, HEO satellites may provide a solution to this problem. Table 7.11 shows the characteristics of HEO systems.

7.6.1 Archimedes [30,31]

The Archimedes satellite constellation is based on multiregional highly inclined elliptical orbits (M-HEO) to provide good coverage of the densely populated northern hemisphere. The disadvantages of the HEO system are that several satellites are needed for continuous coverage, satellite antenna pointing, Doppler compensation, and satellite handover. The M-HEO constellation consists of six satellites in different orbital planes, each having an 8-hr period, with an operational orbit period of 4 hours with an eccentricity of 0.63. The apogee and perigee are 26,786 km and 1,000 km, respectively. During a 24-hour period, one satellite can successively serve three zones spaced 120 degrees apart in longitude. The M-HEO system has the advantage over geostationary satellite systems of allowing a high elevation angle for Earth stations located in a northern latitude.

The difference between the M-HEO orbit and the Molnyia orbits is that the period of the Molnyia is 12 hours, and the number of required satellites is three or four. The M-HEO has lower apogees, and hence shorter path delay and smaller losses.

Each satellite has a 3- to 5-beam coverage pattern. The beams are highly overlapped over the European region and are repointed slowly to compensate for the satellite motion. The services that have been considered in the Archimedes system are not only digital audio broadcasting (DAB), but also personal (to handheld phones and notebook-sized terminals) and mobile (to vehicle-mounted terminals) voice communications. The option to provide interactive multimedia entertainment and information broadcasting is currently considered, but the services have yet to be fully defined [32].

Table 7.11
Characteristics of HEO Systems

Name	Owner	Country	No. of Satellites	Altitude (km)	Latitude	Inclination (deg)	Period (hours)	Weight (kg)	Battery Power (W)	Life Time (yrs)	Launch	User Frequency (MHz)	Feeder Link Frequency (MHz)	Service	Multiple Access
Marathon - Arcos	Informcosmos stock company lim.	Russia	5	GEO (35800)	40.0° E, 90.5° E, 145.5° E, 160.0° W, 135.5° W	0	0	N/A	N/A	7	1996 or 1997	L band 1,631.5–1,660.5/ 1,530.0–1,559.0	C band 6,355.0–6,420.0/ 4,030.0–4,095.0	Voice/data/ fax/ telegraphy/ telex/e-mail/ data	N/A
Marathon - Mayak	Informcosmos stock company lim.	Russia	4	HEO	—	64.0, –84.5	11.96	N/A	N/A	7	1996 or 1997	1,631.5–1,660.5 (up), 1,530–1,559 (down)	6,355–6,420 (up), 4,030–4,095 (down)	Voice/data/ fax/ telegraphy/ telex/e-mail/ data	N/A
Archimedes	ESA	Europe	6	26,800–1,000	—	63.4	8	1050	1470	10	1998	S band or/and L band	Ku band	Voice/data/ fax/DBS	FDMA

References

[1]   Stern, J. A., "Small satellites: worldwide developments and trends in communications and remote sensing," *Proc. Space and Navigation Electronics (SANE) Committee of IEICE,* Nagoya, Japan, Oct. 1996, pp. 23–49.

[2]   Abrishamkar, F., and Z. Siveski, "PCS global mobile satellite," *IEEE Communications Magazine,* Vol. 34, No. 9, Sept. 1996, pp. 132–136.

[3]   Stahl, P., S. Ali, and R. Hoy, "Low earth orbit (LEO) communications constellations and their impact on the commercial space industry—a U.S. perspective," *Proc. AIAA/ ESA Workshop on Int. Cooperation in Satellite Communications,* Noordwijk, The Netherlands, March 27–29, 1995, pp. 255–263.

[4]   Ananasso, F., and F. D. Priscoli, "The role of satellites in personal communication services," *IEEE J. of Selected Areas in Communications,* Vol. 13, No. 2, Feb. 1995, pp. 180–196.

[5]   Hart, N., "Mobile satellite system design," *Satellite Communications—Mobile and Fixed Services,* Miller, M. J., B. Vucetic, and L. Berry, (eds.). MA: Kluwer Academic Publishers, Ch. 3, 1993.

[6]   Corbley, K. P., "INMARSAT—mobile satellite communications from the battlefield and beyond," *Via Satellite,* Nov. 1995, pp. 22–28.

[7]   Sengupta, J. R. "Evolution of the INMARSAT aeronautical system: service, system, and business considerations," *Proc. 4th Int. Mobile Satellite Conf., IMSC'95,* Ottawa, Canada, June 6–8, 1995, pp. 245–249.

[8]   Levin, L. C., and D. C. Nash, "U.S. Domestic and international regulatory issues," *Proc. 3rd Int. Mobile Satellite Conf.,* IMSC'93, Pasadena, CA, June 16–18, 1993, pp. 67–72.

[9]   Pedersen, A., "MSAT wide-area fleet management: end-user requirements and applications," *Proc. 4th Int. Mobile Satellite Conf., IMSC'95,* Ottawa, Canada, June 6–8, 1995, pp. 158–163.

[10]   Johanson, G. A., "MSAT satellite/cellular integration and implications for future systems," *Proc. AIAA/ESA Workshop on Int. Cooperation in Satellite Communications,* Noordwijk, The Netherlands, March 27–29, 1995, pp. 155–161.

[11]   Harrison, S., "MobileSat®—the world's first domestic land mobile satellite system," *Space Communications,* Vol. 13, No. 3, 1995, pp. 249–256.

[12]   Wagg, M., and M. Jansen, "MobileSat® a characteristically Australian MSS," *Proc. 4th Int. Mobile Satellite Conf., IMSC'95,* Ottawa, Canada June 6–8, 1995, pp. 404–408.

[13]   Colcy, J.-N., and L. Vandebrouck, "New applications for the Euteltracs service," *Proc. 4th Int. Mobile Satellite Conf., IMSC'95,* Ottawa, Canada, June 6–8, 1995, pp. 391–396.

[14]   Loisy, C., P. Edin, and F. J. Benedicto, "European mobile satellite services (EMSS) a regional system for Europe," *Proc. 4th Int. Mobile Satellite Conf., IMSC'95,* Ottawa, Canada June 6–8, 1995, pp. 545–550.

[15]   Miracapillo, L., T. Sassorossi, R. Giubilei, "The L-band land mobile payload (LLM) aboard Artemis," *Proc. 16th Int. Communications Satellite Systems Conf.,* AIAA, Washington, DC, Feb. 25–29, 1996, pp. 879–887.

[16]   Tsyrlin, I. S., and V. B. Tamarkin, "Russian mobile satellite communication program," *Proc. 4th Int. Mobile Satellite Conf., IMSC'95,* Ottawa, Canada, June 6–8, 1995, pp. 536–538.

[17]   Yarbrough, P. G., "Operations concept for the world's first commercially licensed low-earth-orbiting mobile satellite service," *Proc. 16th Int. Communications Satellite Systems Conf.*, Feb. 25–29, 1996, Washington, DC, pp. 524–532.

[18]   Yi, B. K., et al., "The GEMnet global data communication," *Proc. 4th Int. Mobile Satellite Conf., IMSC'95*, Ottawa, Canada, June 6–8, 1995, pp. 179–184.

[19]   Brunt, P., "IRIDIUM®—overview and status," *Space Communications*, Vol. 14, 1996, pp. 61–68.

[20]   Hutcheson, J., and M. Laurin, "Network flexibility of the IRIDIUM global mobile satellite system," *Proc. 4th Int. Mobile Satellite Conf., IMSC'95*, Ottawa, Canada, June 6–8, 1995, pp. 503–507.

[21]   Tadano, T. N., "IRIDIUM®—a Lockheed transition to commercial space," *Proc. 4th Int. Mobile Satellite Conf., IMSC'95*, Ottawa, Canada, June 6–8, 1995, pp. 90–95.

[22]   Bell, D., P. Estabrook, and R. Romer, "Global tracking and inventory of military hardware via LEO satellite: a system approach and likely scenario," *Proc. 4th Int. Mobile Satellite Conf., IMSC'95*, Ottawa, Canada, June 6–8, 1995, pp. 204–211.

[23]   Hirshfield, E., "The Globalstar system: breakthroughs in efficiency in microwave and signal processing technology," *Space Communications*, Vol. 14, 1996, pp. 69–82.

[24]   Smith D., "Operations innovations for the 48-satellite Globalstar constellation," *Proc. 16th Int. Communications Satellite Systems Conf.*, AIAA, Washington, DC, Feb. 25–29, 1996, pp. 537–542.

[25]   Schindall, J., "Concept and implementation of the Globalstar mobile satellite system," *Proc. 4th Int. Mobile Satellite Conf., IMSC'95*, Ottawa, Canada, June 6–8, 1995, pp. A-11–A-16.

[26]   Spitzer, C. J., "Odyssey personal communications satellite system," *Proc. 4th Int. Mobile Satellite Conf., IMSC'93*, Pasadena, CA, June 16–18, 1993, pp. 297–302.

[27]   Castiel, D., and J. E. Draim, "The Ellipso™ mobile satellite system," *Proc. 4th Int. Mobile Satellite Conf.*, IMSC'95, Ottawa, Canada, June 6–8, 1995, pp. 409–418.

[28]   Sturza, M. A., "Architecture of the Teledesic satellite system," *Proc. 4th Int. Mobile Satellite Conf., IMSC'95*, Ottawa, Canada, June 6–8, 1995, pp. 212–218.

[29]   Fernandez, R., "Internet in the sky," *Via Satellite*, March 1997, pp. 52–66.

[30]   Galligan, K. P., R. Viola, and C. Paynter, "Evolution of DAB by satellite to meet the multi-media challenge," *Proc. 4th Int. Mobile Satellite Conf., IMSC'95*, Ottawa, Canada, June 6–8, 1995, pp. 433–438.

[31]   Paynter, C., and M. Cuchanski, "System and antenna design considerations for highly elliptical orbits as applied to the proposed Archimedes constellation," *Proc. 4th Int. Mobile Satellite Conf., IMSC'95*, Ottawa, Canada, June 6–8, 1995, pp. 236–241.

[32]   Stojkovic, I., and J. E. Alonso, "Key design issues for user terminals operating with HEO satellites," *Proc. 4th Int. Mobile Satellite Conf., IMSC'95*, Ottawa, Canada, June 6–8, 1995, pp. 370–375.

8

Mobile Earth Stations

8.1 Overview

This chapter is focused mainly on the mobile Earth stations used for the International Maritime Satellite Organization (INMARSAT) system. Mobile Earth stations used in other systems, however (and some terminals that are still in R&D stages), are also introduced. The INMARSAT provides worldwide mobile satellite communication services for maritime, land, and aeronautical users. The INMARSAT mobile satellite communication system consists of a satellite, a gateway Earth station, a network coordination station (NCS), an operations control center (OCC), and a mobile Earth station (MES). A gateway Earth station is called a coast Earth station (CES) in maritime systems and is called a ground Earth station (GES) in aeronautical systems. The INMARSAT-A is the most basic system and has provided analog voice (FM) and telex services since the inauguration of the INMARSAT in 1982. The INMARSAT-C has been providing message (nonvoice) and low-speed data services since 1991, and the INMARSAT-B and -M systems have been providing digital communications since the second generations were introduced in 1993. The M and C systems have been used not only by maritime users but also by land mobile users such as long-haul trucks and trains. This section mainly describes the ship Earth stations of the INMARSAT-A, B, C, aero, and M systems. Table 8.1 lists the main characteristics of these systems.

8.2 INMARSAT-A

8.2.1 System Configuration

An INMARSAT-A terminal, called a ship Earth station (SES), is usually installed on a relatively large ship. As shown in Figure 8.1, a ship Earth station consists

Table 8.1
Main Characteristics of the INMARSAT Systems

System	System-A	System-B	System-C	System-Aero	System-M
Main service for	Large ship	Large ship	Very small ship Land mobile	Aircraft	Small ship Land mobile
Start of service	1982	1993	1991	1990	1993
Fittings (July 1995)	25,124	938	18,642		
Service	Voice/telex	Voice/telex/data	Message/data	Voice/data	Voice/data
G/T	−4 dBK	−4 dBK	−23 dBK	−13 dBK (high) −26 dBK (low)	−10 dBK (sea) −12 dBK (land)
EIRP	36 dBW	33 dBW	16 dBW	25.5 dBW (high) 13.5 dBW (low)	27 dBW
Antenna gain	20–24 dBi	20–24 dBi	0–3 dBi	12 dBi (high) 0 dBi (low)	14 dBi 12 dBi
Antenna type	Parabola	Parabola	Helical, cross-dipole	Phased array	Array, short backfire
Antenna weight	About 110 kg	About 110 kg	About 0.5 kg	About 30 kg	About 40 kg
Terminal weight	About 40 kg	About 40 kg	About 5 kg	About 40 kg	About 15 kg
Voice	FM	O-QPSK (24 Kbps)	Message only	A-QPSK (21 Kbps)	O-QPSK (8 Kbps)

Table 8.1 (Continued)
Main Characteristics of the INMARSAT Systems

System	System-A	System-B	System-C	System-Aero	System-M
Modulation					
Data	BPSK Rx/1.2 Kbps Tx/4.8 Kbps	0-QPSK (24 Kbps) BPSK (6 Kbps)	BPSK Rx/1.2 Kbps Tx/1.2 Kbps	A-BPSK (0.6/1.2/2.4 Kbps)	BPSK (6 Kbps)
Voice CODEC	None	APC-MLQ (16 Kbps)	None	Multipulse-LPC (9.6 Kbps)	IMBE (6.4 Kbps)
Coding					
FEC coding	None	SCPC/Punctured/ Viterbi ($R = 3/4$, $K = 7$) TDM, TDMA/Convol. Viterbi ($R = 1/2$, $K = 7$)	Interleaved, Convolutional/ Viterbi ($R = 1/2$, $K = 7$)	Convolutional/ Viterbi ($R = 1/2$, $K = 7$)	Punctured/ Viterbi ($R = 3/4$, $K = 7$)

APC-MLQ: Adaptive predictive coding with maximum likelihood quantization   IMBE: Improved multi-band excitation

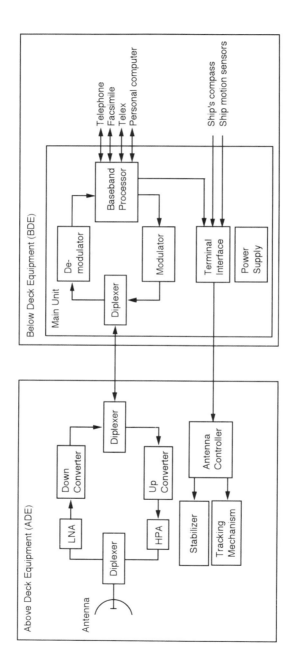

Figure 8.1 Block diagram of basic configuration of a ship Earth station.

of mainly two parts: above-deck equipment (ADE), and below-deck equipment (BDE). Figure 8.2 is a photograph of the ADE covered with a radome that is installed on a ship. Figure 8.3 shows an antenna unit inside a radome, which protects the antenna unit from severe environmental conditions. The ADE consists of an antenna, an antenna mount, a low-noise amplifier (LNA), a high-power amplifier (HPA), a diplexer (DIP), a stabilizer, and an antenna controller. In many cases, parabolic-type antennas have been used because of their good electrical and mechanical characteristics. Such components as the LNA, HPA, DIP, and antenna controller are attached on the antenna mount in order to get a weight distribution suitable for good tracking performance. The specifications of an SES of the system have been defined in the technical requirements for standard-A SESs of the INMARSAT [1]. Table 8.2 lists the main characteristics of an SES for the INMARSAT-A system. The most important requirement is that the figure of merit (G/T) be over -4 dBK. The gain and diameter of an antenna satisfying this requirement are typically about 24 dBi and 1m.

Figure 8.4 is a photograph of the below-deck equipment (BDE) installed in a ship cabin: a telephone handset, a display terminal, a printer, a facsimile,

Figure 8.2 Photograph of ADE installed on a ship. An antenna unit is covered with a radome. (Courtesy of Anritsu.)

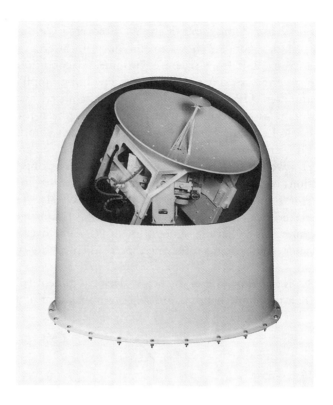

Figure 8.3 An antenna unit inside a radome. (Courtesy of Anritsu.)

a computer, and a main unit. The main unit consists of such components as a diplexer, a modulator, a demodulator, a baseband processor, terminal interfaces, and a power supply. Peripherals such as a telephone, a telex, a facsimile, and a personal computer are connected to baseband processors. Signals from the ship's gyro compass and from several sensors are input to a terminal interface in order to give the antenna controller information about ship motions.

8.2.2 Shipborne Antenna and Tracking Systems

Antennas for the INMARSAT-A and -B ship Earth stations are typically parabolic antennas with gains of 20 to 24 dBi because such an antenna has a simple structure and a high aperture efficiency. Satellite tracking is an essential capability because of ship motions and the small half-power beamwidth: about 10 degrees (see Chapter 4). The four-axis (X-Y-Az-El) stabilizer is the one most commonly used: a fixed horizontal plane is obtained by controlling motion

Table 8.2

Main Characteristics of an SES for the INMARSAT-A System

INMARSAT-A						
Frequency	1636.5–1645.0 MHz	SES → CES		SES: ship Earth station		
	1535.0–1543.5 MHz	SES ← CES		CES: coast Earth station		
Polarization	Right-hand circular					
Antenna gain	Nominal 24 dBi					
EIRP	36 dBW	HPA ~ 40W				
G/T	≧ −4 dBK	LNA noise temperature ~ 100K				

	Type	Service	Access	Link	Information rate	Channel rate	Modulation
Channel types	Request	Channel request	Preassign RA SCPC/FDMA	SES → CES	4.8 Kbps	4.8 Kbps	BPSK
	Assignment	Channel assignment	TDM	SES ← CES	1.2 Kbps	1.2 Kbps	BPSK
	Telex	Telex	TDM	SES ← CES	50 baud	1.2 Kbps	BPSK
			TDMA	SES → CES	50 baud	4.8 Kbps	BPSK
	Voice	Telephone	FDM/FDMA	SES ⇆ CES	Analog	Analog	SCPC/FM

Environmental conditions	Temperature	ADE: −35°C ~ 55°C, BDE : 0°C ~ 45°C	
	ship motion	Roll ±30-degree Cycle (8 sec)	Surge ±0.2G
		Pitch ±10-degree Cycle (6 sec)	Sway ±0.2G
		Yaw ± 8-degree Cycle (50 sec)	Heave ±0.5G
		Turn speed 6°/sec	Speed 30 knot

(ADE : Above-deck equipment)
(BDE : Below-deck equipment)

Figure 8.4 Photograph of BDE installed in a ship cabin. (Courtesy of Anritsu.)

about the X and Y axes, and an antenna system installed on the X-Y plane can be stabilized by controlling Az and El axes. The antenna on the stabilized pedestal, the antenna is directed to the satellite by controlling the Az and El axes. This type of X-Y axis stabilizer needs antenna pedestal control circuits with servo motors to control the axes and needs some sensors (such as an accelerometer, a rate sensor, and a level sensor) to provide information about ship motions. Figure 8.5 shows a block diagram of an antenna pedestal control unit for one axis (the same kind of unit is also used for the other axes). The unit consists of a servo amplifier, an axis driver, and a CPU. A step drives each axis in order to coincide with the angle that is calculated and output by the CPU. The accuracy of the controlled angle is with 0.1 degree.

A flywheel stabilizer has sometimes been used in order to avoid the need for antenna control sensors and electric circuits. This kind of stabilizer makes use of the inertial force generated by one or two rapidly rotating flywheels. Although these four-axis stabilizers are easy to control, stable, and reliable, they are relatively complex, large, and heavy. A three-axis stabilizer was therefore

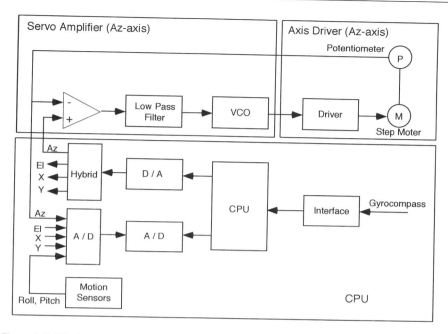

Figure 8.5 Block diagram of an antenna pedestal control circuit.

developed, and it has been used in some recent INMARSAT-A and -B systems [2]. Figure 8.6 shows a schematic configuration of a three-axis-stabilized antenna system for the INMARSAT-A SES.

The present INMARSAT-A and -B SESs have generally used a closed-loop system for satellite tracking because of their simple configuration. In maritime satellite communications in which a ship is using a high-gain antenna, the received signal is very stable because the high-gain antenna has so narrow a half-power beamwidth (as little as 10 degrees) to avoid sea-reflection signals by which sea reflection fading is occurred. The most popular open-loop tracking is a step-track method, which drives an antenna in elevation and azimuth directions alternatively by a step angle of 0.5 degrees in such a way as to keep the received signal level as high as possible.

8.2.3 Communication Channels

As summarized in Table 8.2, the SES has four types of communication links such as a request, an assignment, a telex, and a voice channel. A procedure of voice communications, for example, is explained by a sequence chart of voice communications like that shown in Figure 8.7. When an SES makes a phone call, a request signal is transmitted from an SES to a CES in one of two request

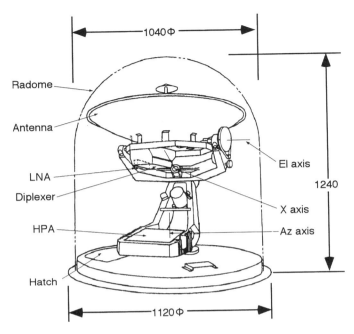

Figure 8.6 Three-axis-stabilized antenna system for the INMARSAT-A ship earth station.

channels in a random access (RA) procedure. Then a RA channel is TDMA/FDMA as there are two frequency channels, each of which has a TDMA random access protocol. On receiving the request signal, the CES transmits a request for a channel assignment to the NCS in an assignment TDM channel. The NCS assigns the channel both to the SES and the CES in an assignment TDM channel. On receiving the channel, the SES and CES establish the voice communications in a single channel per carrier (SCPC) and a frequency division multiplexing (FDMA) channel (SCPC/FDMA). When the voice communication is over, the CES transmits a channel release signal to the NCS in a TDM channel. In the case of telex communications, the sequence is the same except that the time slot is assigned by a CES.

8.2.3.1 Request Channels

There are two frequencies that are preassigned to request channels. A ship will request a channel assignment at random in a request channel with a transmission rate of 4.8 Kbps in a BPSK modulation. This access method is called a preassign random access (RA) single channel per carrier/frequency division multiple access (SCPC/FDMA). Figure 8.8 shows a frame format of a request channel. A length of a frame is 38.5 msec with a transmission speed of a bit rate of

Inmarsat-A Voice

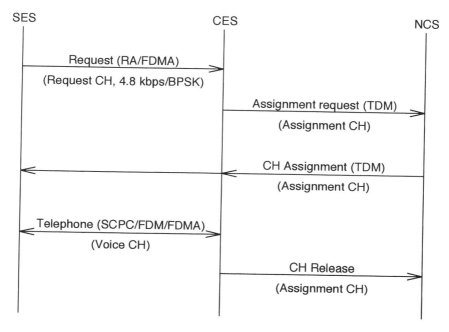

Figure 8.7 Sequence chart of a ship-originated telephone call of the INMARSAT-A system.

4.8 Kbps. As shown in , a request signal contains seven kinds of information: an ID number of a CES, a priority of communication type code, a request type code, a channel type code, a domestic network code, an ID number of an SES, and an ocean area code.

8.2.3.2 Assignment Channels

An assignment signal is transmitted in a time division multiplex (TDM) channel with a BPSK modulation. Figure 8.9 shows a frame format of an assignment channel. A length of a frame is 290 msec with a transmission speed of a bit rate of 1.2 Kbps. As shown in Figure 8.9, this channel is also used to transmit 22 telex channels. An assignment signal contains two frames, one of which is used among an SES, a CES, and an NCS, and the other is used between a CES and an NCS. The first frame contains six kinds of information: an ID number of an SES, a message type code, a channel type code, a time slot, a channel frequency, and a TDM frequency. The second frame contains such information as a priority code and an ID number of a CES.

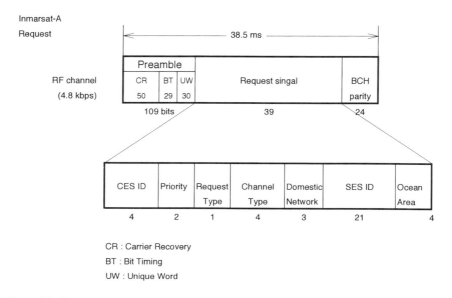

CR : Carrier Recovery
BT : Bit Timing
UW : Unique Word

Figure 8.8 Request channel frame format of the INMARSAT-A SES.

8.2.3.3 Telex Channels

A telex signal from a CES to an SES is transmitted in an assignment channel as described above. A telex signal from an SES to a CES is transmitted in a TDMA channel with a BPSK modulation. Figure 8.10 shows a frame format of a return link telex channel. A time duration of one frame of a bit rate of 4.8-Kbps transmission is 1,740 msec. One frame consists of 22 bursts, and each burst is received by a designated CES with a time interval of 1,740 msec. Each burst contains a preamble and a telex signal of 12 characters, and each burst has a length of 37.7 msec. However, a time slot of 79.1 msec is allocated to the burst in order to have a guard time of about 20 msec between adjacent bursts. A guard time is needed in order to avoid burst conflict due to propagation time differences for the signals from ships in different locations.

8.3 INMARSAT-B

8.3.1 System Configuration

The INMARSAT-B system was introduced in 1993 in order to provide high quality, frequency- and power-saving communications by using the digital communication technologies. The technical requirements of the SES of the system has been defined in the system definition manual (SDM) of the INMAR-

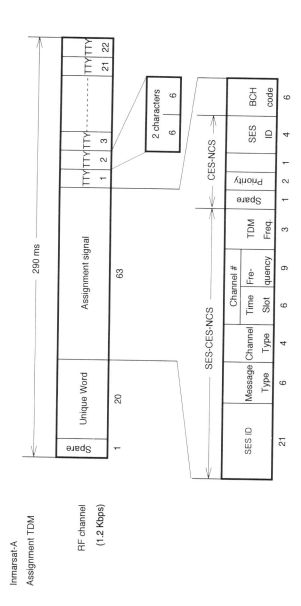

Figure 8.9 Assignment channel format of the INMARSAT-A CES.

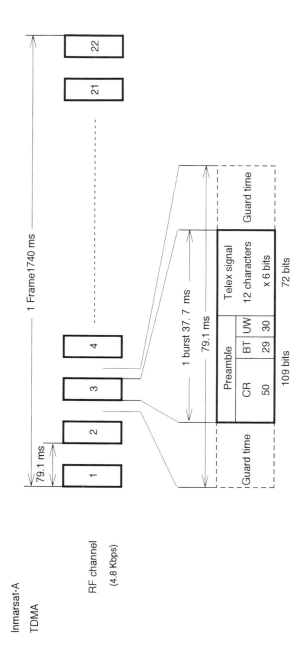

Figure 8.10 Telex channel format from an SES to a CES of the INMARSAT-A system.

SAT [3]. Table 8.3 lists the main characteristics of an SES for INMARSAT-B systems. Except for its using digital communication technologies in modulation and voice codecs, the SES of the INMARSAT-B is basically the same as that of the INMARSAT-A. The required *G/T* is larger than −4 dBK, which is the same as that of the -A system. The effective isotropically radiated power (EIRP) has been reduced from 36 dBW to 33 dBW and can be controlled by 3-dB steps in order to adapt to spot beams of the future third-generation satellites.

The digital modulations that have been used are 24-Kbps offset QPSK (O-QPSK) and 6-Kbps BPSK. Two forward error correction (FEC) methods have been used for TDM/TDMA channels, a convolutional coding and Viterbi decoding (rate $R = 1/2$, constrain length $K = 7$), and punctured coding for SCPC channels ($R = 3/4$, $K = 7$). Voice is transmitted after digital processing by the voice codec, developed by KDD [4], which is named as an adaptive predictive coding with maximum likelihood quantization (APC-MLQ). Figure 8.11 shows a block diagram of voice channels of below-deck equipment for the INMARSAT-B SES.

8.3.2 Antenna and Tracking Systems

The configuration and performance of above-deck equipment are, as mentioned in Section 8.2.2, basically the same as that of the INMARSAT-A SES.

8.3.3 Communication Channels

Table 8.3 shows that the SES has four types of communication channels: request, assignment, telex/low-speed data, and voice/high-speed data. A procedure of voice communications is explained by a sequence chart of voice communications, as shown in Figure 8.12. When the SES makes a phone call, a request signal is transmitted from the SES to the CES in a TDMA channel. On receiving the request signal, the CES transmits a request for a channel assignment to the NCS in an assignment TDM channel. The NCS assigns the channel to both the SES and the CES in an assignment TDM channel. On receiving the channel assignment, the SES and CES establish the voice communications in an SCPC/FDMA scheme. When the communication is over, the CES transmits a channel release signal to the NCS in a TDM channel. In the case of data and telex communications, the sequence is almost the same as that for voice communications.

8.3.3.1 Request Channels

A ship will request a channel assignment in a random access channel with a transmission rate of 24 Kbps in an O-QPSK modulation. Figure 8.13 shows

Table 8.3
Main Characteristics of an SES for the INMARSAT-B System

INMARSAT-B

Frequency	1626.5–1654.5 MHz	SES → CES		SES: ship Earth station	
	1525.0–1545.0 MHz	SES ← CES		CES: coast Earth station	
Polarization	Right-hand circular				
Antenna gain	Nominal 24 dBi	Diameter ~ 80 cm			
EIRP	33, 29, 25 dBW	HPA ~ 30W			
G/T	≧ −4 dBK	LNA noise temperature ~ 100K			

	Type	Service	Link	Information rate	Channel rate	Modulation	FEC
Request	Random Acess	Channel request	SES → CES		24 Kbps	O-QPSK	Convol./Viterbi
	Access						(R = 1/2, K = 7)
Assignment	TDM	Channel assignment	SES ← CES		6 Kbps	O-QPSK	(R = 1/2, K = 7)
	SCPC	Telephone	SES ⇆ CES	16 Kbps	24 Kbps	O-QPSK	Punctured/Viterbi (R = 3/4, K = 7)

(Coding: APC-MLQ)

Table 8.3 (Continued)
Main Characteristics of an SES for the INMARSAT-B System

INMARSAT-B

Voice/data	SCPC	Facsimile	SES ⇆ CES	9.6 Kbps	24 Kbps	O-QPSK	
	SCPC	Data	SES ⇆ CES	9.6 Kbps	24 Kbps	O-QPSK	
	TDM/TDMA	Data		300 bps	6/24 Kbps	BPSK	Convolutional/
	TDM	Telex	SES ⇆ CES	50 baud	6 Kbps	BPSK	Viterbi
	TDMA	Telex		50 baud	24 Kbps	O-QPSK	$(R = 1/2,\ K = 7)$
			SES ← CES				
			SES → CES				

Environmental conditions	Temperature	ADE: −35°C ~ 55°C, BDE : 0°C ~ 45°C		(ADE : Above-deck equipment)
	ship motion	Roll ±30° Cycle (8 sec)	Surge ±0.2G	(BDE : Below-deck equipment)
		Pitch ±10° Cycle (6 sec)	Sway ±0.2G	
		Yaw ± 8° Cycle (50 sec)	Heave ±0.5G	
		Turn speed 6°/sec	Speed 30 knot	

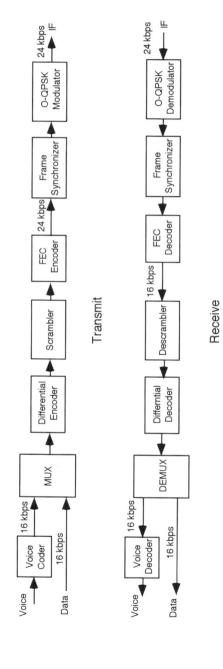

Figure 8.11 Basic block diagram of BDE for the INMARSAT-B SES.

Inmarsat-B Voice

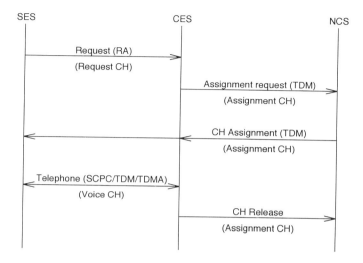

Figure 8.12 Sequence chart of a ship-originated telephone call of the INMARSAT-B system.

Inmarsat-B

Request

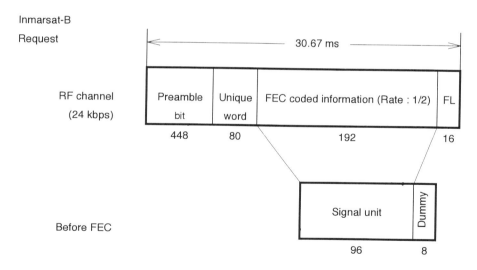

Figure 8.13 Request channel frame format of the INMARSAT-B SES.

a frame format of a request channel. The length of a frame is 30.67 msec, and its transmission bit rate is 24 Kbps. In order to achieve FEC, information data is coded by a convolutional coding with a rate of 1/2 ($R = 1/2$) and constrain length of 7 ($K = 7$).

8.3.3.2 Assignment Channels

From the CES, an assignment signal is transmitted in a TDM channel by a BPSK modulation at a transmission rate of 6 Kbps. Figure 8.14 shows a frame format of an assignment channel. The length of the frame is 264 msec with a channel rate of 6 Kbps.

8.3.3.3 Telex/300-bps Data Channels

A telex/300-bps data channel from a CES to an SES is transmitted in a TDM channel with a channel rate of 6 Kbps with a BPSK modulation. A telex/300-bps data signal from an SES to a CES is transmitted in a TDMA channel with a transmission rate of 24 Kbps with an O-QPSK modulation. Figure 8.15 shows a frame format of a telex/low-speed data channel transmitted from a CES to an SES. The length of one frame is 264 msec, in which the transmission rate is 6 Kbps. One frame consists of eight time slots, and each time slot is used by a low-speed data rate of 300 bps or eight multiplexed telex signals. Figure 8.16 shows a telex/low-speed data channel format in a TDMA channel with a channel rate of 24 Kbps. In a TDMA channel, two kinds of bursts are used for telex and low-speed-data transmission. The length of each burst, including a guard time of about 40 msec, is 148.5 msec.

8.3.3.4 Voice and 9.6-Kbps Data

Voice and 9.6-Kbps-data signals are transmitted in SCPC channels with a channel rate of 24 Kbps in O-QPSK modulation. Figure 8.17(a,b) shows

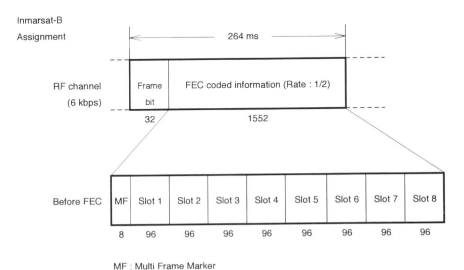

MF : Multi Frame Marker

Figure 8.14 Assignment channel format of the INMARSAT-B CES.

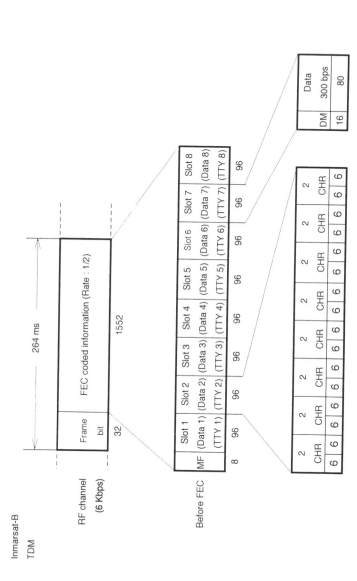

Figure 8.15 Telex channel format from a CES to an SES of the INMARSAT-B system (in a TDM channel with a channel rate of 6 Kbps with a BPSK modulation).

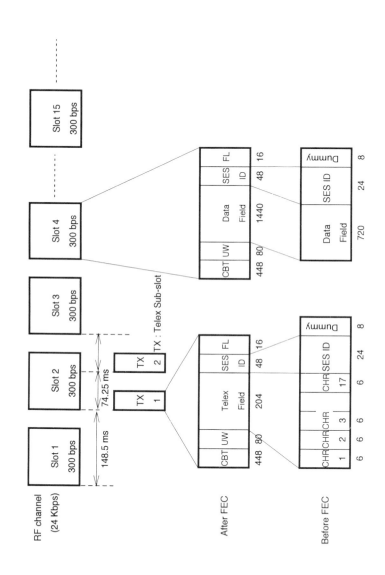

Figure 8.16 Telex channel format from an SES to a CES of the INMARSAT-B system (in a TDMA channel with a channel rate of 24 Kbps with a 0-QPSK modulation).

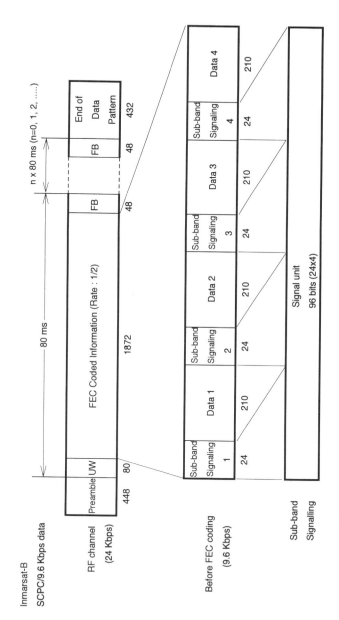

Figure 8.17 (a) 9.6-Kbps data and (b) voice channel formats of the INMARSAT-B SES. These are SCPC channels with a channel rate of 24 Kbps with an O-QPSK modulation.

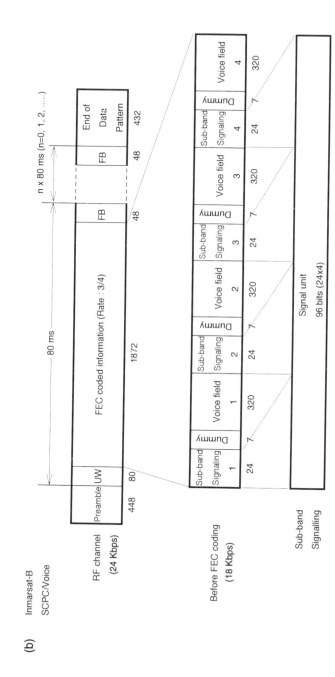

Figure 8.17 (continued)

channel formats of voice and 9.6-Kbps data transmissions. Voice is coded to 16-Kbps digital signals by adaptive predictive coding with maximum likelihood quantization (APC-MLQ) voice codec.

8.4 INMARSAT-C

8.4.1 System Configuration

The INMARSAT-C system was introduced in 1991 to provide data/messaging communications by terminals small enough to be hand carried or fitted to any vessel, aircraft, or vehicle. The Aero-C was introduced several years after the introduction of the basic-C service for maritime and land mobile vehicles. The INMARSAT-C MES has a small, omnidirectional antenna which, because of its light weight and simplicity, can be easily mounted on a vehicle, a vessel, or a hand-carried terminal. The main unit of a terminal is compact and weighs only 3 to 4 kg. Some terminals have message preparation and display facilities incorporated, while some others have standard interfaces so that users can connect their own computer equipment. Hand-carried versions are also available. Figure 8.18 is a photograph of the INMARSAT-C terminal for small ships. The left side is an antenna unit (externally mounted equipment) and the right two units are a main part of the terminal and a computer (internally mounted equipment). A hand-carried version of the INMARSAT-C terminal is shown in Figure 8.39.

A basic technology of data transmission is a store-and-forward technology, in which a land Earth station (LES) stores messages from MESs in a database and distributes them to the designated users. Transmission speed between a satellite and a mobile station is 600 bps, and the signal is coded by a convolutional coding with $R = 1/2$ and $K = 7$. Bit interleaving has been adopted in order to disperse the burst error caused by burst fading, which cannot be neglected because of using an omnidirectional antenna. Table 8.4 lists the main characteristics of an MES for the INMARSAT-C system. The required G/T and EIRP are over -23 dBK and 16 dBW, respectively.

8.4.2 Antenna and Tracking Systems

The antenna has no tracking function, and the antennas usually used for INMARSAT-C are omnidirectional ones such as a quadrifilar helix, a crossed drooping dipole, and a microstrip patch. A quadrifilar antenna has been the most popular type for installation on ships because of its good performance of wide beam coverage under the condition of ship motion (Figure 8.18). A

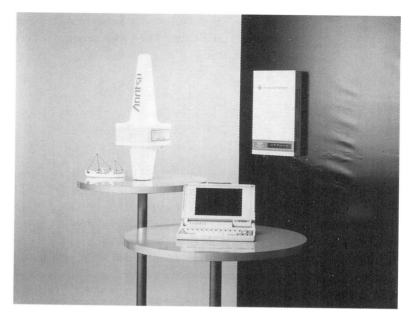

Figure 8.18 The INMARSAT-C terminal for small ships, and a hand-carried version. (Courtesy of Anritsu.)

microstrip antenna is the best one to use in hand-carried or briefcase terminals because of its advantage of having very low-profile characteristics.

8.4.3 Communication Channels

As summarized in Table 8.4, the SES has four types of communication channels: LES TDM, MES message, MES signaling, and NCS common channels. Figure 8.19 shows communication channels and connections between stations.

   1. *LES TDM channel:* Each LES transmits LES TDM signals to mobile Earth stations over at least one channel. These TDM signals contain such signaling information as channel assignments and acknowledgment of message reception.

   2. *MES message channel:* An MES message is transmitted from an MES to an LES in a burst mode. Upon receiving a transmission request from an MES, an LES assigns frequency and timing.

   3. *MES signaling channel:* MES signaling is transmitted in a slotted-ALOHA mode from an MES to an LES or to an NCS. A link from an MES to an LES is used to transmit a channel request,

Table 8.4
Main Characteristics of an MES for the INMARSAT-C System

INMARSAT-C								
Frequency	1626.5–1646.5 MHz	MES → LES		LES : land Earth station				
	1530.0–1545.0 MHz	MES ← LES		MES : mobile Earth station				
				NCS : network coordination station				
Polarization	Right hand circular							
Antenna gain	Nominal 0 dBi	Omni-directional						
EIRP	16 dBW	HPA ~ 20W						
G/T	$\geqq -23$ dBK	LNA noise temperature ~ 100K						

	Type	Service	Link	Access	Information rate	Channel rate	Modulation	Flame length	Coding
Channel types	LES TDM	Signaling	LES → MES	TDM	1,200 bps	600 bps	BPSK	8.64 sec	Convolutional $R = 1/2, K = 7$
	MES message (store and forward)	Message	LES ← MES	Slotted Aloha				$2048\times(N+1)+128$ $N : 0 \sim 4$	
	MES signaling	Signaling	LES ← MES	Slotted Aloha					
			NCS ← MES						
	NCS common	Signaling	NCS → MES	TDM				8.64 sec	

Environmental conditions (for ships)	Temperature	EME : $-35°C \sim 50°C$, IME : $0°C \sim 45°C$
	ship motion	Roll ±30° Cycle (8 sec) Surge ±0.2G
		Pitch ±15° Cycle (6 sec) Sway ±0.2G
		Yaw ± 8° Cycle (50 sec) Heave ±0.5G
		Turn speed 12°/sec Speed 30 knot

EME : externally mounted equipment
IME : internally mounted equipment

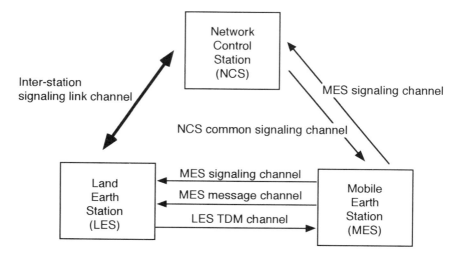

Figure 8.19 Communication channels of the INMARSAT-C system.

acknowledgments of a channel setup, and an acknowledgment of message reception from the LES. A link from an MES to an NCS is used to request a registration of an ocean area where the MES is navigating.

4. *NCS common channel:* Over this communication channel, each NCS is always transmitting such information as registrations of ocean areas and announcements of the reception of messages from an LES.

8.5 INMARSAT-M

8.5.1 System Configuration

The INMARSAT-M system was introduced in 1993 in order to provide high quality, frequency- and power-saving communications with small terminals by using digital communication technologies. Although the INMARSAT-C system cannot provide telephone services, the INMARSAT-M system can provide voice communications with very small and lightweight terminals that are almost the same as those of the INMARSAT-C. The M terminals can be installed in small private cruisers or cars, or carried in a briefcase. These terminals can also be used as solar-powered emergency phones along highways and hand-carried telephone in remote areas.

The technical requirements of the terminal have been defined in the SDM of the INMARSAT [5]. Table 8.5 lists the main characteristics of an

Table 8.5
Main Characteristics of an MES for the INMARSAT-M System

INMARSAT-M

Frequency	1626.5–1646.5 MHz	MES → CES	MES: ship Earth station				
	1525.0–1545.0 MHz	MES ← CES	CES: coast Earth station				
Polarization	Right-hand circular						
Antenna gain	Nominal 24 dBi	Diameter ~ 40 cm					
EIRP	27, 21 dBW	HPA ~ 20W					
G/T	≧ −10 dBK (maritime)	LNA noise temperature ~ 100K					
	≧ −12 dBK (land)						

Channel types	Type	Request	Service	Link	Information rate	Channel rate	Modulation	FEC
	Access	Random Access	Channel request	MES → CES		3 Kbps	BPSK	Convolutional/ Viterbi $(R = 1/2,\ K = 7)$
Assignment	TDM		Channel assignment	MES ← CES		6 Kbps	BPSK	$(R = 1/2,\ K = 7)$
	SCPC		Telephone	MES ⇆ CES	6.4 Kbps (Coding : IMBE)	8 Kbps	O-QPSK	None
Voice/data	SCPC		Facsimile	MES ⇆ CES	2.4 Kbps	8 Kbps	O-QPSK	Punctured $(R = 3/4,\ K = 7)$
	SCPC		Data	MES ⇆ CES	2.4 Kbps	8 Kbps	O-QPSK	$(R = 3/4,\ K = 7)$

Environmental conditions		
Temperature	EME: −25°C ~ 50°C, IME : 0°C ~ 45°C	
ship motion	Roll ±30° Cycle (8 sec)	Surge ±0.2G
	Pitch ±15° Cycle (6 sec)	Sway ±0.2G
	Yaw ± 8° Cycle (50 sec)	Heave ±0.5G
	Turn speed 12°/sec	Speed 30 knot

(EME : externally mounted equipment)
(IME : internally mounted equipment)

MES for the INMARSAT-M system. The required G/T values are over -10 dBK for maritime applications and over -12 dBK for land applications. The EIRP can be set in high-power and low-power modes, respectively, at 27 and 21 dBW. The digital modulations adopted are 8-Kbps O-QPSK and 6/3 (receive/transmit)-Kbps BPSK. In request and assignment channels, a convolutional coding/Viterbi decoding (rate $R = 1/2$, constrain length $K = 7$) has been adopted. In facsimile/data channels, punctured coding/Viterbi decoding ($R = 3/4$, $R = 7$) has been adopted. Voice is transmitted after digital coding by the voice codec of an improved multiband excitation (IMBE). Figure 8.20 shows a photograph of the INMARSAT-M terminal used in a remote area.

8.5.2 Antenna and Tracking Systems

The kinds of antennas used for the M system depend on whether the terminals are used in maritime, land mobile, or hand-carry applications. In the case of maritime applications, the configuration of externally mounted equipment for ships is basically the same as that of the INMARSAT-B SES mentioned in Section 8.2.2. The typical antenna gain is about 15 dBi, and a short backfire (SBF) antenna is one of the favorite M terminal antennas for maritime use. Although the SBF antenna is a compact, simple-configuration, and high-efficiency shipborne antenna, it has a narrow frequency bandwidth of about 3%, which is too narrow to cover the required frequency bandwidth of about 8% [6]. The electrical characteristics of the conventional SBF antenna have

Figure 8.20 Photograph of an INMARSAT-M briefcase-type terminal. (Courtesy of NEC.)

been improved by changing the main reflector from a flat disk to a conical or a step plate and by adding a second small reflector. With the improved SBF antenna, better performance with an aperture efficiency of about 80% and a frequency bandwidth of 20% for VSWRs under 1.5 is obtained [7], and gain is also improved by about 1 dB without changing sidelobe levels [8]. Figure 8.21 is a photograph of an improved SBF antenna for the ETS-V experiment. The antenna has a gain of 15 dBi, is 40 cm in diameter, and weighs about 40 kg including a stabilizer. These antennas generally have two

Figure 8.21 An improved SBF antenna installed on the deck for the ETS-V experiment. The antenna has a gain of 15 dBi, is 40 cm in diameter, and weighs about 40 kg including a stabilizer.

axes (Az-El) stabilized and are simple and compact antennas, suitable for mounting in small ships. A step-track tracking method is the most commonly used, as in the INMARSAT-A and -B systems. Although aperture types of antennas such as the parabolic and SBF types have been very popular in maritime satellite communications, a phased-array antenna has sometimes been used for the M terminal. Figure 8.22 is a photograph of a phased-array antenna with 10 elements, and the configuration of a phased-array antenna is shown in Figure 8.23. Although this antenna uses a two-axis (Az-El) stabilizer, a beam can be scanned within ±30 degrees in the Az directions by electrical scanning in order to avoid gimbals lock [9]. In the case of the Az-El stabilizer, very high speeds and accelerations are required for accurate satellite tracking when elevation angles are near 90 degrees. So, it becomes very difficult to track a satellite when elevation angles are near 90 degrees, and this condition is called gimbals lock. A step-track method is also adopted by this antenna system.

8.5.3 Communication Channels

As summarized in Table 8.5, the MES has three types of communication channels: request, assignment, and voice/data channels. Signaling procedures, including signal format, are basically the same in the B and M systems. A mobile station will request a channel assignment in a random access channel

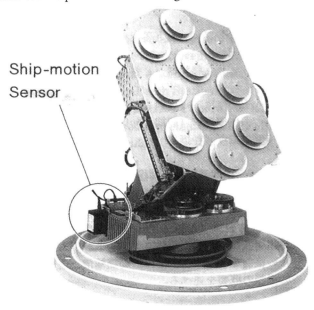

Ship-motion
Sensor

Figure 8.22 A 10-element phased-array antenna system for the INMARSAT-M ship terminal. (Courtesy of JRC.)

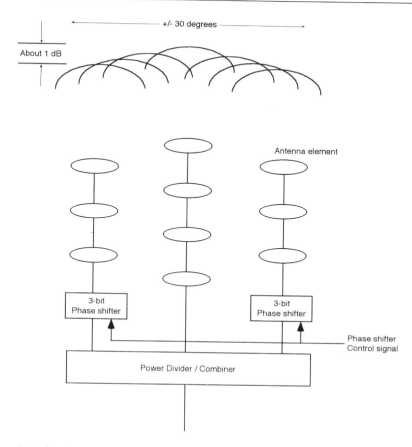

Figure 8.23 Configuration of a phased-array antenna with 10 elements.

in which the transmission rate is 3 Kbps in a BPSK modulation. For FEC, data is coded by convolutional coding with a rate $R = 1/2$ and a constrain length $K = 7$. An assignment signal is transmitted in a TDM channel in a BPSK modulation. Voice communications are transmitted in SCPC channels with 8-Kbps O-QPSK modulation after digital coding by using a voice codec of the improved multiband excitation (IMBE) of 6.4 Kbps. Facsimile and digital data with information rates of 2.4 Kbps are transmitted in SCPC channels with O-QPSK modulation.

8.6 INMARSAT-Aero

8.6.1 System Configuration

Worldwide commercial aeronautical satellite communication services have been provide by the INMARSAT since 1991. The system consists of a space segment, a GES, an NCS, and an aeronautical Earth station (AES). The INMARSAT has defined the system in the aeronautical SDM [10]. The AES portion of the system for commercial aviation has been defined in Aeronautical Radio Incorporated (ARINC) Characteristics 741 [11].

The ARINC 741 categorizes the operational types of the AES into four classes, as shown in Table 8.6. The class 1 type can provide only low-speed data services, which include aeronautical operational control (AOC) and aeronautical administrative communication (AAC) services by using a low-gain antenna (0 dBi). In the future, this type of station will be used for air traffic control (ATC). The class 2 type provides only voice services, mainly for passengers in the cabin (aeronautical passenger communication, APC), with a high-gain antenna (12 dBi). The class 3 type provides both voice and high-speed data services by adding data systems to the class 2 type. The class 4 type combines the features of class 1 and class 3, and is expected to provide all kinds of aeronautical communications with high-gain and low-gain antennas. Table 8.7 lists the main characteristics of an AES for the INMARSAT system. The system has four types of communication channels, as shown in Table 8.7, and Figure 8.24 shows an aeronautical network configuration of these channels.

Figure 8.25 and Figure 8.26 show block diagrams of an AES with low-gain (class 1), and high-gain antennas (class 3). The ARINC 741 Standard describes the characteristics of a satellite data unit (SDU), a radio frequency unit (RFU), a diplexer/low-noise amplifier (DIP/LNA), a high-power amplifier (HPA), and a high-gain antenna (HGA) with its associated beam-steering electronics. Figure 8.27 is a schematic illustration showing the location of AES equipment on a Boeing 747.

8.6.2 Airborne Antenna and Tracking Systems

As mentioned in the previous section, there are two types of airborne antennas for satellite communications: a low-gain antenna and a high-gain antenna, with nominal gains of 0 and 12 dBi, respectively.

8.6.2.1 Low-Gain Antenna Subsystems

A low-gain antenna system consists of an antenna element, a DIP, an LNA, and a C-class HPA. Its gain is about 0 dBi, and its radiation pattern is omnidirectional to cover over 85% of the upper hemisphere above an elevation angle of 5 degrees. The main specifications of low-gain and high-gain antenna subsystems are listed in Table 8.8. Figure 8.28 is a photograph of a low-gain antenna with a C-class HPA and a DIP/LNA.

Table 8.6
Operational Types of AES for the ARINC 741 Standard

Type	Antenna	Voice/Data	Service	User (Air)	User (Ground)
Class 1	Low-gain antenna (0 dBi)	Low-speed data	AOC	Pilot	Airline company
			AAC	Cabin crew	Airline company
			ATC [in the future]	Pilot	Control authority
Class 2	High-gain antenna (12 dBi)	Voice	APC	Passenger	Subscriber
Class 3	High-gain antenna (12 dBi)	Voice & high-speed data	AAC	Cabin crew	Airline company
			APC	Passenger	Subscriber
Class 4	Low-gain antenna (0 dBi) & high-gain antenna (12 dBi)	Voice & high/low-speed data (Class 1 & Class 3)	AOC	Pilot	Airline company
			AAC	Cabin crew	Airline company
			APC	Passenger	Subscriber
			ATC [in the future]	Pilot	Control authority

Note: AOC: aeronautical operational control; AAC: aeronautical administrative communication; ATC: air traffic control (in the future); APC: aeronautical passenger communication.

Table 8.7
Main Characteristics of AES for the INMARSAT System

INMARSAT-Aero

Frequency	1626.5–1660.5 MHz	AES → GES	
	1525.0–1559.0 MHz	AES ← GES	
Polarization	Right-hand circular		
Antenna gain	High-gain antenna	Low-gain antenna	
Antenna type	12 dBi Phased array	0 dBi Helical	
EIRP	25.5 dBW	15.5 dBW	
HPA	60W (A-class)	40W	
G/T	−13 dBK	−26 dBK	
LNA noise temp.	150 K	150 K	

Type/access	Channel name	Service	Link	Information rate	Channel rate
Packet mode/TDM	P channel	Signaling / user data	AES ← GES	300 bps	600 pbs
Random access	R channel	Signaling / user data (≤ 33 bytes)	AES → GES	300 bps	600 bps
Reservation/TDMA	T channel	User data (> 33 bytes)	AES → GES	300 bps	600 bps
Circuit-mode/SCPC	C channel	Voice / signaling	AES ↔ GES	4.8 / 9.6 Kbps	21 Kbps

Channel types

Modulation   A-BPSK (≤ 2.4 Kbps)
            A-QPSK (> 2.4 Kbps)

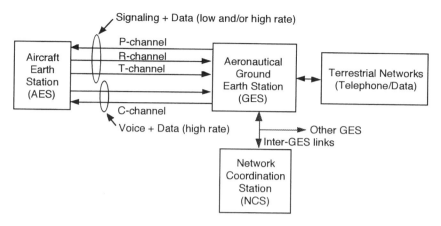

Figure 8.24 Aeronautical network configuration.

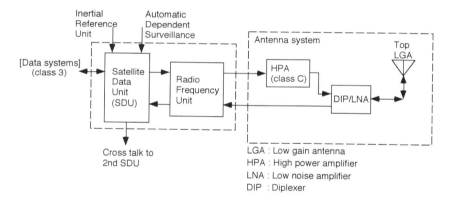

Figure 8.25 Block diagram of an AES with a low-gain antenna (class 1).

8.6.2.2 High-Gain Antenna Subsystems

As shown in Figure 9.27, there are two types of phased-array antennas. The first is a top-mount type, which is installed on the top of a fuselage. The second is a side-mount type, which is installed on both sides (port and starboard) of the fuselage. The top-mount type has the advantage of eliminating keyhole areas where the beam cannot be scanned, but it has the disadvantage of increasing air drag. The conformal type, on the other hand, has the advantage of low air drag, but has the disadvantage of keyholes. A high-gain antenna subsystem consists of a phased-array antenna, a diplexer, a low-noise amplifier, an A-class high-power amplifier, and a beam-steering unit (BSU) that steers the beam to track a satellite. The satellite tracking is carried out by a program tracking method. A BSU steers the beam by controlling digital phase shifters

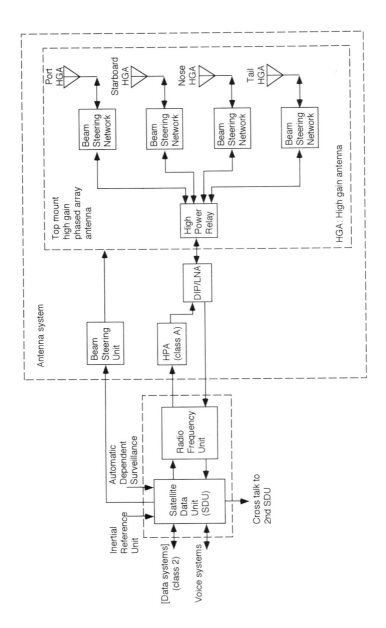

Figure 8.26 Block diagram of an AES with a high-gain antenna (class 2).

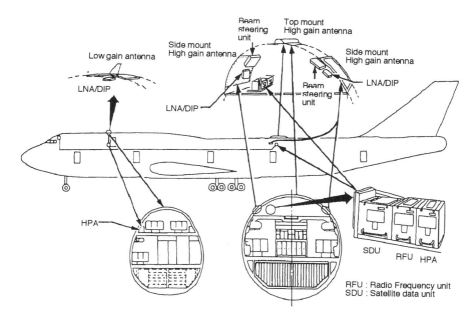

Figure 8.27 Location of AES equipment on the Boeing 747.

Table 8.8
Main Specifications of Low-Gain and High-Gain Antenna Subsystems

	Low-Gain Antenna	High-Gain Antenna
Frequency	1,530.0 ~ 1,559.0 MHz (Receive)	
	1,626.5 ~ 1,660.5 MHz (Transmit)	
Polarization	Right-hand circular	
Axial ratio	< 6 dB	
Figure of merit (*G/T*)	≧ 26 dBK	≧ −13 dBK
Radiation power (EIRP)	≧ 13.5 dBW	≧ 25.5 dBW
Antenna gain	≧ 0 dBi	≧ 12 dBi
Coverage for semisphericity above elevation angle 5 degrees	≧ 85%	≧ 75%
Tracking	None (Omnidirectional)	Program tracking

of the phased-array antenna, and the information used to operate the BSU is calculated from the signals from the inertial navigation system (INS), which gives the position, heading direction, and attitude of the aircraft. To track the satellite, the high-gain antenna has to steer the beam to cover over 75% of the hemisphere above an elevation angle of 5 degrees. When a phased-array

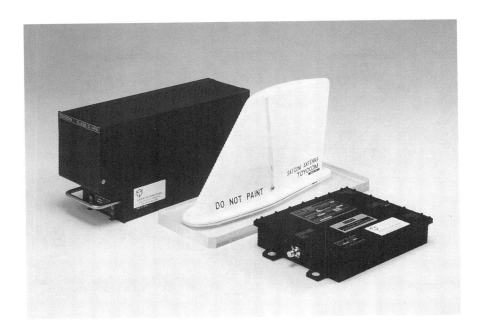

Figure 8.28 Photograph of a low-gain antenna with a C-class HPA and a DIP/LNA. (Courtesy
of TOYOCOM.)

antenna is used, these inevitably are keyhole areas in the fore and aft directions
(Figure 8.29) because it is very hard for a phased-array antenna to scan the
beam to wide angular areas over 60 degrees.

An A-class HPA is used in order to avoid channel intermodulation in
multicarrier operation. The main specifications of a high-gain antenna subsystem are listed in Table 8.8. Figure 8.30 is a photograph of a top-mount type
high-gain antenna with an A-class HPA and a DIP/LNA. A prototype of this
antenna was developed by the Communications Research Laboratory and was
installed on a Japan Air Line's Boeing 747 to carry out the experiments over
the transpacific flight routes between Tokyo and Anchorage in the ETS-V
project [12]. As shown in Figure 8.26, this high-gain phased-array antenna
has four phased-array units , and nose and tail array units have been used to
make keyhole areas narrow enough to keep communication links over 85%
in the upper hemisphere [13]. Figure 8.31 shows a low gain and a top-mount
high gain antenna installed on a Boeing 747.

8.6.3 Communication Channels

Irrespective of its class, each AES is equipped with a capability to receive a
medium-rate forward P-channel transmitted from a GES with transmission

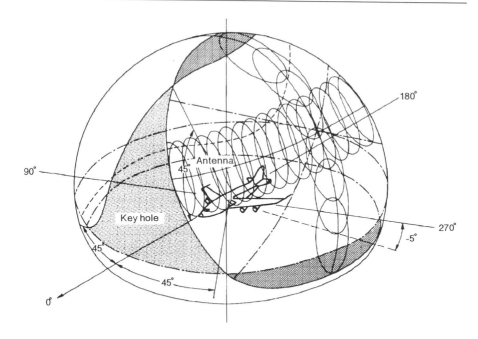

Figure 8.29 Beam-steering coverage and keyholes of a phased-array antenna.

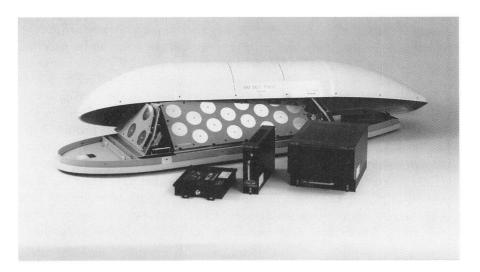

Figure 8.30 Photograph of a high-gain antenna with an HPA, a DIP/LNA, and a BSU. (Courtesy of TOYOCOM.)

Figure 8.31 Photograph of low-gain and top-mount high-gain antennas installed on a Boeing 747.

rates of 600 bps, 1.2 Kbps, 4.8 Kbps and 10.5 Kbps, and carrying signaling messages in packet form. Each AES is also able to transmit a return carrier signal in burst mode (R channel) at a transmission rate that can be switched between 600 bps, 1.2 Kbps, and 10.5 Kbps. At logon, the AES uses a transmission rate of 600 bps for transmission on the R channel. The GES responds on the P channel at the same rate and indicates to the AES the R channel and transmission rate to use for further transactions. The GES determines the transmission rate according to the satellite in use and the signal quality of the logon request. The class 2, 3, and 4 AESs are also equipped with pairs of transmit/receive voice channel equipment (C channel). The number of voice channels is at the discretion of the AES owner/operator. The class 1, 3, and 4 AESs also require additional protocol-oriented capabilities to support data services, which provides error-free delivery of user messages with a high probability of success. These capabilities include provision of the T-channel protocol using transmission bit rates the same as for the R channel. Figures 8.32 and 8.33 show P-channel and C-channel formats, respectively.

Aeronautical satellite channels utilize satellite power and bandwidth efficiently by using FEC. The modulation methods used are aviation quadrature phase shift keying (A-QPSK) and aviation binary phase shift keying (A-BPSK).

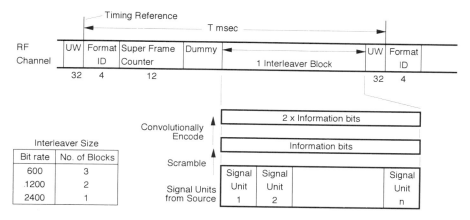

Figure 8.32 P-channel format (600, 1,200, and 2,400 bps).

An A-QPSK is a form of O-QPSK [14], and an A-BPSK is a form of differentially encoded BPSK in which alternate modulation symbols are transmitted in notional in-phase and quadrature channels [15]. The functional blocks of a C channel are shown in Figure 8.34. A pseudonoise (PN) scrambler with a 15-stage shift register is used for scrambling data before FEC coding [16]. The majority of channel types use FEC coding consisting of a convolutional encoder of constrain length $K = 7$ and an eight-level soft decision Viterbi decoder. The FEC coding rate is 1/2. Because of the multipath fading characteristics of the aeronautical transmission path, interleaving is used to preserve the FEC gain. The interleaving has a block size of 192 or 384 bits, resulting in an overall transmission delay of about 30 msec.

8.7 Land Mobile Earth Station

8.7.1 Overview

In the INMARSAT system, M and C Earth stations have been used not only by maritime users but also by land mobile users such as those on trucks and trains. The C Earth station for land mobile users is exactly the same as that for shipboard users because its antenna is omnidirectional without tracking function. As mentioned in the previous section, a quadrifilar antenna is the most popular and suitable for the C Earth station. This type of the C Earth station antenna has been widely mounted on relatively large trucks to provide message/data communications. Although the quadrifilar antenna is a good antenna for trucks, it is too large for installation on small cars.

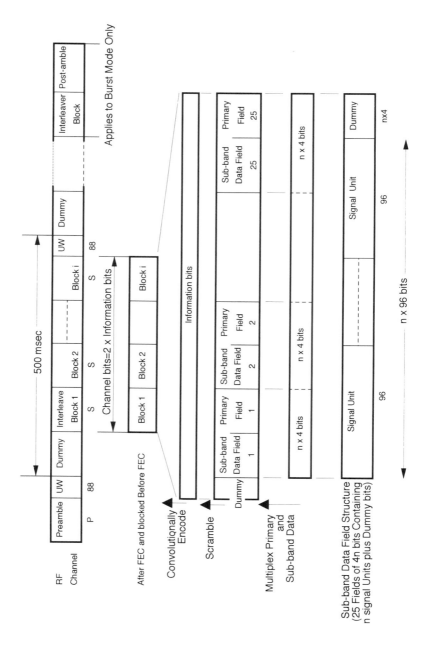

Figure 8.33 C-channel format (general case).

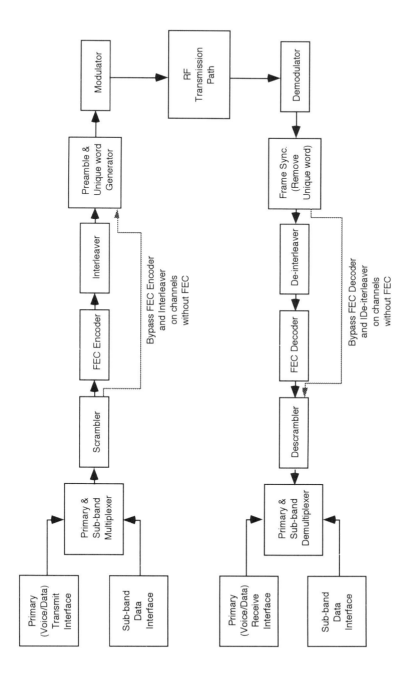

Figure 8.34 Functional blocks of a C channel.

The M Earth station for land mobile stations are basically the same as those for use on ships except that they use different antennas and tracking methods. Antennas for land mobile use have to have a low profile and be very compact, especially the antennas used on private cars. As medium-gain antennas with gains of 12 to 15 dBi, phased-array antennas are considered the best to use on small private cars because of their low profile and high-speed electrical tracking. For satellite tracking, it is very difficult to use closed-loop methods such as step tracking, very popular in SESs because of severer propagation conditions such as fading, blocking, and shadowing. Open-loop tracking methods using new technology such as optical-fiber gyros have been developed.

8.7.2   Antennas and Tracking Systems

8.7.2.1 Antennas

Directional antennas have been expected to provide links for voice signals and high-speed data signals not only for long-haul trucks but also for small private cars. Cost is a very important factor to be taken into consideration when designing such antennas, and phased-array antennas are considered the best kind of antennas for vehicular use because of their low profile, high-speed tracking, and potentially low cost. Phased-array antennas with tracking functions have not yet been used in commercial land mobile satellite communication systems, but many research and development projects have been focused on phased-array antennas and on the tracking method they would use. Mainly in United States, Canada and Japan, several types of phased-array antennas have been developed in MSAT-X, PROSAT and ETS-V experiments [17,18]. Figure 8.35 shows a general block diagram of a land mobile Earth station that has a phased-array antenna with tracking function. In many cases, 19 circular patch elements are used to get system gain of about 13 dBi (G/T is about −13 dBK), and an axial ratio of about 4 dB is obtained in an elevation angle of 45 degrees. A typical antenna is 4 cm in height, 60 cm in diameter and 5 kg in weight. Figure 8.36 is a photograph of a phased-array antenna developed by the CRL [17–19]. The antenna has electromagnetically coupled antenna elements printed on a very thin film (Figure 8.36) and has frequency-dependent 3-bit PIN diode phase shifters [20]. The total number of PIN diodes is half that of conventional phase shifters. These characteristics help keep cost low and reduce tracking error. The decrease in the number of PIN diodes will greatly contribute to the reduction of the cost because the cost of PIN diodes mainly dominates the total cost of phased array antenna. Table 8.9 shows the main characteristics of the phased-array antenna developed for the ETS-V experiments.

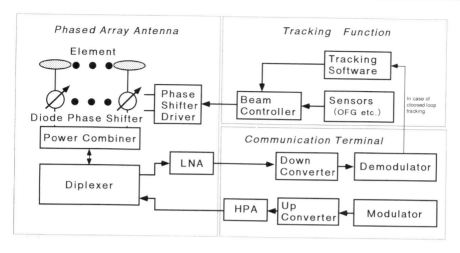

Figure 8.35 Block diagram of the phased-array antenna system.

8.7.2.2 Tracking

Open-loop tracking is, for land mobile satellite communications, more suitable than closed-loop tracking for the following reasons:

1. Because of shadowing and blocking effects, signal levels from/to the satellite are not always stable.

2. A closed-loop method such as a monopulse, a sequential lobbing, or a step-track method requires otherwise unnecessary lowering of signal levels in order to determine the satellite direction under the condition in which a very small link margin is expected.

3. Movement of land vehicles is more complicated than that of ships and aircraft. Land vehicles change direction and speed very quickly.

In land mobile satellite communication systems operated at L-band frequencies, satellite tracking is required only in the azimuth direction. Because the half-power beamwidth of the antenna is about 30 degrees, and everywhere the land vehicle moves is within a radius of a few hundred kilometers, the elevation angle of the satellite is almost constant. In open-loop tracking, the characteristics of sensors are very important because the total communication performance of the mobile Earth station is determined primarily by the tracking capability.

Several kinds of sensors give information on vehicle motion. In the ETS-V experiments, a geomagnetic sensor and an optical-fiber gyro were evaluated.

Figure 8.36 A 19-element phased-array antenna printed on a very thin film. The *G/T* is about
−13 dBK at an elevation angle of 45 degrees, and the performance was evaluated
using the ETS-V satellite and an open-loop tracking method.

As mentioned in Chapter 4, these two are typical kinds of sensors. A geomagnetic
sensor is a very attractive because of its low cost and its ability to determine
absolute direction, but it is affected by environmental conditions such as
buildings, bridges, and power poles. An optical-fiber gyro, on the other hand,
shows excellent performance in determining relative directional changes independent of environmental condition. But it cannot carry out the initial acquisition of the satellite because it cannot determine the absolute direction. In
open-loop tracking, an optical-fiber gyro is usually used to get the information
about vehicle motion, and a geomagnetic sensor gives an absolute direction
that is required to calibrate the accumulative error of the optical-fiber gyro at
appropriate time intervals. Figure 8.37 shows an example of the results obtained

Table 8.9

Main Characteristics of the Phased-Array Antenna Developed for Land Mobile Systems in the ETS-V Program

Frequency	1642.5–1650.0 MHz (Tx)
	1543.0–1548.0 MHz (Rx)
Polarization	Left-handed circular polarization
Antenna	19-element phased array
Element	Circular patch antenna
Gain	12 dBi (El. = 45°)
HPBW	28 degrees
Axial ratio	4 dB
Scanned angle	El. 30 degrees ~ 90 degrees (1-degree step)
	Az. 0 degrees ~ 360 degrees (1-degree step)
System noise	340 K
G/T	−13 dBi (El. = 45 degrees)
EIRP	22 dBW (El. = 45 degrees)
HPA	10 W
Tracking	Open-loop method using optical-fiber gyro
Volume	600 mm (D), 40 mm (H)
Weight	5 kg

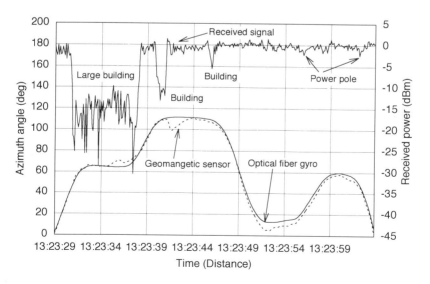

Figure 8.37 Tracking performance of phased-array antenna with an open-loop tracking method using an optical-fiber gyro.

in an experimental evaluation of tracking performance. The phased-array antenna installed on the test van tracked the ETS-V satellite by using an optical-fiber gyro. Although the test van was changing directions, the signal level received from the satellite was found to be almost constant except for the blocking effects caused by buildings and power poles. It was proven that in short time ranges the optical-fiber gyro has excellent performance that is not effected by environmental conditions. The output of a geomagnetic sensor is also shown in the figure and the geomagnetic sensor had a tracking error (relative to that of the optical-fiber gyro) of about 10 degrees.

8.7.2.3 Tracking Algorithm

Open-loop tracking will require two kinds of sensors. The first one will be a geomagnetic sensor that will determine the absolute direction of the vehicle in order to carry out the initial acquisition of the satellite and to calibrate the accumulative error of the second sensor, which will give an accurate relative direction over short time ranges. The following is one example of an open-loop tracking algorithm, and its flow chart is shown in Figure 8.38 [21,22].

1. Before the vehicle starts to move, the initial direction of the antenna beam is measured by using a geomagnetic compass. The accuracy of this direction depends on the performance of the geomagnetic compass, which ordinarily has an accuracy of about 1 degree. This data is also used to set the initial direction of the optical-fiber gyro. After this initial procedure, the antenna begins to track the satellite.

2. Data from the optical-fiber gyro and the geomagnetic sensor are sent to a computer, which processes the data to control the antenna beam directions.

3. If a standard deviation σ of the data from the geomagnetic compass is more than 1 degree, it is estimated that a magnetic disturbance has occurred or that a vehicle has changed its direction. Then the antenna beam is controlled by the optical-fiber gyro. Go to step 5.

4. When the value of standard deviation is within 1 degree, the antenna beam is controlled by the geomagnetic compass and the optical fiber gyro is calibrated by the geomagnetic sensor to cancel out its accumulative errors.

5. Return to step 2.

In step 4 it can be assumed that the geomagnetic compass is not affected by magnetic turbulence and is reliable for determining vehicle direction. Therefore, the angle information output from a geomagnetic compass can be used

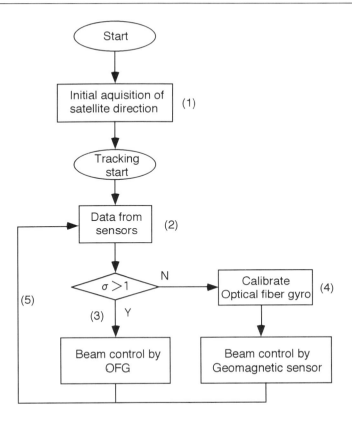

Figure 8.38 Flow chart of an open-loop tracking algorithm.

as a "true" azimuth angle to cancel out the accumulative errors of the optical-fiber gyro.

8.8 Hand-Carried Terminals

Hand-carried terminals are used mainly at the INMARSAT-C and -M Earth stations. As described before, the INMARSAT-C Earth station cannot provide voice communication services and can transmit and receive only low-speed data and messages. The INMARSAT-M Earth station, on the other hand, can provide digital voice and data communication services. Figure 8.39 is a photograph of a hand-carried INMARSAT-C Earth station. Antennas and receiver are assembled in one unit (left), and data are input by a keyboard and are output on the display of a conventional personal computer (right). The weight of the station is about 4 kg excluding the personal computer. An antenna

Figure 8.39 Hand-carried INMARSAT-C Earth station. A receiver with a flat antenna weighs about 4 kg and has a compact size (320 mm (W) by 210 mm (H) by 62 mm (D)). (Courtesy of Toshiba.)

has an array of four patch-antenna elements, and its gain is about 8 dBi. Because of its wide beamwidth, the antenna can be directed to the satellite by hand. This terminal can be used outdoors for about 50 minutes under battery operation.

The hand-carried INMARSAT-M Earth station is becoming very popular for providing outdoor voice and data communications for relief agencies, news reporters, and oil/mining companies. Figure 8.40 is a photograph of a hand-carried INMARSAT-M Earth station.

Before the introduction of the INMARSAT-C system, several small terminals were reported [23,24]. In the ETS-V program [25], for example, a briefcase-size message communication terminal was developed for the experiments [26]. Figure 8.41 is a photograph of an attaché-case-size message communication terminal for the ETS-V experiments. Its main characteristics are listed in Table 8.10. The hardware configuration of a very small terminal must be simplified, even though such simplification reduces such basic performance as frequency stability and transmission power. To communicate when using a low transmission power, a very narrowband transmission technique should be employed in order to reduce signal transmission power. A burst-mode

Figure 8.40 Hand-carried INMARSAT-M Earth station. (Courtesy of NEC.)

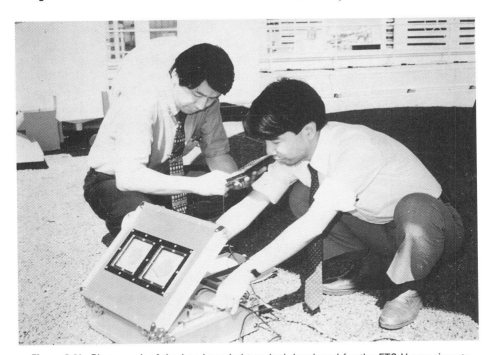

Figure 8.41 Photograph of the hand-carried terminal developed for the ETS-V experiments.

transmission scheme is also effective for using power efficiently. A 100-bps message terminal using only 1W was developed. The key technology of this 100-bps transmission system is how to achieve an initial acquisition and how to keep a synchronization of a received signal. Because a signal has frequency deviation caused by frequency unsuitability of oscillators both in the onboard

Table 8.10
Main Characteristics of the Attaché-Case-Size Message Communication Terminal
Developed for the ETS-V Experiments

Frequency	1644.7–1644.8 MHz (Tx)
	1543.2–1543.3 MHz (Rx)
Antennas	Two patch antennas for Rx and Tx.
	Gain Tx : 7.5 dBi, Rx : 7.0 dBi
Polarization	Left-handed circular polarization
Tracking	Hand operation
G/T	−17 dBK
EIRP	6.5 dBW (HPA : 1W)
Modulation	Digital FM, 100 bps, burst mode
Frequency stability	approximately 1×10^{-6}
Weight	13 kg

transponder and the Earth station. The frequency compensation function is implemented only in the base station terminal in order to keep the hand-carried terminal as small as possible. The terminal has only one local oscillator, which is commonly used for both transmission and reception. It has no frequency compensation function itself. The base station can detect the frequency deviation of the signal from the terminal by using an FFT technique [27] and control its local oscillators for reception and transmission to communicate with the hand-carried terminal.

References

[1] Technical Requirements for INMARSAT Standard-A Ship Earth Station, July 1981.

[2] Hoshikawa, T., et al., "INMARSAT Ship Earth Station Type RSS401A," *Anritsu Technical Journal*, No. 56, Sept. 1988.

[3] INMARSAT-B System Definition Manual, Nov. 1991.

[4] Yatuzuka, Y., and S. Iizuka, "A Variable Rate Coding by APC with Maximum Likelihood Quantization from 4.8 Kbps to 16 Kbps," *Proc. of ICASSP'86*, April 1986, pp. 3070–30744.

[5] INMARSAT-M System Definition Manual, Nov. 1991.

[6] Ehrenspeck, H. W., "The Backfire Antenna, a New Type of Directional Line Source," *Proc. IRE*, Vol. 48, Jan. 1960, pp. 109–110.

[7] Ohmori, S., et al., "An Improvement in Electrical Characteristics of a Short Backfire Antenna," *IEEE Trans. Antennas & Propagation*, Vol. AP-31, No. 4, July 1983, pp. 644–646.

[8] Shiokawa, T., and Y. Karasawa, "Compact Antenna Systems for INMARSAT Standard-B Ship Earth Stations," *IEEE 3rd Int. Conf. on Satellite Systems for Mobile Communication and Navigation*, London, U.K., June 1983.

[9] Eguchi, K., "Small Antenna System for INMARSAT-M Ship Earth Station," *Symposium Papers for RTCM Annual Assembly Meeting*, Florida, 1992.

[10] INMARSAT Aeronautical System Definition Manual, Module I, 1990.

[11] Aeronautical Radio Inc., ARINC Characteristic 741: Aviation Satellite Communication Systems, 1990.

[12] Taira, S., M. Tanaka, and S. Ohmori, "High Gain Airborne Antenna for Satellite Communications," *IEEE Trans. on Aerospace and Electronic Systems*, Vol. 27, No. 2, March 1991, pp. 354–360.

[13] Ohmori, S., et al., "Experiments on Aeronautical Satellite Communications Using ETS-V Satellite," *IEEE Trans. on Aerospace and Electronic Systems*, Vol. 28, No. 3, July 1992, pp. 788–796.

[14] Fang, R.J.F., "Quaternary Transmission Over Satellite Channels with Cascaded Nonlinear Elements and Adjacent Channel Interference," *IEEE Trans. Communications*, Vol. COM-29, No. 5, May 1981, pp. 567–581.

[15] Winters, J. H., "Differential Detection with Intersymbol Interference and Frequency Uncertainty," *IEEE Trans. Communications*, Vol. COM-32, No. 1, Jan. 1984, pp. 25–33.

[16] CCIR Report 384-3, Annex III, Section 3, Module 1.

[17] Woo, K., et al., "Performance of a Family of Omni and Steered Antennas for Mobile Satellite Applications," *Proc. Int. Mobile Satellite Conf.*, Ottawa, Canada, 1990, pp. 540–546.

[18] Nishikawa, N., K. Sato, and M. Fujino, "Phased Array Antenna for Land Mobile Satellite Communications," *Trans. IEICE*, J72-B-II, No. 7, July 1989, pp. 323–329.

[19] Ohmori, S., et al., "A Phased Array Tracking Antenna for Vehicles," *Proc. Int. Mobile Satellite Conf.*, Ottawa, Canada, June 1990, pp. 519–522.

[20] Ohmori, S., S. Taira, and A. Austin, "Tracking Error of Phased Array Antenna," *IEEE Trans. Antennas and Propagation*, AP-39, No. 1, Jan. 1991, pp. 80–82.

[21] Yamamoto, S., et al., "An Antenna Tracking Method for Land-mobile Satellite Communications Systems," *Electronics and Communications in Japan*, Part 1, Vol. 78, No. 9, 1995, pp. 91–102.

[22] Tanaka, K., et al., "Antenna and Tracking System for Land Vehicles on Satellite Communications," *IEEE Proc. Vehicular Technology Conf.*, Denver, CO, May 1992.

[23] Fang, R. J., et al., "Design of an MSAT-X Mobile Transceiver and Related Base and Gateway Stations," *COMSAT Tech. Rev.*, Vol. 17, No. 2, 1987, pp. 421–466.

[24] Rossiter, P., D. Reveler, and L. Tibbo, "L-band Briefcase Terminal Network Operation," *Proc. of Int. Mobile Satellite Conf.*, Ottawa, Canada, 1990, pp. 279–284.

[25] Hase, Y., S. Ohmori, and N. Kadowaki, "Experimental Mobile Satellite System (EMSS) using ETS-V Satellite," *IEEE Denshi Tokyo*, No. 25, 1986, pp. 10–12.

[26] Kadowaki, N., et al., "The Attach-Case-Size Message Communication System via Satellite Links," *IEEE Trans. Vehicular Tech.*, Vol. 41, No. 4, Nov. 1992.

[27] Kadowaki, N., Y. Hase, and S. Ohmori, "ETS-V/EMSS Experiments on Message Communications with Handheld Terminal," *Proc. ICC'89*, 1989, pp. 211–215.

9

Radio Navigation

9.1 Overview

Radio navigation is rather a special application of mobile *communications,* while its prominent usefulness and chance-making potential will tempt everyone to study its essentials for a while here. The original meaning of *navigation* was to determine a ship's position onboard, while in the following we take this word for determining the positions of satellite users in general, without respect to where the work of position determination actually takes place.

Determining the user positions by satellites requires, first of all, determining the positions of the satellites themselves—this work is called *tracking and orbit determination,* and is carried out using range and/or Doppler frequency measurements (and sometimes angle measurements) made at satellite tracking stations. How the satellite orbits are determined is, however, out of our interest here, so we will simply assume the satellite positions as already known to proper accuracy. Then, measurements made over the satellite-user link will determine the user's position. Thus the user position determination traces back to the tracking stations' positions via the satellite communication link.

What kind of measurements are then possible over the satellite-user link? Antennas carried by mobile communication users cannot be as large-apertured as to make precise angle measurements, and it is more or less the same for satellite-carried antennas. So, possible measurements will be either range or Doppler frequency, and choices from these will specify the principle of the navigation system. In fact, existing satellite navigation systems fall into the categories of range navigation, range difference navigation, and Doppler navigation.

Discussing the details of particular navigation systems is not our intention here, since many existing references already do it. Our approach will be rather generic—to study the key points of position fix mechanisms for each category, with particular stress on what is different and what is common to each category. Simple analogy will clarify the positioning mechanisms without using mathematical equations, so that the readers will enjoy surveying the essentials of radionavigation concepts.

9.2 Range Navigation

Among our three categories, the range navigation is probably the easiest to understand because it is purely geometrical in its positioning principle. A typical concept is illustrated in Figure 9.1; two satellites S_1 and S_2 are at known positions, and each satellite measures the distance to a ground user U. Given the distances r_1 and r_2, the user U is imagined as being tied to the satellites with ropes of lengths r_1 and r_2, with U being pulled so as to keep each rope stretched to its length. This makes a rigid triangle $S_1 S_2 U$, which is free to rotate around the fixed axis $S_1 - S_2$. The triangle's vertex U then plots a circle, which will cross the ground surface generally at two points, U and U'. The unwanted U' will be eliminated by the directivity of the satellite antenna or by some a-priori position information so that the user's ground position is correctly fixed at U. Obviously, greater separations between S_1 and S_2 lead to better accuracies of position fix.

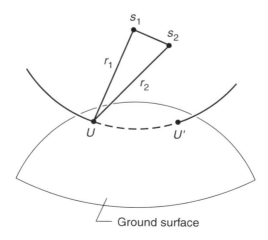

Figure 9.1 Range navigation using two satellites.

Problem 9.1

If U is near to being under the line connecting S_1 and S_2, what kind of difficulty will arise in the position fix?

Measuring a distance between two points needs a timing signal traveling from one point to the other and then back so as to count its round-trip travel time. In practice, distance measurements in our stattelite system will proceed as follows: A communication hub station generates a timing signal and sends it to the user via one satellite, say S_1. The user receives it and sends it back to the hub station via S_1, and to the hub station via S_2 as well. The hub-station measures the round-trip travel times for the thus received two timing-signals, and this becomes equivalent to measuring r_1 and r_2 since the positions of S_1 and S_2 are known. Position determination is then processed at the hub station, and its result is sent to the user as a message. Because of this scheme of signal transmission and data processing, this system is suitable for surveying a number of users altogether at the hub station. In practice, the users will be called one by one in turn by the hub station, since the position determination operates for one user at a time. So, each user will have a preregistered address of its own, so that the timing signals and position messages should be delivered to the addressed user. After theoretical studies [1] and experimental proofs [2], systems of this category using geostationary satellites have been put into practice for surveying land vehicles with accuracies of a few hundred meters—see [3,4] for details.

Position fix can be three-dimensional if three satellites measure each the distance to the user. Our "rope analogy" provides an image of this case, as illustrated in Figure 9.2, in that a lamp is suspended from the ceiling by three

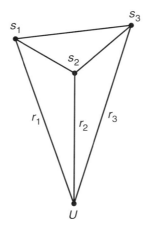

Figure 9.2 Range navigation using three satellites.

ropes of lengths r_1, r_2, and r_3. If the suspension points on the ceiling (S_1, S_2, S_3) are not forming a straight line, determining the rope lengths will immediately fix the user's position. Though this mechanism of position fix may appear quite reasonable, it would not attract practical interest because it makes for a complex system with too many signal transmission routes.

9.3 Range-Difference Navigation

The range navigation was to serve, in practice, a closed society of preregistered users. If the positioning service is to be made wide open to public, the signal transmission must be one-way from satellite to user. Then it follows that the absolute distance measurement between satellite and user is impossible, and to solve this problem is the task of our next category, the range-difference navigation.

9.3.1 Positioning Using Three Satellites

Let us start with a simple case of using three satellites to fix a ground user's position. Let S_1, S_2, and S_3 be the satellites, respectively at distances r_1, r_2, and r_3 from the ground user. Each satellite emits its own message to users, while the message emissions are synchronized between the satellites (how to synchronize them will be discussed later). These messages will inform the users of the present positions of the satellites. The users receive the messages from the three satellites and measure their times of arrival against the users' own clocks. Users' clocks will be more or less fast or slow; so, what is correctly measured at the user end is only the time intervals between the times of arrival of the satellite messages. This measurement provides, when multiplied by the speed of light, the differential distances to the satellites from the user: $r_1 - r_2$ and $r_2 - r_3$. Then we have our problem to solve: Given the satellite positions and the measured differential satellite distances, find the user's position. The position fix procedure for this case becomes a little bit complicated, though we try to interpret it again with our rope analogy.

Imagine a device as illustrated in Figure 9.3. Its ceiling, made of thin plate, has three small holes S_1, S_2, and S_3, and they represent the satellite positions. The floor is curved round so as to represent the ground service area. Thus the ceiling holes together with the floor represent the satellite-ground geometry—with a scaling factor, of course. Through each hole S_i, a rope denoted by r_i passes, sliding smoothly. The lower ends of the ropes are knotted together, while their upper ends are wound by a windlass. The below-ceiling parts of the lengths of ropes r_1, r_2, and r_3 are denoted by $\bar{r}_1$, $\bar{r}_2$, and $\bar{r}_3$, respectively.

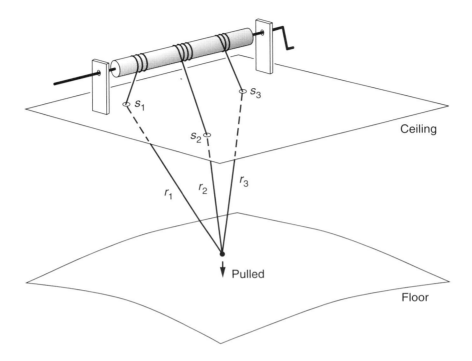

Figure 9.3 Range-difference navigation using three satellites.

Our device is operated as follows: At first the ropes are hanging separately, not yet being knotted together. Adjust the lengths $\bar{r}_1$, $\bar{r}_2$, and $\bar{r}_3$, by trimming the ropes so that the relative lengths $\bar{r}_1 - \bar{r}_2$ and $\bar{r}_2 - \bar{r}_3$ become equal to the differential satellite distances $r_1 - r_2$ and $r_2 - r_3$ that have been measured. Absolute rope lengths are out of our interest here; we are only concerned with their relative lengths. Then, knot the rope-ends together and pull the knot adequately so that every rope becomes stretched taut, and this fixes the knot to a certain position. Now wind the handle to lift the knot up and down, while keeping the ropes taut. Note that the winding of the handle makes $\bar{r}_1$, $\bar{r}_2$, and $\bar{r}_3$ vary equally, and so the relative lengths $\bar{r}_1 - \bar{r}_2$ and $\bar{r}_2 - \bar{r}_3$ will not change. Thus the knot will plot a curve as it moves up and down, and on this curve somewhere the user must exist. The knot can go down and down until it meets the floor at one particular point, and this is where the user's position is fixed, since the user has been assumed to be on the ground.

Good position fix requires a good arrangement of the satellites. If any two of S_1, S_2, and S_3 are close to each other as viewed from the user, small errors in $\bar{r}_1$, $\bar{r}_2$, and $\bar{r}_3$ will cause a big error to the position fixing. So, the satellites must form a constellation of a wide-appearing triangle to ground users.

Problem 9.2

Let S_1, S_2, and S_3 form a horizontal triangle. If $\bar{r}_1 - \bar{r}_2 = \bar{r}_2 - \bar{r}_3 = 0$, where is the user on the ground?

What we have observed explains exactly the positioning principle of the global positioning system (GPS) when it fixes a ground position using three satellites. Synchronizing the message emissions between the GPS satellites is made as follows: Each satellite carries a precise clock of atomic frequency standard that controls the frequency and timing of the message emissions. These messages arrive at ground monitoring stations so as to provide satellite tracking data, and this makes possible the orbit determination of the satellites. A thus-made orbit determination has a by-product of determining the offset and drift of each satellite's clock. The ground monitoring station will then adjust the satellite clocks so as to maintain all satellite clocks in synchronized manner. For the details of the satellite and ground equipment and their actual performance, see, for example, [5,6].

We should remember here that, in a range-navigation system, it was the hub station that identified each user among a number of users—this relation is inverted in GPS. A number of GPS satellites are revolving around the Earth, from which three satellites must be selected by the user so as to obtain a good triangle for positioning. Thus the user must identify each satellite from others, and this is done by identifying the PN-code that carries the satellite message, because each satellite uses a uniquely assigned PN-code.

We must examine here a particular case of three-satellite range-difference navigation: It is the case that the satellites form a straight line as illustrated in Figure 9.4. In this case, the knot can swing around the axis S_1-S_2-S_3 to complicate our discussion; so we assume for the moment that the knot's motion is confined in one plane, say, in the vertical plane, and forget about the floor. Note here that stretching all the ropes is not always possible—this is clarified in Figure 9.5. When the ropes are wound up enough, the rope r_2 will become slack; this is because the knot is driven up quickly by the ropes r_1 and r_3, which are bent inwards when passing through the holes. (Imagine, for clear understanding, that the knot comes up close to the ceiling.) When unwound enough, on the contrary, either r_1 or r_3 will become slack, because r_1 and r_3 will try—in vain—to drive the knot down quicker than r_2 actually drives. Hence, at one particular position of the turning handle, the three ropes will become equally stretched, with none being slack. So we will wind and unwind the handle until this "equal stretch" of the ropes is found. Then (going back to Figure 9.4), fix the handle and recollect the floor. The knot can still swing around the axis S_1-S_3 so as to plot a circle; this situation is identical to the range navigation illustrated in Figure 9.1, so that the user position is fixed to

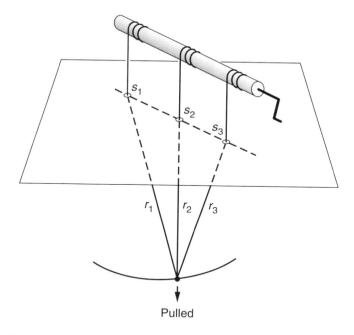

Figure 9.4 Particular case of three-satellite, range-difference navigation.

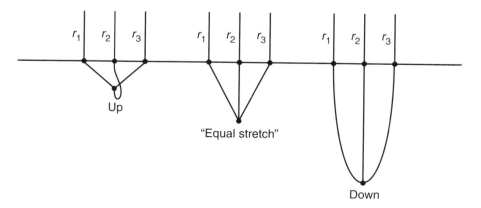

Figure 9.5 Finding *equal stretch* of the ropes.

two candidates on the ground. This mechanism of position fix is close to what happens if the three satellites are geostationary satellites arranged at not very wide separations. With geostationary satellites being used, the work of generating the synchronized messages can all be done at the ground hub station, thanks to the uninterrupted geostationary communication links. The satellites then need to simply have through-repeaters, thus allowing a low-cost navigation system. Such a system has been applied to navigating ships—see [7] for its operational details.

9.3.2 Positioning Using Four Satellites

Now add the fourth satellite S_4, placed at a distance r_4 from the ground user, so that we have four satellites emitting synchronized navigation messages— thus using four satellites to fix a three-dimensional position is the normal operation mode of GPS. Correspondingly, our device becomes as illustrated in Figure 9.6, where the floor has been removed. So that the four small holes can represent the satellite positions, the ceiling will probably need to be curved. The satellite S_4 is placed inside the triangle $S_1 S_2 S_3$ in this figure, with a proper reason that will become clear later.

Assume that the user has measured the differential satellite-distances $r_1 - r_2$, $r_2 - r_3$, and $r_3 - r_4$, and that we have adjusted the relative rope lengths in the same way as we did in the three-satellite case. It is important to remember here that three ropes were necessary and sufficient for fixing the knot position. If we have four ropes, generally one rope will be left slack. Then, what will happen in our device of Figure 9.6 if we wind and unwind the handle? When the ropes are wound up enough, the ropes r_1, r_2, and r_3 will drive up the knot quickly so that the rope r_4 will become slack, by the same mechansim as we

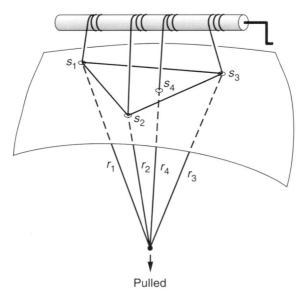

Figure 9.6 Range-difference navigation using four satellites.

have observed in Figure 9.5. When unwound enough, on the contrary, either one of r_1, r_2, or r_3 will become slack. Therefore, at one particular position of the winding handle, the four ropes come into "equal stretch," and at this very moment the knot position marks the user's position. This explains why and how the position can be fixed in three dimensions by GPS. The actual procedure of the positioning may seem quite complex (see for example, [8] for detail), while understanding its basic principle is not at all difficult with our rope analogy.

Good geometry of the four satellites is essential. If any two satellites come close to each other, the performance of the position fix simply reduces to that of three-satellite position fix, thus reducing to two-dimensional fix on the ground. If the satellites form a straight line, the positioning performance becomes the same as that of Figure 9.4, thus reducing again to two-dimensional fix on the ground. Let us then examine an unfavorable geometry that is particular to four-satellite positioning: Satellites S_1, S_2, S_3, and S_4 lie on a horizontal circumference, and the user rests on the vertical line that passes through the circumference center. In this case the four ropes will continue to be in equal stretch, however hard we wind and unwind the handle. This means that the user's position is indefinite along the vertical line, and so the positioning reduces to two-dimensional fix. This example suggests that the user-satellite geometry becomes unfavorable when the saellites lie in the same plane with a certain kind of symmetry.

Problem 9.3

Suppose that the four satellites form a rectangle. If the user rests in a plane of symmetry of this rectangle, then in what manner will the positioning become indefinite?

The ideal satellite-user geometry develops from that of Figure 9.6: Make a wide equilateral triangle $S_1 S_2 S_3$ while elevating S_4 from the plane of this triangle, and place the user near to the triangle's center. Consequently, the user on (or near) the ground will see the three satellites at low elevations with fully diverted azimuths and one satellite at a high elevation—see [5,6].

If five or even more satellites are usable, our device of Figure 9.6 must have the same number of ropes. Then we have to find out the "equal stretch" for all those ropes, thus leading to an increased precision of position fix.

Finally, let us operate our device once more in a slightly different manner, so as to observe another aspect of the performance of a GPS navigation receiver. Assume, that we have finished the adjustment of the relative rope lengths but have not yet knotted the rope-ends. Wind up the ropes until the end of the shortest rope (assume this to be r_1 here) reaches the ceiling. Then start unwinding the ropes, while counting the unwound length by the number of turns of the handle. The rope-ends are knotted together when sufficiently unwound. Counting the unwound length continues until the moment that the knot reaches the floor (three-satellite positioning), or until the ropes come to their equal stretch (four-satellite positioning). At this very moment of position fix, the unwound length determines r_1, the absolute distance for the satellite S_1, and thus determines all the satellite distances at the same time. Note that the position fix and the range finding yielded in a quite parallel manner, with neither being a by-product of the other. So, in this context, a GPS navigation receiver may be thought of as performing a kind of "ranging" to the satellites, even though this receiver does not transmit any timing signals to the satellites.

9.4 Doppler Navigation

One must have noticed while studying the range navigation and the range-difference navigation, that the position fix is finished in one moment and that the satellites may be in any orbits if the satellite geometry appears favorable to the user at the moment of the position fix. These characteristics both become absent to our last category of Doppler navigation, and what makes this difference is that the position fix by Doppler is "dynamic" as we will see, while that by range/range difference was "geometrical."

The basic positioning concept is illustrated in Figure 9.7, where a satellite flies over a user U, passing through the points S_1, S_2, and S_3 at the times t_1, t_2, and t_3, respectively, and emitting a continuous beacon signal of frequency f all the way. This signal is received at U, with a Doppler shift Δf in its frequency. Since the Doppler shift is due to the satellite motion relative to the user, the satellite must fly in a low-altitude orbit so as to generate a sufficient

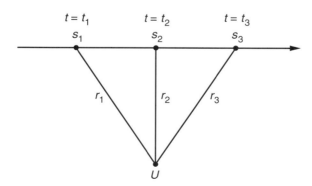

Figure 9.7 Doppler navigation.

Δf for its measurement. Measuring the Doppler shift then determines the rate of change in the satellite distance r by

$$dr/dt = -c\Delta f/f$$

with c the velocity of light. Then, by integrating the measured Δf with respect to time, from t_1 through t_2, we know how much the distance has changed during the interval, as

$$\frac{c}{f}\int_{t_1}^{t_2} \Delta f\, dt = \int_{t_1}^{t_2} (-dr/dt)\, dt = r_1 - r_2$$

The distance change $r_2 - r_3$ over the time period of t_2 through t_3 is known in the same way. Meanwhile the satellite signal informs the orbital elements so that the user should know the satellite positions at $t = t_1$, t_2, and t_3. Now assume, that S_1, S_2, and S_3 in Figure 9.7 are separate satellites; then our positioning problem becomes equivalent to the range-difference position fix by three satellites. Note here that the orbital arc from S_1 through S_3 is nearly a line because the orbit is low in its altitude, so only its short arc is visible to the ground user. Consequently, the position fix must refer to the special case that we saw in Figure 9.4, where the position fix had two candidates on the ground. In practice, however, the measured Doppler shift is modulated by the user's motion due to the Earth's rotation, and this enables us to identify the true position out of the two candidates. If the user was, unfortunately, near to being under the satellite path, position fix is indefinite along the direction perpendicular to the satellite path; in this case, the user will have to wait for another satellite coming in a better path. The ground user position is thus fixed by one satellite with one-way signal transmission while taking a finite period of time. This is the principle of Transit/Navy Navigational Satellite System (NNSS) which operated in the pre-GPS period—see [9] for its early historical background and [6] for a concise explanation. Readers are now encouraged once more to see Figures 9.7, 9.3, and 9.6 in this order so as to review how the navigation concept using one-way signal transmission evolves from a single-satellite two-dimensional fix to a multiple-satellite three-dimensional fix.

The Doppler navigation has a still-useful variation. In this variation, the user emits a beacon signal while the satellite receives it and observes its Doppler shift. As the satellite flies over the user, observed Doppler shifts appear as the curves illustrated in Figure 9.8. If the curve was like Figure 9.8(a), the user

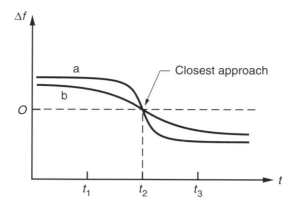

Figure 9.8 Doppler curves.

was close to the satellite path, while Figure 9.8(b) indicates a distant user. This will determine the distance from the satellite path to the user, while the curve's shape determines t_2, the time of the satellite's closest approach to the user. Hence the user's position is fixed on the ground in two dimensions. That there appears two candidates is the same here.

Problem 9.4

Assume a satellite flying at a velocity v along a linear path, and a user at a distance d from the satellite path. Derive, as a function of time, the Doppler curve observed by the satellite and examine how it depends on d.

The measured Doppler-shift is not time-integrated, since doing so would force the satellite to concentrate on one user's signal, thus having to ignore other users that may exist in the same area of view of the satellite. Practically, the satellite, or its control station, has a spectrum analyzer to observe the Doppler curves. If there are a number of users emitting their beacons, the same number of Doppler curves come out from the analyzer. These curves will be different from each other owing to the users' different positions and different beacon frequencies, so that each curve is separately identified. In addition, the beacon signals may carry slow-rate data including their identifications and messages to the control station. Such a system is suitable for surveying the positions of widely distributed objects—in fact, oceanographic buoys, migrating birds and animals (with tiny beacon transmitters attached), and so on are surveyed by the ARGOS system, while COSPAS-SARSAT is surveying marine and aviation distress signals for search and rescue; for these particular systems, see [6,10,11].

References

[1] Rothblatt, M. A., *Radiodetermination Satellite Services and Standards*, Norwood, MA: Artech House, 1987.

[2] Morikawa, E., et al., "Communications and Radio Determination System Using Two Geostationary Satellites," *IEEE Trans. AES*, Vol. 31, No. 2, 1995, pp. 784–794.

[3] Ames, W. G., "A Description of QUALCOMM Automatic Satellite Position Reporting (QASPR) for Mobile Communications," *Proc. of 2nd Int. Mobile Satellite Conf.*, Ottawa, Canada, 1990, pp. 285–290.

[4] Colcy, J. N., and R. Steinhauser, "EUTELTRACS, The European Experience on Mobile Satellite Service," *Proc. of the 3rd Int. Mobile Satellite Conf.*, Pasadena, CA, June 1993, pp. 261–266.

[5] Kaplan, E. D., (ed), *Understanding GPS, Principles and Applications*, Norwood, MA: Artech House, 1996.

[6] Canadian GPS Associates, Wells, D., (ed), *Guide to GPS Positioning*, Canadian Institute of Surveying and Mapping, Ottawa, Canada, 1986.

[7] Ott, L., "The Starfix Satellite Navigation System," *Proc. of IEE 4th Int. Conf. on Satellite Systems for Mobile Communications and Navigation*, London, U.K., Oct. 1988, pp. 208–213.

[8] Hoffman-Wellenhof, B., H. Lichtenegger, and J. Collins, *Global Positioning System— Theory and Practice*, Springer-Verlag, 1994.

[9] Weiffenbach, G. C., "The Genesis of Transit," *IEEE Trans. AES*, Vol. 22, No. 4, 1986, pp. 474–482.

[10] "Argos User's Guide," Service Argos, CNES, Toulouse, France.

[11] "Introduction to the COSPAS-SARSAT System," COSPAS-SARSAT Secretariat, INMARSAT, London, U.K.

10

Mobile Broadcasting

This chapter describes mobile reception of satellite broadcasting services, especially satellite digital audio broadcasting (DAB) services to mobile vehicles and to portable and fixed receivers. Section 10.1 describes system requirements of satellite DAB services and Section 10.2 presents examples of DAB system plans in several countries.

10.1 System Requirements for Digital Audio Broadcasting Services

New digital broadcasting services of CD-quality music programs from satellites to listeners in mobile vehicles and in homes, which is called DAB internationally, digital audio radio services (DARS) in the United States, and digital radio services (DRS) in Canada, have recently been attracting the attention of many people. Actually, digital sound broadcasting services have already been provided in several countries; for example, NICAM 728, which was developed for use with phase alternation each line (PAL), the Digital Satellite Radio (DSR) system, and the Astra Digital Radio (ADR) system for delivering digital stereophonic sound programs at a data rate of 192 Kbps via Astra satellites. In Japan, Music Bird Inc. and Zipang & Sky Communications Inc. are providing digital pulse code modulation (PCM) sound broadcasting programs through eight and six channels, respectively, using the Ku-band transponders on the JCSAT-2 satellites. However, these systems are designed to be received only by fixed receivers, and they may not provide reliable reception in mobile environments.

The International Telecommunication Union's (ITU's) World Administrative Radio Conference for Dealing with Frequency Allocations in Certain

Parts of the Spectrum (Malaga-Torremolinos, in 1992) (WARC-92) allocated the following frequency bands to the satellite sound broadcasting service and complementary terrestrial broadcasting service for the provision of digital audio broadcasting:

- 1,452 to 1,492 MHz worldwide, except for the following specific countries;
- 2,310 to 2,360 MHz for the United States and India;
- 2,535 to 2,655 MHz for Japan, Republic of Korea, China, Russian Federation, Ukraine, Belarus, Singapore, Thailand, Pakistan, Bangladesh, and Sri Lanka.

The worldwide allocation is the L band (1.5 GHz), but the United States and a group of countries mainly in Asia (but also including the Russian Federation) selected the S band.

The ITU-R (ITU Radiocommunications Sector) has considered and recommended the following system conditions for DAB services to vehicular, portable, and fixed receivers in the frequency range 1,400 to 2,700 MHz [1]. The system:

- Can provide a range of audio qualities up to high-quality stereophonic two-channel/multichannel sound with subjective quality indistinguishable from high-quality consumer digital recorded media ("CD quality") to vehicular, portable, and fixed receivers;
- Uses technically feasible and mature source and channel coding, modulation, and advanced digital signal-processing technologies;
- Has better spectrum and power efficiency as well as better performance in multipath and shadowing environments than conventional analog systems;
- Uses satellite and terrestrial systems complementarily for better power and spectrum efficiency through the implementation of hybrid and mixed satellite/terrestrial DAB services;
- Can provide enhanced facilities for program-related data, such as service identification, program labeling, program delivery control, and copyright control;
- Can provide value-added services with different data capacities, such as business data, paging, still picture/graphics, future integrated services digital broadcasting (ISDB), and low-bit-rate video/audio multiplexing;

- Uses common signal processing in receivers for any satellite and terrestrial DAB applications and allows receivers and antennas to be manufactured at low cost through mass production;

- Enables satellite-based DAB systems to provide full coverage for subnational, national, or supranational service areas.

Recently, the ITU-R has been considering recommendations on sound broadcasting via one of two proposed systems. One is the Eureka 147 system, which has been developed by a European consortium established in 1987 and is referred to as Digital System A by the ITU. The project office of the Eureka 147 consortium is managed by the Deutsche Forschungsanstalt fur Luft und Raumfahrte (DLR), based in Cologne, Germany. The other system, Digital System B, was proposed jointly by the Voice of America and the Jet Propulsion Laboratory in the United States. Evaluation tests of the two systems in several propagation environments are being conducted. Both these systems are briefly described in Section 10.1.2. Moreover, studies on several methods of providing DAB systems that take the above requirements into account are being studied in various parts of the world.

10.1.1 Source Coding

The DAB system is designed to provide high-quality digital audio broadcasting for reception by mobile vehicles and by portable and fixed receivers. To transmit high-quality audio signals, which have frequency components ranging from 20 Hz to 20 kHz and a dynamic range wider than 90 dB, the data rate must exceed 2×768 Kbps for a stereophonic sound program sampled at 48 kHz. To broadcast several music programs from satellite transponders, which have limited power and bandwidth, the source coding must use a high degree of compression while preserving the original audio quality. Recently, a number of highly efficient compression algorithms have been developed, and low-cost chips are now available. Below, we present several source-coding techniques for digital audio broadcast.

Efficient source coding can help to increase the spectral efficiency of the overall DAB system by significantly reducing the bit rate. Two main coding techniques have been developed. One is subband coding, where a sound source signal is split into several subband frequency components, and the degree of quantization in each subband varies according to the signal energy. The other is transform coding, which splits frequency components using a transform windowing, related to a bit-allocation technique aided by psychoacoustics rules. Although both these methods can theoretically reduce the bit rate to around

100 Kbps, they can be severely limited by the uncertainty due to the filter banks or by the transform windowing. Examples of DAB source-coding techniques are shown in Table 10.1.

In Europe, the Eureka 147 project started in 1987 to promote research on audio source coding for future DAB services. It uses MUSICAM (Masking pattern-adapted Universal Subband Integrated Coding And Multiplexing), which is an extension of MASCAM. The original MASCAM, which divides the broadband audio spectrum into a number of subband signals with a suitable filter band and quantizes each subband with a bit allocation based on the masking thresholds, was developed by Institut fur Rundfunktechnik (IRT) in Germany. It was improved for different applications by CETT (France), IRT (Germany), and Philips (Netherlands), and renamed MUSICAM. The compression rate of MUSICAM is from 1/4 to 1/12. The apt-X 100 was developed by Audio Processing Technology (APT), a subsidiary of Solid State Logic in the United Kingdom, using subband adaptive differential pulse code modulation (ADPCM) techniques with a compression rate of 1/4.

In the United States, AT&T Bell Laboratory developed ASPEC in cooperation with Thomson and FhG (Germany). Dolby Laboratory developed AC-2 using an adaptive transform coder (ATC) based on the modified discrete cosine transform (MDCT). Scientific Atlanta developed spectrum-efficient digital audio technology (SEDAT) using ATC and based on an understanding of the human auditory system. Moreover, several digital audio coding techniques have been developed for Advanced Television systems.

In Japan, ATAC was proposed by Fujitsu, Victor, NEC, and Sony. NHK has developed subband coding with ADPCM in low-frequency bands using psychoacoustic characteristics with 128 Kbps per monophonic channel, and Sony developed adaptive transform acoustic coding (ATRAC) for their Mini Disc system, which can transmit coded data in a continuous stream or as packets.

The design of source coding and decoding algorithms should be determined based on a trade-off between several opposing factors, as follows:

- Bit rate;
- Quality of the decoded audio signal;
- Time delay for coding and decoding procedures;
- Complexity of coding and decoding circuits;
- Compatibility with other sampling frequencies.

10.1.2 Transmission Schemes

Several transmission schemes were studied, mainly by European, U.S., Canadian, and Japanese groups. In this section, we present briefly the ITU-R's

Table 10.1
DAB Source Coding Techniques

Method	Subband Coding With ADPCM in Low Frequency Bands	MUSICAM (Masking Pattern-Adapted Universal Subband Integrated Coding and Multiplexing)	PASC (Precision Adaptive Subband Coding)	SB-ADPCM	apt-X100	ATRAC (Adaptive Transform Acoustic Coding)	AC-2	SEDAT (Spectrum Efficient Digital Audio Technology)
Start of development	1990	1989	1991	1988	1988	1991	1990	1991
Developer	NHK (Japan)	IRT (Germany), CCETT (France), Philips (Netherlands), Matsushita (Japan)	Philips (Netherlands)	PTT (Switzerland)	Audio Processing Technology (APT), a subsidiary of Solid State Logic (UK)	Sony (Japan)	Dolby (U.S.)	Scientific Atlanta, Inc. (U.S.)
Coding method	Divide into 32 subbands	32 subbands	32 subbands	4 subbands	4 subbands, ADPCM	Subbands (QMF) + transform (MDCT) coding	MDCT + adaptive transform coder (ATC)	DCT + ATC
Compression rate	1/6 – 1/8	1/4–1/12	1/4	1/4	1/4	N/A	N/A	N/A

Digital Systems A and B. In Europe, Digital System A has been adopted as a standard (ETSI standard ETS 300401) for broadcasting satellite service (BSS)/broadcasting service (BS) (sound) to vehicular, portable, and fixed receivers.

10.1.2.1 Digital System A (Eureka 147) [2]

Figure 10.1 shows the conceptual block diagram of Digital System A, which was developed by the Eureka 147 DAB consortium and was actively supported by the European Broadcasting Union (EBU) to introduce DAB services in Europe in 1995. Since 1988, the system has been successfully demonstrated and extensively tested in Europe, Canada, the United States, and other countries [3–6].

The audio source coding method is International Standards Organization (ISO)/International Electro-technical Commission (IEC) MPEG-Audio Layer II, whose encoding consists of the basic mapping of the PCM audio signals at a sampling rate of 48 kHz into 32 subbands, fixed segmentation of data into blocks, a psychoacoustical model for the adaptive bit allocation, and quantization using block companding and frame coding. The subband coding compression system is the MUSICAM system. Available bit rates for a monophonic sound signal are 32, 48, 56, 64, 80, 96, 112, 128, 160, and 192 Kbps with 2 Kbps for program-associated data. Stereophonic sound signals are conveyed as two cophased monophonic signals for better bit error performance. Decoding in the receiver is simple and straightforward, requiring only demultiplexing, expanding, and inverse-filtering operations. In mode III, which is applicable to satellite and mixed and hybrid satellite/terrestrial DAB, the frame duration T is 24 msec. The fast information channel includes multiplex configuration information, service information, and other data. The main service channel includes program services (audio data and program-associated data) and spare capacity.

The transmitted signal is built around a conceptual frame format with time-division-multiplexed components consisting of a synchronization channel, a fast information channel, and a main service channel, as shown in Figure 10.2. In the synchronization channel, the first null symbol is transmitted to provide a coarse synchronization, followed by a fixed reference symbol to provide fine synchronization, automatic gain control, automatic frequency control, and phase reference functions in the receiver. It is also used to carry encoded transmitter identification codes. The main service channel includes both sound and data channels that are defined in the fast information channel. The total duration T is either 96, 48, or 24 msec, depending on the transmission mode. The transmission modes of the Eureka 147 system are shown in Table 10.2 [7].

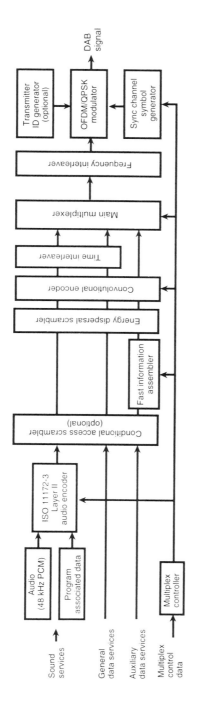

Figure 10.1 Block diagram of the transmission part of Digital System A (Eureka 147).

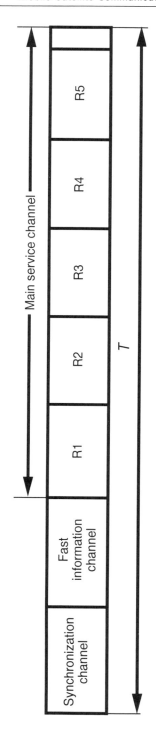

Figure 10.2 Multiplex frame structure of Digital System A (Eureka 147).

Table 10.2
Transmission Modes of the Eureka 147 DAB System

Parameter		Mode I	Mode II	Mode III	Mode IV
Guard interval duration	μsec	246	62	31	123
Frame duration	msec	96	24	24	48
Number of carriers		1,536	384	192	768
Carrier spacing	kHz	1	4	8	2
Total symbol duration	μsec	1,246	312	156	623
Symbols per frame		76	76	153	76
Frequency range up to		300 MHz	1.5 GHz	3.0 GHz	1.5 GHz

Four different operating modes are used, since the propagation conditions vary with frequency. Mode I is most suitable for a terrestrial single-frequency network (SFN), in which all transmitters covering a particular area broadcast the same set of sound programs on the same radiofrequency (RF) channels, operating at frequencies below 300 MHz. All the SFN transmitters need to be synchronized, and receivers need to recognize the signals transmitted from the various transmitters with different time delays. Mode II is applicable to local and regional services requiring one terrestrial transmitter, and to hybrid satellite/terrestrial transmission at frequencies below 1.5 GHz. Mode III is most suitable for satellite and complementary terrestrial transmission at all frequencies up to 3.0 GHz. Mode IV has recently been introduced for transmitters providing seamless coverage of large areas by SFNs in the L band.

The error-correction methods for the fast information channel and the main service channel are both convolutional encoding with a constraint length of 7, variable coding rates, and interleaving both in frequency and time. For the audio signals, the unequal error-protection method, where greater error protection is given to some important bits than to others, is used. The average coding rate, which is defined as the ratio of the number of source-encoded bits to the number of encoded bits after convolutional encoding, takes a value from 1/3 to 3/4. The fast information channel is encoded at a constant rate of 1/3. Time interleaving is applied to the convolutionally encoded data at an interleaving depth of 16.

The modulation scheme is coded orthogonal frequency division multiplexing (COFDM), which was presented in Chapter 6 of this book. The multiplexed signal is transmitted via several hundred or thousand closely spaced RF carriers having a total bandwidth of about 1.5 MHz. COFDM can reduce impairment due to multipath fading, since the symbol duration is designed to be larger than the delay spread of the transmission channel, so an echo shorter than the guard interval will not cause intersymbol interference. Each carrier is modulated by differentially encoded PSK. Under frequency-selective multipath fading, some of the carriers are enhanced while others are suppressed. Therefore, frequency interleaving due to rearrangement of the bit stream can also reduce the effect of frequency-selective fading.

10.1.2.2 Digital System B [2]

Figure 10.3 shows a functional block diagram of the receiver in Digital System B. The optional functions for mitigating propagation problems are shown by dashed blocks. Carrier recovery and symbol recovery are performed by a QPSK Costas loop and a matched filter with timing provided by a symbol-tracking loop, respectively. After the frame synchronization has been established, the signal is deinterleaved, decoded, and demultiplexed. If the Reed-Solomon decoder cannot remove all the errors in a data block, the data block is marked as a bad block. This indication is used at the audio decoder to control audio output muting, and can be used by the time- or signal-diversity combiner to select the better signal. Time diversity is optionally implemented to transmit a delayed version of a data stream multiplexed together with the original. In the receiver, one of these two data streams, whichever has the fewest errors, is selected for output. Signal diversity requires the independent processing of the signal, or of different frequency signals, up to the diversity combiner. Then, the diversity combiner performs the time alignment and selects the most error-free data stream.

Figure 10.4 shows a block diagram of Digital System B. The transmitter performs multiplexing for all analog audio and digital data sources to be combined onto one carrier, forward error correction, and QPSK modulation. A number of audio encoders are provided to handle the required number of limited-bandwidth monaural, limited- and full-bandwidth stereo, and full-bandwidth five-channel surround sound channels. The bit rate ranges from a minimum of 16 Kbps for limited-bandwidth monaural to approximately 320 Kbps for five channels. Audio encoder data rates are limited to multiples of 16 Kbps. All audio channels and data channels are multiplexed into a composite serial data stream. The output data rate ranges from a minimum of 32 Kbps to a maximum in the range from 1 to 10 Mbps. Error-correction encoding is a convolutional encoding with a rate of 1/2 and a constraint length of 7, preceded by Reed-Solomon encoding with a rate of 140/160. A block time interleaver is used with the block length proportional to the composite data rate to provide an interleaver frame time on the order of 200 msec at any data rate. The modulation scheme is QPSK modulation, and FDM is used.

The receiver performs the demodulation, decoding, and demultiplexing functions, as well as the digital-to-analog conversion of the selected audio

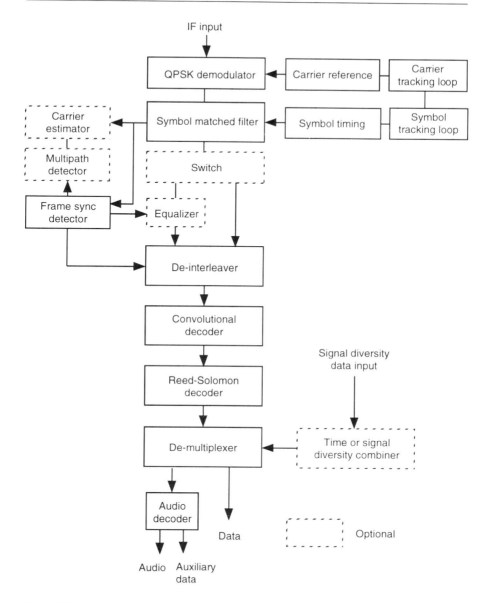

Figure 10.3 Functional block diagram of a receiver in Digital System B.

signal. In mobile reception environments, where there are propagation problems with signal blockage due to trees, buildings, and other obstacles, the receiver has functions needed for time or signal diversity, or equalization. Carrier demodulation is performed with a phase-locked coherent QPSK demodulator,

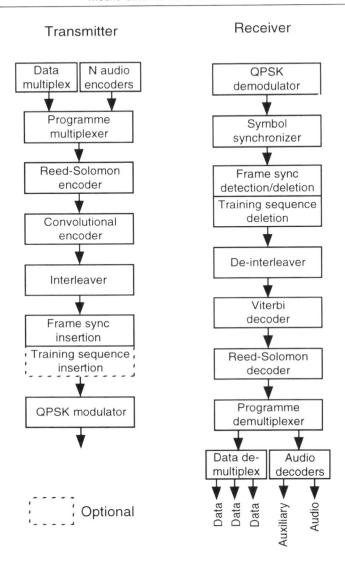

Figure 10.4 Block diagram of Digital System B.

and symbols are detected by a matched filter with timing provided by a symbol-tracking loop. A Viterbi decoder, followed by a Reed-Solomon decoder, reduces the error rate. If the Reed-Solomon decoder cannot remove all the errors in a data block, the data block is marked as bad. This indication is used by the diversity combiner to select the better signal, as well as by the audio decoder to control audio output muting.

10.1.3 System Configuration

In providing satellite DAB services to mobile vehicles with high reliability, it is necessary to minimize service outages (which are caused by multipath fading, Doppler shifts, and shadowing and blockage due to buildings, trees, and terrain) by using several outage mitigation methods, such as satellite spatial diversity and a downlink margin for overcoming shadowing by vegetation.

The shadowing and blockage mainly depend on the satellite elevation angle. The higher the elevation angle, the less frequent shadowing becomes. In northern countries, such as northern Europe and Canada, the satellite elevation angles are low, so a large downlink margin will be required to overcome fading. To reduce this margin requirement, higher elevation angles can be achieved by using highly inclined elliptical orbits (HEO) instead of geostationary orbits (GEO). In Europe, HEO DAB systems are being studied as part of the Archimedes program, in which elevation angles of 55 degrees or more are possible.

Various solutions to ease this problem have been proposed. Satellite and terrestrial DAB systems that cooperate to provide reliable DAB service using the same frequency band have been considered. Such systems are presented below, and compared in Table 10.3.

10.1.3.1 Single Satellite System

The concept of satellite DAB systems is similar to the conventional satellite broadcasting systems via individual satellites. With the conventional digital modulation schemes and omnidirectional antennas having gain of about 5 dBi, a fading margin of about 15 dB is required to reduce shadowing and multipath Rice or Rayleigh frequency-selective fading in urban and suburban areas. The frequency-interleaving technique of COFDM or a spatial-diversity technique can reduce the fading margin to 10 dB.

10.1.3.2 Hybrid Systems

Terrestrial-based signal repeaters, called *gap fillers,* retransmit the satellite-transmitted signals at low power on the same frequency to an area where the satellite signals cannot be received satisfactorily due to blockage from high buildings, thick foliage, or terrain features. This concept was proposed by EBU, and it is used in the Eureka 147 DAB system. While the listener's receiver receives signals from both the satellite and the gap filler at slightly different times, the two sets of signals add together constructively to enhance the signal strength. This can reduce the fading margin from 10 dB to 5 dB.

Table 10.3
Comparison of Satellite DAB Systems

	Single Satellite	Hybrid	Mixed	Mixed + Hybrid
System concept	DAB via a single satellite	Terrestrial signal repeaters, gap fillers, retransmit the satellite-transmitted signals at low power on the same frequency to an area where the satellite signals cannot be received satisfactorily due to blockage	Both satellite and terrestrial systems use the same frequency band, but the terrestrial systems operate in the frequency blocks that are not used in the satellite system	Mixed DAB systems together with gap fillers are operated
Characteristics	Simple and relatively low cost for development	This can solve the blockage problems, and receivers don't need to switch the channels for satellite and terrestrial services	This can broadcast different programs for satellite and terrestrial services, and requires less initial cost, if only terrestrial services start	Medium characteristics between hybrid and mixed systems, and most frequency-efficient system
Link margin	10 dB	5 dB	10 dB	5 dB
Suitable usage	Cover a medium-size country via a satellite	Cover an area where frequent blockage due to buildings and terrain occurs (in urban areas)	Area where large cities are scattered	Area that is suitable to mixed system with frequent shadowing
Problems	Frequency blockage and shadowing in urban areas	Large cost for construction of gap fillers, if there are too many frequent blockages	Switching between satellite and terrestrial channels is required	Large cost for construction of gap fillers

10.1.3.3 Mixed Systems

In the mixed DAB system, which was proposed by Canada, both satellite and terrestrial systems use the same frequency band, the same transmission format, and the same modulation schemes, but the terrestrial system operates in the frequency blocks that are not used in the satellite system. Receivers are designed to receive both satellite and terrestrial DAB signals with omnidirectional antennas.

For example, after local terrestrial DAB services have been introduced in urban areas in the first stage, the satellite-based DAB services with the mixed system configuration can be easily introduced. Conversely, after global satellite DAB services have been introduced, local terrestrial DAB services can be introduced in the complementary mixed configuration.

10.1.3.4 Mixed + Hybrid Systems

This system uses the mixed DAB system together with several gap fillers, which transmit the DAB signals at the same frequencies as those of the satellite system. The use of terrestrial retransmission stations, which transmit the terrestrial-based signals, provides a countermeasure against shadowing by buildings. The isolation distance between neighboring countries, which is necessary to prevent mutual interference of broadcasting signals, can be shorter than in the mixed system.

Before introducing a DAB system, one must evaluate it from economical, operational, and future developmental aspects. In the Eureka system, an SFN concept has also been proposed. In the SFN, all transmitters covering a particular area broadcast the same set of sound programs on the same RF channels. All the SFN transmitters need to be synchronized, and receivers need to recognize the signals transmitted from the various transmitters with different time delays.

10.1.4 Receiver

Several configurations of receivers can be considered for the above four system configurations and two modulation schemes, COFDM and QPSK. For the single-satellite system and the hybrid terrestrial/satellite system, the same config-urations of receivers can be used. Since, in the mixed terrestrial/satellite system, receivers have to compare the powers of satellite-based and terrestrial DAB signals and switch to the stronger signal, the receivers for the mixed system need a relatively complicated configuration. Moreover, the mixed + hybrid system requires a more complicated configuration of receivers.

Figure 10.5 shows the configuration of a mobile vehicle's receiver, where both FM and AM receivers, and a cassette tape player (or CD player or mini disc player), are installed. Space-diversity antennas are installed for continuous

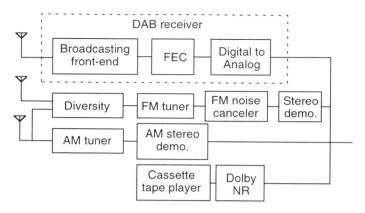

Figure 10.5 Configuration of a mobile vehicle's receiver.

reception in a moving vehicle, and an FM noise canceller is installed to reduce ignition noise produced in a car. Since the satellite DAB receiver needs to be small and inexpensive, mobile antennas should be azimuthally nondirectional. Several antennas for mobile satellite communications can also be used in the satellite DAB systems.

10.2 Digital Audio Broadcasting (DAB)

In this section, we survey various satellite DAB systems proposed in several countries.

10.2.1 United States

In the United States, Satellite CD Radio Inc. was the first to apply to the Federal Communications Commission (FCC) for a license to provide DAB services, in 1990. Subsequently, the FCC selected one of the three satellite and terrestrial DAB frequency allocations successfully designated at ITU's WARC'92. Of the five companies that followed CD Radio's lead in filing an application for a DAB license, three remain: American Mobile Radio Corporation (AMRC), Digital Satellite Broadcasting Corporation (DSBC), and Primosphere L.P. Most DAB systems that have been studied in the United States operate with in-band/on-channel (IBOC) systems, which allow simultaneous operation of digital and traditional analog broadcasting on the same FM radio bands [8].

Table 10.4 shows terrestrial DAB systems developed or proposed in the United States. The Acorn DAB system, which was opposed by USA Digital

Table 10.4
Terrestrial DAB Systems Developed or Proposed in the United States

Systems	Acorn DAB	Digital FM-S	LinCom DAB	MFM	Stanford Telecom
Developers	USA Digital Radio	Synetcom Digital	LinCom Corporation	Mercury Digital Communications	Stanford Telecommunications
Source coding	MUSICAM	N/A	N/A	N/A	Dolby AC-2
Channel coding	CPVDM	FM-S	DMSK	MFM	D-SCPC
Bandwidth	In-band	In-band	In-band	In-band	Out-of-band
Compatibility	Yes	Yes	No	No	No

Radio, combines a DAB signal that is −30 dB lower in power with the conventional analog FM signal. The modulation scheme of the DAB signal is coded polyvector digital modulation (CPDM). Within the FM bandwidth of 200 kHz, more than 21 carriers are assigned and each carrier is modulated by QPSK. The band efficiency is 1.6 bps/Hz and the total transmission capacity is 320 Kbps. However, the 200-kHz bandwidth is not wide enough to compensate for frequency-selective fading by using frequency interleaving, and 320 Kbps is not enough to use forward error control.

Digital FM-S (subcarriers) multiplexes DAB signals from four carriers onto the baseband spectrum of conventional analog FM. As a multipath mitigation method, it uses adaptive antennas with diversity capability. A high-compression source-coding algorithm is required and mobile reception may not be possible.

LinCom DAB, which was proposed by LinCom Corporation, uses the same bandwidth as but is not compatible with conventional FM, and the compression ratio is 1/10. The modulation scheme is DMSK with a convolutional encoding rate of 7/8, resulting in bandwidth efficiency of 1 bps/Hz. The total transmission rate is 177 Kbps with frequency-diversity capability.

MFM, which was proposed by Mercury Digital Communications, uses the same bandwidth as conventional FM. The modulation scheme is multifrequency modulation (MFM), in which 192 orthogonal carriers are modulated by trellis-coded 8-PSK. The MFM signal is transmitted at a power 16 dB lower than the analog FM signal. As in Acorn DAB, the bandwidth of 200 kHz is not wide enough for frequency interleaving.

Stanford Telecom DAB was proposed by Stanford Telecommunications to CCIR (presently ITU-R) in 1990, and is currently known by ITU as Digital

System B. The source coding is Dolby AC-2. Within the bandwidth of 3.5 MHz, 12- channel data is dynamically assigned to 12 carriers. Each carrier is modulated by QPSK at 537 Kbps, and frequency hopping is performed in every 2-msec period. The FEC is convolutional encoding with a rate of 1/2 and Viterbi decoding. Adaptive equalization is used at the receiver.

A number of DAB systems have been proposed, and a number of tests and demonstrations have been carried out. Since 1992, the Eureka 147 system has been tested along with several in-band proposals where the digital audio signals are transmitted in the same band as the current analog service. Two concepts have been proposed: in-band/on-channel (IBOC) and in-band/adjacent-channel (IBAC).

The following in-band systems were tested by the Electronics Industry Association (EIA) [7]:

- USADR-AM (0.54 to 1.7 MHz)—an IBOC system;
- AT&T (FM band)—an IBAC system;
- AT&T Amati (FM band)—an IBOC system;
- USADR-FM1 (FM band)—an IBOC system;
- USADR-FM2 (FM band)—an IBOC system.

The experimental results show that the in-band digital systems may cause/suffer intolerably high interference to/from analog systems, particularly in a multipath environment.

Finally, let us consider one example of U.S.-based DAB systems: WorldStar, which plans to provide low-cost service and portable receivers to customers in the developing nations of Africa, Asia, and Central and South America using the ITU's L-band DAB allocation that was avoided by the United States.

10.2.1.1 WorldStar [9]

WorldSpace Inc., the satellite radio company based in Washington, DC, is preparing a worldwide DAB system using three geostationary WorldStar satellites called AfriStar (to be located at longitude of 21 degrees east), CaribStar (95 degrees west), and AsiaStar (105 degrees east). This system will provide DAB services equivalent to AM mono, FM mono, FM stereo, and CD-quality stereo to small portable radios using layer 3 of MPEG 2 at bit rates of 16, 32, 64, and 128 Kbps, respectively, in the 1,467 to 1,492 MHz frequency band (L band). Uplinks are accessed by VSAT-sized broadcaster's Earth terminals with FDMA/QPSK schemes at a prime rate of 16.056 Kbps. Up to 288 uplink prime-rate carriers can be transmitted to the satellite in its global uplink beam.

The downlink scheme is multiple channels per carrier (MCPC) with TDM. Each satellite has three downlink spot beams, each supporting 96 prime-rate channels, which can be combined to create program channels at bit rates up to 128 Kbps. The WorldStar system is very similar to Digital System B, recommended by the ITU-R.

Candidates for the receiver's antenna are a small compact patch antenna with gain of 4 to 6 dBi with no vertical pointing, a phased array antenna with gain of 10 to 12 dBi, and a rod-shaped helical antenna. A detachable antenna is mounted on a small tripod on the ground or mounted on a window frame. For mobile reception, antennas with gain of about 4 dBi are mounted on the vehicle. To mitigate multipath interference, spatial diversity array antennas mounted at various locations on the vehicle are used to choose the maximum signal component. The small broadcaster's Earth terminals have 2.4m-diameter X-band parabolic antennas with 25W power amplifiers, which can transmit a 128-Kbps program channel signal.

Forward error correction (FEC) is provided by a concatenated coding scheme, which includes a (255, 223) Reed-Solomon block coding, block interleaving, and 1/2 convolutional encoding, for a low bit error rate. FEC encoding increases the prime rate to 36.826 Kbps with a control word and a preamble code. Onboard the satellite, the uplink FDMA/QPSK signals are received, demultiplexed, demodulated, routed to one or more of the three downlink beams, and then time division multiplexed with 1-sec-long frames. The downlink MCPC/TDM has a rate of 1.767688 megasymbols per second.

The WorldStar system will also provide multimedia services, such as large database downloads to personal computers for business applications, map and text information for travelers, and color images to augment audio programs for advertising and entertainment.

10.2.2 Canada [10]

In Canada, existing AM and FM radio services are being replaced by terrestrial DAB services using the Eureka 147 system in the 1,452 to 1,492-MHz frequency band (L band). Terrestrial DAB services will be provided in small- to medium-sized service areas, such as cities and their suburbs, while satellite DAB services can cover larger areas as well as remote or sparsely populated areas and highways. Use of a common transmission format for both types of services makes possible a common receiver, and the mixed satellite and terrestrial approach is being seriously considered in Canada and in other countries. Canada has been testing DAB systems since 1990, and the first licensed commercial terrestrial DAB station was scheduled to start in early 1996. The commercial introduction of satellite DAB is expected in about 2003 to 2007.

The source coding for CD-quality audio uses layer II from the ISO/IEC 11172-3 standard, also known as MPEG I, at 1,152 Kbps. A 1/2 FEC uses 2,304 Kbps in a given channel bandwidth of 1.536 MHz, which contains five CD-quality stereo programs. A wideband multicarrier modulation scheme implemented as COFDM in conjunction with frequency and time interleaving is used.

When designing satellite DAB systems that have high reliability for all reception conditions, including rural, suburban, and urban environments, important factors to consider are propagation anomalies due to multipath fading, shadowing, and Doppler effects. It is not realistic to assign a high downlink margin in order to overcome frequent signal attenuation caused by shadowing or blockage in urban areas. The Canadian approach is a hybrid satellite service that uses terrestrial on-channel repeaters to fill in holes in the coverage due to shadowing, especially in urban areas. Therefore, the downlink margin for their satellite DAB service is specified based on suburban reception. Table 10.5 shows an example of the Canadian satellite DAB downlink budget for a GEO satellite [10]. The satellite receiver assumes a circularly polarized omnidirectional antenna having gain of 4 dBi and a noise figure of 2 dB. This budget suggests that 62.5 dBW of equivalent isotropically radiated power (EIRP) is required for a fading margin of 6 dB. Covering Canada with eight satellite beams requires spot beams as small as 1.4 degrees, which is possible with a 10m parabolic reflector; this would provide 41 dB of gain. Therefore, the transmitting power per beam is 24.5 dBW or 280W, because each beam provides two DAB channels, one of which is a spare for future growth. On the other hand, the terrestrial receiving antenna for dual-mode receivers assumes a vertically polarized antenna with gain of 0 dBi. The coexistence of these two

Table 10.5
Canadian Satellite DAB Downlink Budget

Operating frequency	1,472.0 MHz
Data rate	1,152.0 Kbps
Required E_b/N_0	8.0 dB
Theoretical downlink C/N_0	68.6 dBHz
Margin	3.9 dB
Required downlink C/N_0	72.5 dBHz
Satellite EIRP per channel	62.5 dBW
Propagation loss	163.0 dB
Receiver antenna gain	4.0 dBi
Coupling and filter losses	1.0 dB
Receiver noise figure	2.0 dB
Receiver G/T	−21.8 dB/K

services in the same band requires the ability to select any of the 1.536-MHz-wide channel blocks in the 40-MHz band and to reject strong neighboring channel blocks. Such a system has been developed and tested by the Communications Research Centre (CRC) [11].

10.2.3 Europe

In Europe, terrestrial DAB services have succeeded with Eureka 147. The United Kingdom, Sweden, and Denmark have already begun operating terrestrial DAB stations, and many other countries are conducting DAB pilot programs [3–6,12]. The world's first official DAB services were provided by BBC Radio in the United Kingdom, and by the Swedish Broadcasting Corporation in Sweden, on September 27, 1995.

The provision of satellite DAB services by a highly elliptically orbiting HEO satellite, Archimedes, is being considered because when geostationary satellites are used, counties in northern latitudes suffer more frequent shadowing and blockage due to the low angle of elevation. The Archimedes satellite constellation[13] is based on multiregional highly inclined elliptical orbits (M-HEO). It consists of six satellites in different orbital planes, each having an 8-hour period, with an operation orbit period of 4 hours. The apogee and perigee altitudes are 26,786 km and 1,000 km, respectively. In Europe, elevation angles of 55 degrees or more are possible. A handover of transmissions from one satellite to another is required every 4 hours.

As mentioned in the previous section, in their first implementations, DAB systems will provide CD-quality audio programming using the COFDM, Eureka 147, and Jessi AE13 standards in the L band (1,452 to 1,492 MHz), which was assigned in WARC'92. However, a number of broadcasters are considering lower bit rate services. Some of them are expecting to provide a number of multimedia applications to a user's personal computer with megabit-per-second rates, which are intermediate between the kilobits per second rate of cellular services and the 100-Mbps rate of optical fiber. Using the telephone or cellular network, broadcasts with some interactive services can also be provided.

Kozamernik reported recent activities of DAB services in Europe in *EBU Technical Review* [7]. Here, we will briefly summarize his report.

In Belgium, a national DAB has not been established, since Belgium has several different cultural communities such as Flemish, French, and German. However, the Flemish language public broadcaster, BRTN, is planning to start preoperational DAB services at the end of 1997. Denmark is setting up a national DAB platform, and the National Telecom Agency has already decided the frequency allocation for DAB services. Since September 1994, several experiments on DAB services have been carried out.

The Finnish national DAB platform is working, and several experimental transmissions began in February 1994. The Finnish manufacturing industry is developing a combined terrestrial/satellite DAB receiver in cooperation with the European Space Agency (ESA). In France, the France DAB Club, which was established in 1991, has conducted experiments in Paris in the VHF and L bands, and has decided in favor of the L band. Radio France has established a Working Party to study DAB services, and has also signed an agreement with TDF to provide DAB services in all major metropolitan areas and motorways.

The German national DAB platform was established in 1991, and it plans to provide DAB services in time for the Berlin Internationale Funkausstellung fair in 1997, and to achieve complete coverage of Germany by the year 2000. In Hungary, a DAB experiment using one transmitter was conducted in 1995, and regular DAB services will start in 1997. In Italy, several experiments in the VHF band are being carried out at a DAB testbed in the Aosta Valley. In the Netherlands, tests in the VHF band are being carried out by NOZEMA in Haarlem, Hilversum, and Rotterdam to provide 40% coverage of the Dutch population. Recently, DAB was the subject of governmental auditing.

In Norway, after the installation of the first DAB transmitter in Oslo in April 1994, a fourth transmitter will soon be installed to evaluate the services in mountainous areas. In Sweden, after the first DAB experiments started in 1992 in Stockholm, there are now four DAB transmitters operating. An official DAB service started in Stockholm on September 27, 1995.

In Switzerland, Swiss Telecom PTT has been operating two test SFNs in the VHF and L bands since 1993 and 1994, respectively. The official introduction will be in 1997. In the United Kingdom, the national DAB Forum was established in 1992. Official service was started by the BBC on September 27, 1995. Initially, five transmitters will cover southeast England, which has 20% of the U.K. population. Within two and one-half years, 27 transmitters will cover about 60% of the U.K. population.

10.2.4 Japan

In Japan, while several digital PCM sound broadcasting services are being provided by Ku-band transponders on the JCSAT-2 satellites, these systems were designed to be received by a fixed receiver. For experiments in mobile reception of satellite DAB services, there are plans to launch a satellite called Engineering Test Satellite VIII (ETS-VIII) [14].

In Japan, the S band (2,535 to 2,655 MHz) was assigned for satellite DAB services in WARC-92. The ETS-VIII's experimental system has an S-band digital audio broadcasting system as well as an S-band mobile satellite communication experimental system, which is being developed to achieve mobile-to-mobile direct communications between handheld terminals using large onboard antennas, onboard signal processing, and so forth. The two systems share several onboard subsystems, such as 10m-diameter antennas and a beam-forming network (BFN). Figure 10.6 shows the configuration of the ETS-VIII satellite, Figure 10.7 shows the antenna coverage of an EIRP of 65 dBW, and Table 10.6 shows the specifications of ETS-VIII's satellite DAB system. The frequency from the hub to the satellite is 30 GHz (Ka band) and the frequency of the satellite-to-mobile receivers is 2,535 to 2,540 MHz (S band). The modulation scheme is COFDM with each carrier modulated by QPSK, based on the terrestrial/satellite hybrid system configuration. The maximum data rate is 256 Kbps, and six channels are available. The mobile receiver is designed to have an antenna gain of 7 dBi and a figure of merit, G/T, of −18.8 dB/K. The launch mass of ETS-VIII will be around 2,500 kg,

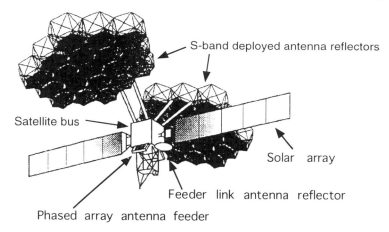

Figure 10.6 Configuration of ETS-VIII.

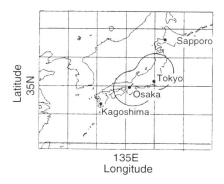

Figure 10.7 ETS-VIII's antenna coverage with an EIRP of 65 dBW.

Table 10.6
Specifications of the DAB System of the ETS-VIII System

Frequency	
Hub-to-satellite	Ka band (30 GHz)
Satellite-to-mobile receiver	S band (2,535–2,540 MHz)
Modulation	COFDM/QPSK
Data rate	256 Kbps (maximum)
Number of channels	6
Required BER	10^{-6}
Satellite EIRP	65 dBW
Terminal	
Antenna gain	7 dBi
G/T	−18.8 dB/K

its geostationary position will be longitude 135 degrees east, and the mission life will be three years.

10.2.5 Other Countries

The DAB services in other countries are presented here, based on Kozamernik's report [7].

In Australia, terrestrial demonstrations of DAB were conducted during 1994 in Sydney and Canberra with the Eureka 147, and in June 1994, the first L-band satellite DAB test was carried out using the Australia Optus B3 satellite. The national telecommunications carrier, TELSTRA, was scheduled to carry out experiments in Sydney and Melbourne by the end of 1996.

In China, terrestrial DAB tests started in 1995, in cooperation with the European Commission and the German national DAB platform, using the Eureka 147. In India, DAB services will be implemented in three phases. In the first phase, due to commence in 1998, a limited terrestrial DAB service will be provided in four metropolitan cities. In the second phase, independent local services will be gradually added to a number of FM stations until the year 2003, when satellite DAB service will start.

In Mexico, a terrestrial DAB test was successfully conducted with Eureka 147 in the L band in Mexico City in 1993. Then, in July 1995, an L-band satellite test was conducted via the Solidaridad 2 satellite.

References

[1]   Recommedation ITU-R BO.789-2, "Service for digital sound broadcasting to vehicular, portable, and fixed receivers for broadcasting-satellite service (sound) in the frequency range 1400-2700 MHz," *ITU-R Recommendations*, 1995 BO Series Fascicle, Broadcasting-Satellite Service (Sound and Television), Geneva, Switzerland, 1995.

[2]   Recommendation ITU-R BO.1130-1, "Systems for digital sound broadcasting to vehicular, portable, and fixed receivers for broadcasting-satellite service (sound) bands in the frequency range 1400-2700 MHz," *ITU-R Recommendations*, 1995 BO Series Fascicle, Broadcasting-Satellite Service (Sound and Television), Geneva, Switzerland, 1995.

[3]   Lau, A., M. Pausch, and W. Wutschner, "First results of field tests with the DAB single frequency network in Bavaria," *EBU Technical Review*, Autumn 1994, pp. 4–27.

[4]   Erkkila, V., and M. Jokisalo, "DAB field trials in Finland," *EBU Technical Review*, Autumn 1994, pp. 28–35.

[5]   Maddocks, M.C.D., I. R. Pullen, and J. A. Green, "Field trials with a high-power VHF single frequency network for DAB: Measurement techniques and network performance," *EBU Technical Review*, Autumn 1994, pp. 36–55.

[6]   Cominetti, M., "The RAI plans for DAB field tests," *EBU Technical Review*, Winter 1994, pp. 50–57.

[7]   Kozamernik, F., "Digital audio broadcasting—radio now and for the future," *EBU Technical Review*, Autumn 1995, pp. 2–27.

[8]   Kunz, C., "DARS wars: the government strikes back," *Via Satellite*, Vol. 12, No. 4, April 1997, pp. 48–56.

[9]   Campanella, S. J., "The worldspace satellite-to-radio multimedia broadcast system: a technical overview," *Proc. 16th Int. Communications Satellite Systems Conf.*, AIAA, Washington, DC, February 25–29, 1996, pp. 826–831.

[10] Paiement, R. V., R. Voyer, and D. Prendergast, "Digital radio broadcasting using the mixed satellite/terrestrial approach: an application study," *Proc. 4th Int. Mobile Satellite Conf.*, IMSC'95, Ottawa, Canada, June 6–8, 1995, pp. 439–444.

[11] Wight, J. S., "Versatile high dynamic range receiver front end for digital radio broadcast and M-SAT reception," *Proc. IEEE MTT-S Symp. on Technologies for Wireless Applications*, Feb. 1995, pp. 161–166.

[12] O'Leary, T., "Terrestrial digital audio broadcasting in Europe," *EBU Technical Review*, Spring 1993, pp. 19–26.

[13] Galligan, K. P., R. Viola, and C. Paynter, "Evolution of DAB by satellite to meet the multimedia challenge," *Proc. 4th Int. Mobile Satellite Conf.*, IMSC'95, Ottawa, Canada, June 6–8, 1995, pp. 433–438.

[14] Kitahara, H., et al., "Japan's geostationary mobile communication test satellite," *Proc 2nd Ka-band Utilization Conf. and Int. Workshop on SCGII*, Florence, Italy, Sept. 24–26, 1996, pp. 77–82.

11

Future Systems—Towards Intelligent Satellites

11.1 Overview

The concept of a future advanced mobile satellite communication system is shown in Figure 11.1. The advanced mobile satellite communications will provide several kinds of services not only for aircraft, ships, and land vehicles, but also for portable very small aperture terminals (VSATs), and even for people walking who are carrying a handy terminal. Furthermore, communica-

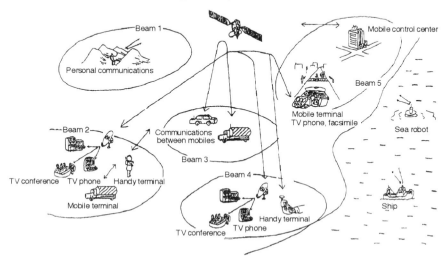

Figure 11.1 Concept of a future advanced mobile satellite communication system.

425

tion services will be expanded not only to voice and data, but also to pictures, movies, and TV conferences.

A handy terminal will be achieved by the following:

1. Technical innovations to achieve higher performance and higher efficiency in devices such as antennas, amplifiers, and circuit components;

2. Improvements in modulation and coding technologies to achieve high quality with low carrier to noise power ratio;

3. Use of higher frequency bands such as Ka and millimeter waves to get sufficient frequency bandwidth and to make devices small;

4. Development of large and intelligent satellites having high-power transmission, large onboard antennas, and high performance.

The *intelligent satellite* will be the most important concept of future mobile satellite communication systems [1]. The future advanced services will be achieved by "intelligent satellites" that have intelligent functions such as onboard signal processing, beam interconnection, and scanning spot beams. These intelligent functions will enable very small handy terminals to establish communication channels by direct access to satellites without any gateway Earth stations (one-hop communications).

The size of a terminal will depend on services covering the range from very low-speed to high-speed transmissions, as shown in Table 11.1.

If the uplink channel quality (C/N_0)up is given, then the aperture antenna size onboard satellite A_s is given by the following equation (see (3.18) and (3.19)):

$$A_s \eta_s = \frac{(C/N_0)_{up} d^2 \lambda^2 N_{0S} L}{P_E A_E \eta_E} \tag{11.1}$$

where

$(C/N_0)_{up}$ is the required carrier to noise power ratio in an uplink;

d is the distance between satellite and Earth station;

λ is the wavelength;

N_{0S} is the receiving noise power density of satellite;

L is the rain attenuation, feeder loss, and so on;

A_E, η_E is the antenna aperture area of Earth station and its efficiency.

Table 11.1
Example of Services in Future Advanced Mobile Satellite Communications

Service	Transmission Rate	Antenna Diameter of Earth Station	Earth Station Type
Paging	20 bps–500 bps	2 cm	Handy
Messages	200 bps–5 Kbps	2–10 cm	Handy
SCADA	200 bps–5 Kbps	about 10 cm	Handy, portable
Low-speed voice	about 2.4 Kbps	about 10 cm	Handy
Facsimile	about 10 Kbps	10–30 cm	Mobile, portable
High-speed voice	10 Kbps–30 Kbps	10–30 cm	Mobile, portable
Still pictures	2 Kbps–20 Kbps	10–30 cm	Mobile, portable
Compressed moving pictures	2 Kbps–300 Kbps	over 30 cm	Portable
TV conference	200 Kbps–1M bps	over 30 cm	Portable
Computer network	2 Kbps–1 Mbps	over 30 cm	Portable

Note: SCADA: Supervisory control and data acquisition; bps: bit per second.

It is known that using higher frequency bands can lead to small terminals, because the received power is proportional to the square of the frequency assuming the same transmission power and the same size of satellite antennas. In deriving (11.1), only the (C/N_0) required in the uplink is taken into consideration assuming that the same (C/N_0) is obtained in the downlink by enough output power of the transponder. Table 11.2 shows the sizes of antennas onboard the satellite, calculated by the above equation in the Ka band (30/20 GHz), in order to provide services with transmission speeds of 2.4, 24, and 300 Kbps. However, we assume that the efficiency of an onboard antenna is 35% and the margin for rain attenuation is 10 dB. For example, a rain margin of 10 dB gives a channel operation rate of 99.8% throughout the year in Tokyo. The maximum required onboard antenna size is found to be about 5m. The half-power beamwidth and footprint radius are also shown in the same table. We found that the footprint radius becomes as small as a few hundred kilometers. This shows that multibeam antennas and relevant beam interconnection onboard the satellite are essential technologies for the future mobile satellite communication systems.

11.2 Intelligent Satellites

11.2.1 Definition

The *intelligent satellite* is a recent term in satellite communications that has not yet been formally defined. *Intelligence* means the ability to learn or under-

Table 11.2
Antenna Sizes Onboard a Satellite for Wide Variety of Services

Service	Low-Speed Voice	High-Speed Voice	TV Conference
Transmission rate	2.4 Kbps	24 Kbps	300 Kbps
Uplink required C/N_0 ((E_b/N_0)up = 11 dB)	44.8 dBHz	54.8 dBHz	65.8 dBHz
Antenna diameter of Earth station (efficiency = 50%)	10 cm	15 cm	30 cm
Transmission power of Earth station	0.1W	0.3W	2W
Rain attenuation	10 dB	10 dB	10 dB
Feeder loss	6 dB	6 dB	6 dB
Receiving noise temperature of satellite	800K	800K	800K
Antenna diameter of satellite	4.0m	4.8m	3.4m
Half-power beamwidth	0.6 deg	0.5 deg	0.7 deg
Radius of a footprint (f = 30 GHz, efficiency = 35%)	250 km	190 km	310 km

stand from experience and to respond quickly and successfully to a new situation (*Webster's New World Dictionary*). In some senses, the present satellites can be called intelligent satellites because, for example, they can automatically control their attitudes or antenna directions based on output signals from appropriate onboard sensors. This is the first intelligent function, which is performed by the procedures based on the signal level. The second intelligent function is to program the next steps by logical procedures based on the symbol level. For example, beam switching is an intelligent function because switching is carried out by extracting information about beam destinations from transmission signals.

Intelligent satellites are required to have such functions as self-judgment, logical operation, memory, reference to memory, inference, self-execution and self-requirement.

11.2.2 Technologies in Intelligent Satellites

Technologies for intelligent functions of satellites for mobile satellite communications are described in this section. The basic intelligent functions are performed by a regenerative transponder and multibeam antennas onboard the satellite.

11.2.2.1 Regenerative Transponders

A key feature of a regenerative transponder is to have baseband signal processing onboard the satellite, which provides such functions as switching of beam interconnections, rate conversion of data transmission, and scanning spot beams of onboard antennas. These functions have more advantages compared to a conventional transponder in providing communication services and networks for a large number of compact terminals for mobiles and individuals. Figure 11.2 compares regenerative and conventional transponders. A conventional transponder only transmits a signal to an Earth station after frequency converting and amplifying the uplink signals. So, a conventional transponder is sometimes called a bent-pipe or transparent transponder. On the other hand, a regenerative transponder demodulates uplink signals to baseband levels, and retransmits them after modulating them for downlink channels.

From the standpoint of link budget, a regenerative transponder can reduce the effective isotropically radiated power (EIRP) of Earth stations. As shown in Chapter 3 (Example 3.7 and Figure 3.10), the total (C/N_0) when using a transparent transponder is determined by the performance of the downlink, which is constrained by the limited power onboard the satellite. In order to provide the required total (C/N_0), the (C/N_0) in the uplink must be about 20 dB more than that in the downlink. On the other hand, in using a regenera-

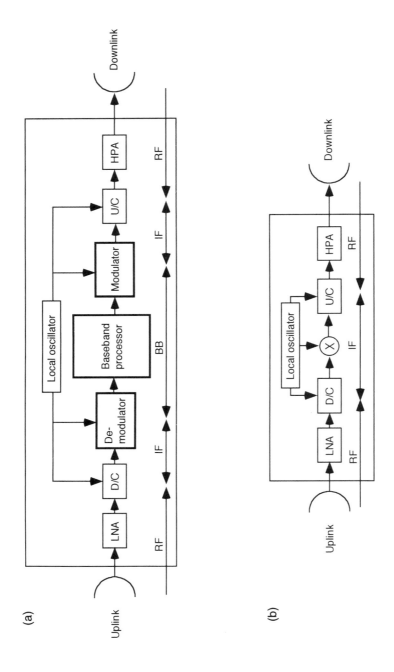

Figure 11.2 Configurations of (a) regenerative and (b) transparent transponders.

tive transponder, the channel qualities of uplinks and downlinks are determined independently, and both links can have almost the same (C/N_0). The reduction of transmission power would make Earth terminals compact. Another advantage of a regenerative transponder is that it can provide different multiple access formats in uplinks and downlinks. In general, frequency division multiple access (FDMA) is suitable in the uplink for mobiles, because each mobile transmits a signal less often and its data rate is not so high speed, so FDMA does not need as much power and can have a simple, less expensive configuration compared to time division multiple access (TDMA). On the other hand, time division multiplexing (TDM) is preferable in the downlink, because the high-power amplifier (HPA) can be operated in the saturation region to give the maximum transmission EIRP to reduce the G/T of mobile terminals.

11.2.2.2 Multibeam Antennas

There are two main advantages in using multibeam antennas onboard a satellite. The first is that high-power flux density can be obtained on the Earth's surface, which enables mobile users to use smaller terminals with a compact antenna and low transmission power. The second one is that frequencies can be reused several times more in the same frequency band, as in the case of satellite 1. This results in the efficient use of limited frequency bands in order to achieve a large capacity for users. Figure 11.3 shows the basic concept of a multibeam antenna (satellite 3) compared with a single-beam antenna (satellite 1 and satellite 2). We assume that satellite 1 has an antenna with a diameter of d which makes a footprint of diameter a on the Earth surface. When satellite 2

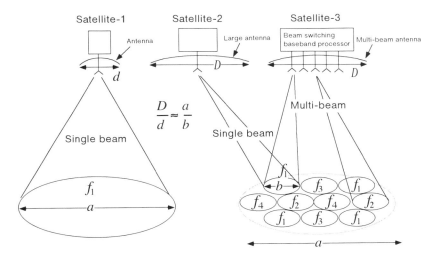

Figure 11.3 Multibeam antenna compared with single-beam antenna.

has a larger antenna with a diameter of $D(D \geq d)$, and still has a single beam and the same frequency, the diameter b of the footprint becomes narrower than in the former case. The value of b becomes $b = -(d/D)$, which can be obtained from the fact that half the power beamwidth of an antenna is nearly proportional to (λ/D), as shown in (4.7). Satellite 2 makes possible smaller user terminals with low-gain antennas having aperture sizes about (b/a) times smaller, and can provide transmission capability about (a/b) times greater than that of the satellite 1. Although satellite 2 can achieve better performance than satellite 1, we recommend satellite 3, whose antenna has many beams, in order to cover the required areas. It is clear that the number of beams can be obtained from the value of $(D/d)^2$ when frequency bands are the same as for satellite 2. This type of antenna, which is called a multibeam or a spot-beam antenna, has been used in some of the present commercial and experimental satellites such as MSAT, ITALSAT, and COMETS, as shown in the following section. When these beams are fixed, namely their footprints do not vary with time, they are called a *fixed* multibeam. When a multibeam (spot beam) can be scanned or hopped within some areas of the required service areas, this multibeam is called a scanning spot beam or a hopping beam. The scanning beam gives a satellite the ability to hop or scan beams to collect and deliver signals from one designated area to another. This performance can be achieved by switching to interconnect beams in the baseband or intermediate frequency levels, and by using onboard memory for transmission at the time the beam is scanning the designated area.

Onboard switching can be achieved by the following technologies:

1. *Filter bank:* Beam switching by filter banks can be achieved by a very simple configuration. This method is suitable for a small number of multibeams. Signals transmitted in a beam by a user are interconnected by frequency filtering to a designated beam. This method was chosen for the COMETS satellite, which is introduced in Section 11.3.3.

2. *Microwave switch matrix:* The signals are simply interconnected from an uplink beam to a downlink beam by changing switches, which are operated in the radio frequency (RF) or intermediate frequency (IF) levels. This method has been chosen for the ACTS satellite, which is introduced in Section 11.3.2.

3. *Baseband switch:* The uplink signals are downconverted to baseband signals, which are easier for digital signal processing. The beam switching is carried out on baseband levels by digital signal processors. This type of beam switching has been experimentally tested on the ACTS and COMETS satellites.

These onboard signal processing functions are generally very sophisticated ones and they need very advanced technologies.

11.2.2.3 Other Functions

The use of a signal processor and memory allows the following intelligent functions to be installed onboard the satellite:

- Signal error correction;
- Channel control for efficient operability;
- Power and gain control for efficient traffic control;
- Protocol conversion for communications between different networks;
- Translation between different languages;
- Voice and data storage onboard the satellite;
- Programmable modulation and protocols by the large onboard memory, digital signal processor, and software that can be downloaded by the ground station;
- Information about communication charges;
- Intersatellite communication to reduce delay time.

11.3 Examples of Intelligent Satellites

11.3.1 ITALSAT

ITALSAT is the first intelligent satellite to have an onboard regenerative transponder. ITALSAT-1 was launched in 1991, and ITALSAT-2 is currently under development. The mission of the ITALSAT satellite is research, development, and demonstration of digital communications in the Ka band via a satellite that has functions of onboard switching, onboard regeneration, and multibeam interconnections. As shown in Figure 11.4, the ITALSAT satellite has six spot beams generated by two 2m onboard antennas in the Ka band. The configuration of the satellite transponders is also shown in Figure 11.5. The transmission power of the traveling wave tube amplifier (TWTA) is 20W and the noise figure of the receiver is 6 dB. The EIRP and *G/T* are 57 dBW and 17 dBK, respectively, for a transmission rate of 147 Mbps. Beam interconnection between six beams is carried out by satellite-switching time division multiple access (SS-TDMA). The signal is regenerated by a QPSK demodulator, and it is input to a QPSK modulator after interconnection to a designated beam by a 6 by 6 baseband switch matrix [2].

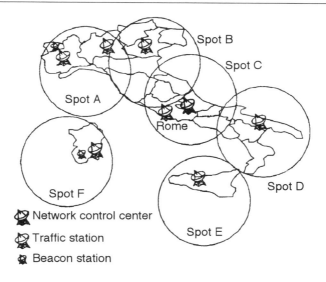

Figure 11.4 Footprints of ITALSAT spot beams.

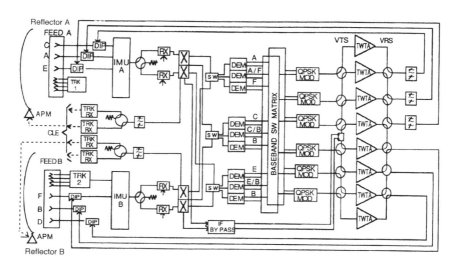

Figure 11.5 Configuration of transponders of ITALSAT.

11.3.2 ACTS [3]

The advanced communications technology satellite (ACTS), launched by NASA in 1993, has the mission of furnishing the technology necessary to establish very small aperture terminal (VSAT) digital networks, which provide on-demand, full-mesh connectivity 1.544-Mbps services with only a single hop.

Utilizing onboard switching and processing, each individual voice and data circuit may be separately routed to any location in the network.

ACTS has two separate Ka-band antennas, each with horizontal and vertical polarization subreflectors, for transmitting and receiving signals. The uplink and downlink transmissions take place within two 900-MHz frequency bands centered at 19.7 GHz and 29.42 GHz, respectively. The offset Casegrain antenna system provides two hopping spot-beam families (east and west) plus three fixed spot beams with the coverage shown in Figure 11.6. Hopping beams offer certain advantages relative to fixed beams. In a period equal to a single TDMA frame, 1 msec in the case of ACTS, a beam can hop to many locations, dwelling long enough at each to pick up the offered traffic. By adaptively varying each dwell time, the system's capacity is efficiently matched to a nonuniform demand for traffic [4]. However, where predicable, large, nearly uniform trunked traffic exists, a fixed system can provide a less complex solution.

As shown in Figure 11.7, ACTS has two operating modes: baseband processor and 3 by 3 microwave switch matrix. The baseband processor using baseband-switched TDMA performs three functions. After the TDMA traffic bursts coming from the ground terminals have been sampled during the beam dwelling period during a receiving frame, it routes the contents of the individual uplink channels from the appropriate demodulators to the input memory. During the next frame, it formats these signals into a new configuration of downlink TDMA traffic bursts and stores them in output memories. In the third frame, it transmits the reformatted traffic bursts, in either the selected coded or uncoded format, to the destination stations in the appropriate downlink hopping beam dwelling. The baseband switch interconnects the two hopping beams and operates based on onboard stored baseband-switched TDMA.

No onboard demodulation, storage, or message traffic rearrangement is possible with the microwave switch matrix. The traffic bursts are simply connected from a particular uplink beam to a destination downlink beam during their passage through the satellite. The traffic burst time length, the microwave switch matrix time schedule, and the uplink and downlink beam pointing schedule all must be coordinated to accomplish unimpeded traffic flow between the network terminals. The microwave switch matrix operation is based on SS-TDMA and interconnects the three fixed beams. The switch and the multibeam antenna make possible wideband Ka-band communications (900 MHz) with the ground terminals. Table 11.3 shows a summary of burst rates used in the ACTS baseband processor and microwave switch.

In the ACTS program, several VSAT terminals have been developed for high data rate services. These terminals are used mainly for fixed services. One

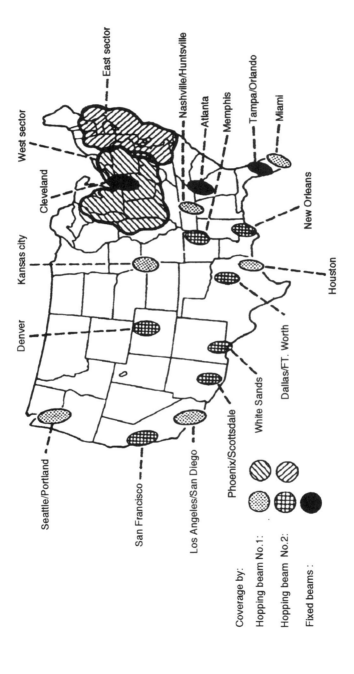

Figure 11.6 ACTS multibeam antenna coverage [3].

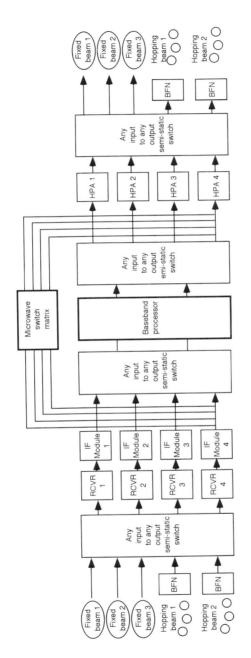

Figure 11.7 ACTS multibeam communication system block diagram.

Table 11.3
Outline of Mission Payload of ACTS

Antenna	Spot beam and hopping beam antennas.
	    Diameter: 3.3m (20 GHz) / 2.2m (30 GHz), linear polarization
	Mechanically steerable antenna, linear polarization
	    Diameter: 1m

Beam	Three fixed beams Two hopping spot beam families (east and west) One steerable beam (steering rate 1 deg/minute)
Frequency	Ka-band: 28.97–29.87 GHz (uplink) 19.25–20.15 GHz (downlink)
Transponder	46W TWTA × 4

Beam switching	3 × 3 Microwave switch matrix
	    Uplink: nominal 220 Mbps
	    Downlink: nominal 220 Mbps
	Baseband processor
	    Uplink: 110 Mbps (single CH TDMA), 27.5 Mbps (two CH
	        FDMA/TDMA)
	    Downlink: 110 Mpbs (single and two channels)

of the most interesting experiments in the ACTS program is the aeronautical terminal experiment (Aero-X) carried out by NASA and the Jet Propulsion Laboratory, where 4.8-Kbps and 9.6-Kbps duplex voice links were established between a NASA jet aircraft and the ACTS ground station via the ACTS satellite. In Ka-band mobile satellite communications, a vehicle antenna is considered one of the most changeable key technologies. In the ACTS Aero-X experiments, three types of monolithic microwave integrated circuit (MMIC) array antennas have been developed and evaluated. Their main characteristics are shown in Table 11.4 [5].

11.3.3 COMETS

COMETS is a joint R&D satellite project of the Communications Research Laboratory (CRL) and the National Space Development Agency (NASDA) of Japan [6]. The COMETS satellite is scheduled to be launched in August 1997 by NASDA. One of its main missions is to develop basic technologies of advanced mobile satellite communication systems. Table 11.5 outlines the mission payload of COMETS related to advanced mobile satellite communications.

The frequencies used in the COMETS system are 47/44 GHz for millimeter-wave communications, and 31/21 GHz for Ka-band communications. As

Table 11.4
Characteristics of MMIC Ka-Band Array Antennas for ACTS Aero-X Experiments [5]

MMIC Array	Type 1	Type 2	Type 3
Type	Transmit	Receive	Receive
Frequency	29.6	19.9	19.9
No. of elements	32	23	16
Array configuration	Tile	Brick	Brick
Element configuration	Squared grid	Triangular grid	Square grid
Element spacing (cm)	0.82 (0.8λ)	0.81 (0.53λ)	0.84 (0.55λ)
Radiating elements	Aperture coupled Circular patch	Dielectrically loaded Circular waveguide	Printed circuit Endfire dipole
Polarization	Linear	Linear	Linear
Phase shifter bits	4	4	3
Scanning range (degrees)	±30	±60	±60
EIRP at boresight (dBW)	23.4	N/A	N/A
G/T at boresight (dBK)	N/A	−16.6	−16.1
Cooling	Thermoelectic and fan	None	Fan

Table 11.5
Outline of Mission Payload of the COMETS Satellite

Antenna	Spot beam antenna shared with millimeter wave and Ka band Diameter: 2m, Circular polarization, antenna pointing system
Beam	Two Ka-band beams (Tokyo and Nagoya beams) One millimeter-wave beam (Tokyo beam)
Frequency	Ka band: 30.75–30.85 GHz (uplink) 20.98–21.07 GHz (downlink) Millimeter wave: 46.87–46.90 GHz (uplink) 43.75–43.78 GHz (downlink)
Transponder	Ka band: 2 (20W and 10W SSPA) Millimeter wave: 1 (20W TWTA)
Beam switching	IF repeater: 2 × 2 matrix beam interconnecting by IF filter bank Wideband filter (6 MHz) and narrowband filter (500 kHz) Regenerative transponder: 8 ch SCPC (uplink)/TDM (downlink) Beam interconnection by baseband switching

shown in Figure 11.8, the payload consists of a multibeam antenna, millimeter-wave and Ka-band transmitters and receivers, an IF filter bank, and regenerative modems having 2 by 2 matrix beam interconnections. The transponder consists of 20W and 10W solid-state power amplifiers (SSPAs) for Ka-band communications, 20W TWTA for millimeter-wave communications, and LNAs. An HEMT-LNA with a very low noise figure of 2.5 dB has been developed for millimeter-wave missions. The multibeam antenna has three beams: two adjacent Ka-band beams (Tokyo and Nagoya beams), and one millimeter-wave beam (Tokyo beam), as shown in Figure 11.9. The satellite has two methods of switching beams. The first is an IF filter bank. As shown in Figure 11.10, signals from an uplink beam are divided by the IF filter bank into three frequency subbands, and each subband interconnects between Tokyo-Tokyo, Tokyo-Nagoya, and Nagoya-Nagoya. Each frequency band is preassigned to each beam. This is a very simple and flexible method, suitable for a small number of multibeam interconnections, and has the advantage of transmitting various types of signals because of the transparency of the transponder. In using a transparent transponder, it is desirable to optimize the transponder gain and frequency bandwidth in order to increase the satellite's transmitting power (EIRP) and reduce the noise power.

The second method of interconnecting beams is carried out by a digital polyphase FFT filter bank in a regenerative transponder. Figure 11.11 shows the basic diagram of a regenerative transponder. Eight-channel SCPC signals are received by one regenerative modem and a single TDM signal is transmitted. The SCPC signal transmission is 24 Kbps or 4.8 Kbps with BPSK modulation. Eight-channel SCPC signals are demultiplexed by a digital polyphase FFT filter and demodulated discretely. A link controller manages such functions as channel setup, and SCPS and TDM channel assignment. This control is achieved on the satellite by using one SCPC packet signal channel and one TDM signal time slot. This method is suitable for a large number of beam interconnections and for mobile satellite communications, because regenerative processing enables transmission power to be reduced.

In the development of mobile terminals for Ka-band and millimeter-wave bands, a mobile antenna is one of the most leading edge technologies. In the COMETS project, three types of mobile antennas have been developed. The first is a mechanically steered waveguide slot array antenna. Figure 11.12 shows a photograph of a waveguide slot array antenna for the Ka band. The gain is 23.6 dBi for receiving and 26.4 dBi for transmitting. The half-power beamwidths (HPBWs) in elevation are 17 degrees and 19.5 degrees for transmitting and receiving, respectively, which are wider than those in azimuth to enable the satellite to be tracked only in azimuth directions by a mechanical tracking method [7]. The second type is an active phased-array antenna, which

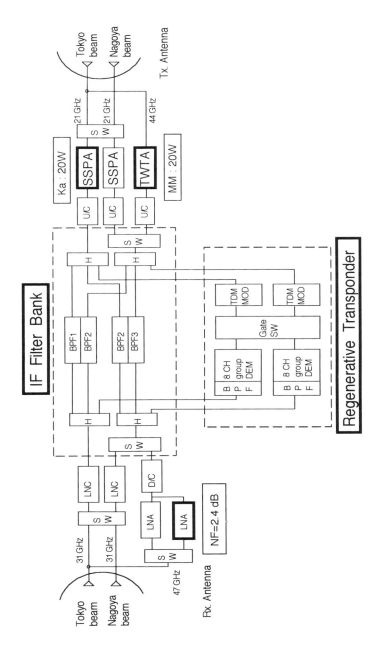

Figure 11.8 Block diagram of a transponder of the COMETS satellite.

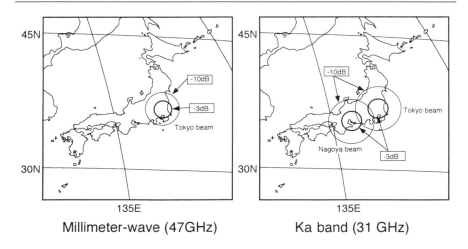

Figure 11.9 Footprints of the COMETS antenna, two beams in the Ka band and one beam in the millimeter-wave band.

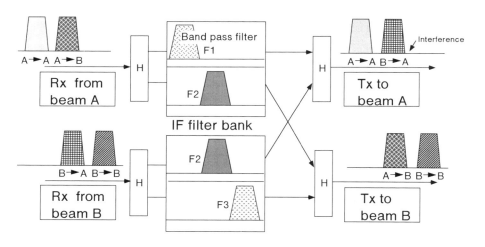

Figure 11.10 Concept of interconnecting beams by IF filter bank.

is under development. The third one is a torus antenna for the millimeter-wave frequency band, which has a torus main reflector and a primary waveguide feed horn with a subreflector. The horn antenna makes the frequency bandwidth wide enough to cover both Tx and Rx frequencies [7]. Figure 11.13 shows a configuration and a photograph of a torus antenna for the millimeter-wave band. Table 11.6 shows characteristics of three types of antennas for the Ka band and for the millimeter waves, which have been developed in the COMETS project.

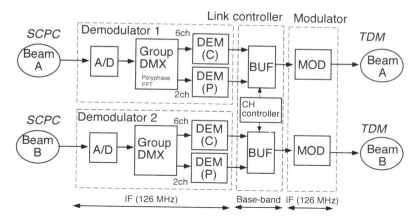

Figure 11.11 Basic block diagram of a regenerative transponder.

Figure 11.12 Photograph of a waveguide slot array antenna for the Ka band.

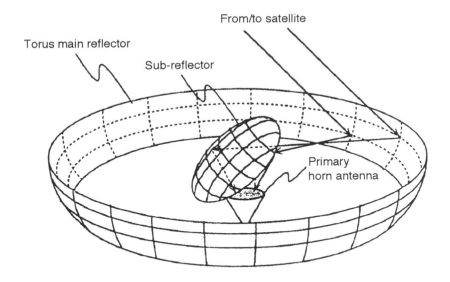

Figure 11.13 Configuration and photograph of torus antenna for the millimeter-wave band.

Table 11.6

Three Types of Antennas for the Ka Band and Millimeter-Wave Band That Have Been Developed in the COMETS Project

Antenna Type	Waveguide Slot Array		Active Phased Array	Torus Reflector	
Function	Two sets for Tx & Rx		Rx only	Tx & Rx	
Frequency (GHz)	Transmit	Receive	Receive	Transmit	Receive
	30.8	21	21	46.9	43.8
	Ka band			Millimeter-wave band	
Radiating Elements	Cross slot		Circular patch	Waveguide aperture	
No. of elements	256	144	169	1	
Gain (dBi)	24.4	21.9	21	26	25
HPBW in elevation (degrees)	17	19.5	10	10	11
Polarization			Circular		
Tracking	Mechanical		Electrical	Mechanical	

References

[1] "Report on Technologies of Future Intelligent Satellites in Next Generation," Key Technology Center (Japan), Feb. 1994.

[2] Mastracci, C., and G. Cedrone, "ITALSAT, New Technologies Orbital Demonstration," *Proc. AIAA 14th Int. Communication Satellite Systems Conf. and Exhibit*, ITA, 1992, pp. 842–848.

[3] Richard, G., and S. Ronald, "Advanced Communications Technology Satellite (ACTS)," *Int. Conf. on Communications (ICC'89)*, 1989, pp. 1566–1577.

[4] Regier, F. A., "The ACTS Multibeam Antenna," *IEEE Trans. MTT*, Vol. 40, No. 6, June 1992, pp. 1159–1164.

[5] Raquet, C., et al., "Ka-band MMIC Array for ACTS Aeronautical Terminal Experiments (Aero-X)," *IMSC'95, 4th Int. Mobile Satellite Conf.*, Ottawa, Canada, 1995, pp. 312–317.

[6] Ohmori, S. et al., "Advanced Mobile Satellite Communications using COMETS Satellite in MM-wave and Ka-band," *3rd Int. Mobile Satellite Conf.*, Pasadena, CA, June 1993, pp. 549–553.

[7] Hase, Y., M. Tanaka, and H. Saito, "Mobile Antennas for COMETS Advanced Mobile Satcom Experiment," *IMSC'95, 4th Int. Mobile Satellite Conf.*, Ottawa, Canada, 1995.

About the Authors

Dr. Shingo Ohmori received B.E., M.E., and Ph.D. degrees in electrical engineering from the University of Thohoku, Japan, in 1973, 1975, and 1978, respectively.

Since 1978, he has been with the Communications Research Laboratory (CRL) of the Ministry of Posts and Telecommunications, Japan and has been engaged in research on mobile satellite communications, especially on antenna and propagation.

During 1983–1984, he was a visiting research associate at the Electro-Science Laboratory of Ohio State University, Columbus, Ohio. He led the ETS-V satellite experiments as a chief of the Satellite Communications Section in Kashima Space Research Center of CRL from 1988 to 1991, and from 1991 to 1993 in the COMETS satellite program as a chief of the Advanced Satellite Communications Section in Space Communications Division of CRL.

He was a director of Kashima Space Research Center of CRL from 1995 to 1996. He is now a director of the Space Communications Division of CRL. He also has a professorship of the University of Electronics and Communications in Tokyo.

He is a co-author of *Mobile Antenna Systems Handbook* (Artech House, 1994).

He was awarded the Excellent Research Prize from the Minister of Science and Technology Agency of Japan in 1985 and the Excellent Research Achievements Prize of the IEICE in 1993.

Dr. Ohmori is a member of the Institute of Electronics, Information and Communication Engineers (IEICE), Japan, and a senior member of the IEEE.

Dr. Hiromitsu Wakana received B.Sc., M. Sc. and Ph.D. degrees in applied physics from Waseda University, Japan, in 1976, 1978, and 1984, respectively. Since 1982, he has been with the CRL of the Ministry of Posts and Telecommunications, Japan, and has been engaged in experimental research on satellite communication systems using the CS-series, ETS-V, ETS-VI and COMETS satellites. During 1989–1990, he was a visiting research associate of Aussat Pty. Ltd., Australia. During 1990–1995, he led mobile satellite communication experiments by ETS-V and intersatellite communication experiments by ETS-VI in the Kashima Space Research Center as a chief of the Satellite Communications Section. Since 1996, he has been a visiting associate professor of the University of Electro-Communications in Tokyo. Since 1997, he has been a project manager leading the COMETS project of the Ka-band and millimeter-wave mobile satellite communications and the 21-GHz satellite broadcasting in the Kashima Space Research Center, CRL.

Dr. Wakana is a member of the Institute of Electronics, Information and Communication Engineers (IEICE), Japan, and the IEEE.

Dr. Seiichiro Kawase received B.E. (1972) and M.E. (1975) degrees from Tokyo Institute of Technology, both in mechanical engineering, and a D.E. (1994) from Tokyo University in electronics. In 1975, he joined the CRL of the Ministry of Posts and Telecommunications to work on the orbital dynamics of experimental telecommunications satellites (CSE and BSE). He worked for Telesat Japan (1982–1983) to support the first commercial version of telecommunications satellites (CS-2 and BS-2) in their orbital dynamics.

After a visit to European Space Operations Center, Germany (1984–1985), he started his work again at CRL, where he is in charge of developing the techniques for monitoring and controlling the orbital motions of the increasingly crowded geostationary satellites. Presently he is head of the Space Systems Section, Kashima Space Communications Center.

Index

Recent Titles in the Artech House
Mobile Communications Series

John Walker, Series Editor

Mobile Telecommunications: Standards, Regulation, and Applications, Rudi Bekkers and Jan Smits

Practical Wireless Data Modem Design, Jonathon Y. C. Cheah

Radio Propagation in Cellular Networks, Nathan Blaunstein

Resource Allocation in Hierarchical Cellular Systems, Lauro Ortigoza-Guerrero and A. Hamid Aghvami

Signal Processing Applications in CDMA Communications, Hui Liu

Spread Spectrum CDMA Systems for Wireless Communications, Savo G. Glisic and Branka Vucetic

Understanding Cellular Radio, William Webb

Understanding Digital PCS: The TDMA Standard, Cameron Kelly Coursey

Understanding GPS: Principles and Applications, Elliott D. Kaplan, editor

Universal Wireless Personal Communications, Ramjee Prasad

Wideband CDMA for Third Generation Mobile Communications, Tero Ojanperä and Ramjee Prasad, editors

Wireless Technician's Handbook, Andrew Miceli

For further information on these and other Artech House titles, including previously considered out-of-print books now available through our In-Print-Forever® (IPF®) program, contact:

Artech House
685 Canton Street
Norwood, MA 02062
Phone: 781-769-9750
Fax: 781-769-6334
e-mail: artech@artechhouse.com

Artech House
46 Gillingham Street
London SW1V 1AH UK
Phone: +44 (0)20 7596-8750
Fax: +44 (0)20 7630-0166
e-mail: artech-uk@artechhouse.com

Find us on the World Wide Web at:
www.artechhouse.com